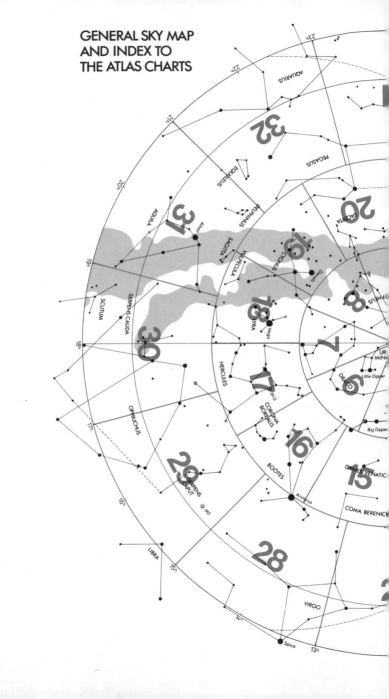

GENERAL SKY MAP
AND INDEX TO
THE ATLAS CHARTS

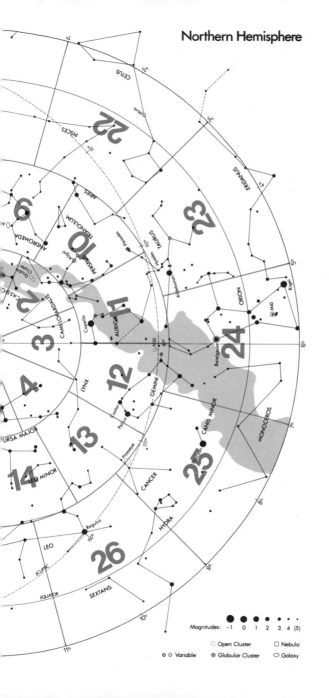

Northern Hemisphere

Magnitudes: -1 0 1 2 3 4 (5)

○ Open Cluster □ Nebula
⊙ ○ Variable ⊕ Globular Cluster ○ Galaxy

THE PETERSON FIELD GUIDE SERIES
Edited by Roger Tory Peterson

Frontispiece (overleaf) — Cluster of galaxies in Virgo, including the elliptical galaxies M86 and M84 at center and bottom, respectively. North is at right. (The Kitt Peak National Observatory)

A Field Guide to
Stars
and Planets

Donald H. Menzel
and
Jay M. Pasachoff

Monthly Sky Maps
and Atlas Charts by
Wil Tirion

Second Edition

Sponsored by the National Audubon Society,
the National Wildlife Federation,
and the Roger Tory Peterson Institute

HOUGHTON MIFFLIN COMPANY · BOSTON

Library of Congress Cataloging in Publication Data

Menzel, Donald Howard, 1901–1976
A field guide to the stars and planets.

(The Peterson field guide series)
Bibliography: p. 457
Includes index.
1. Astronomy—Observers' manuals.   I. Pasachoff,
Jay M.   II. Title.
QB64.M4 1983   523   83-8392
ISBN 0-395-34641-X
ISBN 0-395-34835-8 (pbk.)

Printed in the United States of America

A 13 12 11 10 9 8 7

Editor's Note

Although we expect never to publish a guide to the creatures of outer space, it is inevitable that the Field Guide Series should include this volume on recognition of the stars and planets.

Among all the inhabitants of the planet Earth, we alone have systematically considered the heavenly bodies. We have given names to the constellations and charted their relative positions and movements. Although we formerly believed that we alone were able to navigate by celestial means, now we know through the experiments of E.G. Franz Sauer and others that nocturnal bird migrants apparently take their direction by means of an innate ability to read the night sky.

In recent years, and particularly since the development of rocket-launched satellites, more people than ever before have become aware of space and want to know what is out there. The tiny dots of light in the night sky may be obscured by city fog, but on clear nights they cannot fail to stir the inquiring mind.

The first step in astronomy, as in zoology, is to put names to things, to identify them. This *Field Guide* will facilitate the process and is equally usable for the observer depending on the naked eye, the binocular, or a small astronomical telescope. Unlike most of the books in the Field Guide Series, which tend to be regional in scope or at least confined to a single continent, it may be used at any point on the earth's surface and on any day of the year. In line with the general policy of the other *Field Guides,* emphasis has been put on new and simplified techniques of recognition.

Much has happened in the science of astronomy during the last 20 years since the first edition of *A Field Guide to the Stars and Planets* was published (1964). The original text by Donald H. Menzel, the late director of the Harvard College Observatory, has been completely revised and brought in line with present knowledge by Jay M. Pasachoff, Director of the Hopkins Observatory at Williams College. The maps and charts of the earlier edition have been replaced by monthly sky maps and atlas charts prepared with great precision by Wil Tirion. In addition, the color photographs are new.

The book should present no problems to beginners interested in finding their way around in the heavens, but at the same time the completeness of its charts and tables should make it a useful tool for serious amateurs and even for professionals.

It is a joy to thumb through the book while relaxing in an arm-chair, but inasmuch as it is basically a Field Guide, put it to practical use. Use it on clear nights to interpret the free show put on by the heavens.

Roger Tory Peterson

Acknowledgments

This *Field Guide to the Stars and Planets* presents tours of the stars, the planets, the sun, the moon, and other objects in the heavens. The information it contains is on a level suitable for novices as well as for those who are already quite knowledgeable in astronomy. I am very pleased to have had the complete cooperation of the celestial cartographer Wil Tirion. He not only drew a complete set of Atlas Charts covering the whole sky but also prepared monthly star maps to a new design and executed supplemental charts for variable stars.

I appreciate the assistance of Cathryn Baskin for her work on the preparation of descriptive material to accompany the Atlas Charts and on other phases of the book. Robert Murphy of Scientia, Inc., has not only prepared Graphic Timetables showing the positions of the planets but also designed Graphic Timetables to show which of the brightest stars, clusters, variable stars, nebulae, and galaxies are suitable for viewing in various seasons. Robert A. Victor of the Abrams Planetarium of Michigan State University expertly prepared descriptive material to accompany the Graphic Timetables. The fruits of Ewen Whitaker's lengthy study of the moon show in the material he has prepared to accompany the moon maps. I thank them all.

I appreciate the special cooperation of Leif Robinson, William Shawcross, Roger W. Sinnott, Alan Hirshfeld, Dennis di Cicco, and others at *Sky & Telescope,* and thank them for their permission to reprint some of the tables that they have carefully prepared as part of their *Sky Catalogue 2000.0.*

I thank Dennis di Cicco and Ben Mayer for their helpful advice on many phases of observing, for reading large sections of the text, and for the photographs they provided.

Charles Case and the National Geographic Society were kind enough to allow me to use the special moon maps beautifully prepared in a cooperative effort of the National Geographic Society and the U.S. Geological Survey.

I thank many people who assisted with various phases of the work, including especially Naomi Pasachoff for her work on constellations, Janet Akyüz Mattei (American Association of Variable Star Observers) for providing information on variable stars, and Ewen Whitaker (Lunar and Planetary Laboratory, University of Arizona) for providing descriptive information and photographs of the moon. I also thank Brian Marsden (Harvard-Smithsonian Center for Astrophysics) and John Bortle for comments on minor bodies of the solar system, Peter Millman (Herzberg Institute of Astrophysics, National Research Council of Canada) for material on meteor showers, Robert A. Victor for comments on observing, Robert D. Miller for calculations, Jean Meeus for information about eclipses, and George Lovi for information about mapping.

A Field Guide is obviously based on the work of many people carried out over an extended period of time. Users wanting to go beyond this book to do more observing would naturally refer to such magnificent sources as *Sky & Telescope* and *Astronomy* magazines, Robert Burnham's *Celestial Handbook,* and such additional books as *Norton's Star Atlas,* all of which were also consulted in compiling information for this Field Guide. These books and other additional sources are listed in the bibliography.

Susan Shepard, Wendell Severinghaus, Susan Welsch, Karen Kowitz, and Jon Riecke have assisted me in Williamstown with the preparation of the manuscript, figures, and tables.

Nancy Pasachoff Kutner prepared the Index.

I thank my father, Samuel S. Pasachoff, for his expert reading of the manuscript, and my mother, Anne T. Pasachoff, for her assistance. My wife, Naomi, and my children, Eloise and Deborah, have been my inspiration. All of them have also helped me read proof.

It is a pleasure for me to be associated with my late professor, Donald H. Menzel, on the second edition of this Field Guide. I only regret that he did not live to participate. I learned so much from him on a series of eclipse expeditions, and so much from his example as a scientist dealing with other aspects of astronomy, that I am forever in his debt. I thank also Florence K. Menzel, who has been kind and helpful to me since my student days, and whom my entire family values as a friend.

I hope that you all enjoy the *Field Guide to the Stars and Planets.* It would be nice if it were error-free, but no book is. I do hope that you will write me with your comments, to point out errors, or with observing suggestions that are not in this book.

Jay M. Pasachoff
Williams College
Hopkins Observatory
Williamstown, Massachusetts 01267

Contents

Illustrations

Front and rear endpapers: visual key to Atlas Charts

Color Plates: 82 color photographs grouped after p. 215

Black-and-white Plates: 160 black-and-white photographs, charts, and graphs, distributed throughout the book
10 Moon Maps in Chapter 8

Line Drawings: 72 Monthly Sky Maps in Chapter 3
52 Atlas Charts in Chapter 7
11 special charts

Tables

Appendix Tables

Introduction:
How to Use This Book

This *Field Guide to the Stars and Planets* will show you around the sky. We will try to make it easy for you to identify what you see; at the same time, we will try to demonstrate the excitement of our current understanding of the universe.

General organization. We begin by describing in Chapter 1 a framework for observing the heavens. We describe how to tell stars from planets, how to identify the brightest stars, and how to find a few of the most prominent groupings of stars in the sky. Then, in Chapter 2, we give you a brief tour around the heavens, season by season. This chapter can be used together with the Monthly Sky Maps that follow in Chapter 3. No special knowledge or equipment is needed to use these maps. For observers in the northern hemisphere, a set of four maps appears for each month: two maps — one with constellation outlines and one without — for use when facing north, and a similar pair for use when facing south. For observers in the southern hemisphere, a set of two maps shows stars with constellation outlines.

In Chapter 4, we describe the types of objects that are relatively constant in their places in the sky, including stars, nebulae, and galaxies. We discuss astronomers' current understanding of these objects, including the stages in the life of a star. We also provide information about the times of year when a selection of the most interesting double and variable stars, star clusters, nebulae, and galaxies are visible. In this chapter, and in the associated section of color plates, we include spectacular photographs of some of the most beautiful objects. The celestial objects that *move* with respect to the stars — the moon, the sun, the planets, comets, meteors, and asteroids — have their own chapters later on.

Next, in Chapter 5, we describe the constellations and the classical myths associated with them. A list of the current constellations (Appendix Table A-1) appears on p. 420.

Two types of objects of special interest to those observing the sky are double and variable stars, so Chapter 6 is devoted to them. This chapter includes charts and tables that will enable you to find many examples of these stars.

Chapter 7 is an Atlas of the entire sky, broken down into 52 charts, drawn by Wil Tirion. These charts are a special feature of this Field Guide. All of the brightest stars and constellations are shown, as on the Monthly Sky Maps; however, the Atlas Charts provide a more detailed look at each region of the sky. Although many of the objects on the Atlas Charts can be seen with the naked eye or with binoculars, you will find the charts even more interesting if you have access to a small or medium-sized telescope. Each chart shows not only stars but also nebulae, galaxies, and a wide variety of other celestial objects. Descriptions of these objects and photographs of some of them accompany the Atlas Charts. A list of nonstellar, deep-sky objects, the Messier Catalogue, precedes the charts, along with a table showing the region of sky covered by each chart. A visual key to the Atlas Charts appears on the endpapers of this book.

To use the Atlas, first locate an object of interest, using either the Monthly Sky Maps or the celestial coordinates listed in the Messier Catalogue or in other tables in this guide. Then turn to the Atlas Chart where your object is shown. Alternatively, you may choose to survey the whole area shown on a chart with a telescope.

As part of the introduction to the Atlas Charts, we explain the symbols used on the charts and the names used for stars and other types of astronomical objects. We also briefly explain the system of celestial coordinates — right ascension and declination — used to indicate the locations of objects in the sky. The apparent position of objects in the sky changes slightly over the years because the earth wobbles as it spins; we have compensated for this by drawing the charts and calculating the tables in this book for the year 2000, rather than using the 1950 positions now shown in most books.

Though the positions of distant objects in the universe change only slightly in the sky, the positions of the moon, most of the planets, and the sun change quite drastically in the course of the year. The path the sun follows through the sky in the course of a year is called the *ecliptic;* it is indicated by a dotted line on the Monthly Sky Maps and the Atlas Charts. The moon and planets never stray far from this line.

Chapter 8 describes the moon and includes 10 maps of its craters and other features of its surface. Chapter 9 explains how to locate the planets in the sky and how to predict when they will be visible. Chapter 10 describes what each planet is like and what you may see if you observe it with binoculars or with a telescope.

Chapter 11 describes comets, with special attention to Halley's Comet and its 1985–1986 appearance. Chapter 12 discusses meteors and asteroids, and lists meteor showers that you can see in the sky at different times of year. Meteors usually flash across the sky and asteroids move too slowly for their motion to be apparent. Lights that appear to move slowly and steadily across the sky are usually airplanes or — particularly in the couple of hours after sunset or before sunrise — artificial satellites in orbit around the earth.

In Chapter 13, we turn from the nighttime sky to the daytime sky and discuss the major object that is visible all day — the sun. We discuss not only the everyday sun and how to observe it, but also how to observe at a total solar eclipse and why a total eclipse is so glorious. We also describe annular eclipses, like the one visible from the southeastern United States on May 30, 1984, and how to observe them.

Finally, in Chapter 14, we discuss technical aspects of the positions of objects in the sky, ways to tell time by the sun and the stars, and calendars.

At the end of the book, we present some information on telescopes, a glossary, suggestions for additional reading, an extensive set of tables, and an index.

Illustrations. Rendering the sky in a book is a difficult problem because it requires stretching and squashing the curve of the sky onto a flat page. We have solved this problem in a new and improved way in collaboration with our celestial cartographer, Wil Tirion. Our Monthly Sky Maps are presented in a special projection that makes the maps easy to use while distorting the shapes of the constellations as little as possible.

In addition to the 72 Monthly Sky Maps and the 52 Atlas Charts by Wil Tirion, 10 detailed maps of the moon's surface, prepared in a collaborative effort of the National Geographic Society and the U.S. Geological Survey, enhance this revised edition of *A Field Guide to the Stars and Planets.* A number of Graphic Timetables have also been provided to help you determine when stars, planets, and other celestial objects will be visible above the horizon. The 82 color plates at the center of the book, showing some of the most interesting celestial objects, are another new feature of this revised edition. All photographs in this guide are oriented with north at the top (unless otherwise indicated), to make it easy for you to compare them with the Atlas Charts in Chapter 7.

Some observing hints. Your eyes have to be dark-adapted to see the sky well. This may take 5 to 15 minutes after you go outside from a lighted room. As you watch the sky during this time, more and more stars will become visible. To maintain your adaptation to darkness, cover the front of your flashlight with red plastic.

Telescopes. The observing suggestions in this guide are designed to help you locate interesting objects in the sky, whether or not you have a telescope. When we mention a "small telescope" in this guide, we are referring to one with a lens less than 4 in. (10 cm) in diameter; a "medium-sized telescope" has a lens about 4–10 in. (10–25 cm) in diameter. If you are interested in purchasing a telescope, you will find a list of some telescopes that are popular with amateurs on p. 463, along with a list of telescope manufacturers.

How to use this book. If you want to survey the stars and constellations, use Figs. 3 and 4 in Chapter 1 and the Monthly Sky Maps in Chapter 3. The seasonal tours in Chapter 2 illustrate how the constellations seem to move across the sky as the earth rotates around the sun. The Graphic Timetable of the Brightest Stars (Fig. 2) in Chapter 1 will show you when the brightest stars visible from midnorthern latitudes will be above the horizon.

If you see a bright object in the sky and want to identify it, the first step is to determine whether it is a star or a planet (see p. 5). Then check the Graphic Timetable of the Brightest Stars in Chapter 1 or the Graphic Timetables of the Planets in Chapter 9, to see which bright stars or planets are visible on your date of observation. Or you can refer to the Monthly Sky Maps in Chapter 3. You can plot the positions of the brightest planets on the Monthly Sky Maps in Chapter 3 (or on the Atlas Charts in Chapter 7) using Appendix Table A-8, which shows the planets' longitudes along the ecliptic.

If you are using binoculars or a telescope and want to look at a variety of interesting objects, such as double and variable stars, star clusters, nebulae, or galaxies, use the Graphic Timetables in Chapter 4 to find out which ones will be visible on your date of observation. Then turn to the Atlas Charts in Chapter 7, where observing notes supplement detailed charts of each region of the sky. Each chart is oriented with north at the top. Remember that binoculars give a right-side-up image but most telescopes give inverted images, compared with the way celestial objects appear to the naked eye.

If you want to observe the planets with a telescope, refer to the information in Chapter 10. If you are able to see the moons of Jupiter, you can use the table on p. 370 to identify them.

If you want to see a meteor, use the table of meteor showers in Chapter 12.

If you want to know how to observe the sun safely, or when and where the next solar eclipse will occur, refer to Chapter 13.

If you want to know how to tell time by the sun and the stars, refer to Chapter 14.

1

A First Look at the Sky

Finding your way around the sky is like finding your way around a large city — it is easy if you are familiar with the streets and have navigated there before, but otherwise it takes some time to become familiar with routes and shortcuts. In this first chapter of *A Field Guide to the Stars and Planets,* we will assume that you are new to observing the heavens. We will start from scratch and show you some of the basic ways that you can orient yourself when observing. Our focus here will be on some of the most prominent stars and constellations that you can observe with the naked eye or binoculars.

Before you begin to observe the nighttime sky, it will be helpful for you to determine which way north, south, east, and west are. If you don't know the compass directions for the place where you are observing, though, we describe how to find them with the aid of the Big Dipper and the North Star, Polaris (see p. 11).

One of the first things you will notice when you start to study the sky is that stars and other objects are of different brightnesses. Perhaps the easiest way to determine what is what in the sky is to take advantage of this fact. Except for the moon, the brightest objects in the nighttime sky are some of the planets. The planets change their position slightly from night to night with respect to the stars in the background; in Chapter 9 we show you how to locate the planets on any given night.

Three characteristics will tell you quickly if an object is a planet.

1. ***Brightness.*** Some of the planets simply appear too bright to be stars. Venus, the brightest planet, is an example. It can never be very far away from the sun in the sky, so whenever an extremely bright dot of light — the "evening star" — appears in the sky towards the west after sunset, or in the morning sky towards the east before sunrise — the "morning star" — it is probably Venus. It is often the first bright object visible, before any stars appear in the sky. Mercury also appears in these areas of the sky around sunrise and sunset, but it never looks as bright as Venus nor gets as far from the sun as Venus does. Mercury appears only during twilight and Venus never remains visible through the night.

Whenever a very bright yellowish-white point of light appears in the sky in the middle of the night, it is probably Jupiter. Unlike Mercury and Venus, Jupiter is not always near the sun in the sky; it can appear high in the sky at midnight. Mars and Saturn can also appear far from the sun in the sky, rising well after sunset;

Mars rarely outshines Jupiter, though, and the brightness of Saturn never equals that of Jupiter or Venus. Mars can often be distinguished by the fact that it has a slight but distinct reddish tinge. Saturn, on the other hand, appears to be yellowish. The other planets are too faint to be seen with the naked eye.

2. *Twinkling.* Planets usually seem to shine steadily, while stars twinkle. Twinkling is an effect of turbulence in the earth's atmosphere: the atmosphere bends the starlight passing through it, and, as small regions of the atmosphere move about, the intensity of a star's light varies slightly but rapidly. Observations with a telescope would also reveal that a star appears to move around slightly. The reason why stars twinkle and planets do not is that stars are so far away that they look like points even when viewed through large telescopes; planets, though, are close enough to earth that their telescopic images are tiny disks. The light from different parts of a planet's disk averages out and makes the planet appear relatively steady in both brightness and position.

If the atmosphere is especially turbulent, or if you are looking through an especially large amount of atmosphere (when you are looking at an object low in the sky, for example, making your line of sight pass obliquely through the atmosphere), even planets can seem to twinkle. Under these conditions, the object you are observing may even seem to change in color — when Venus is low in the western sky, it is not uncommon to see it change from greenish to reddish and back again.

3. *Location.* All the planets always appear close to an imaginary line across the sky, so objects located far from that line cannot be planets. The line is called the *ecliptic,* and it is followed (more or less) not only by the planets but also by the moon. (The ecliptic is actually the path followed by the sun across the background of stars in the course of the year.) Since the earth is but one of the planets, and since all the planets orbit the sun in approximately the same plane, from our point of view the planets and sun must follow roughly the same line across the sky. The moon orbits the earth at only a slight angle to the plane of the planets, so it too always appears close to the ecliptic. The location of the ecliptic is plotted as a dotted line on the Monthly Sky Maps in Chapter 3, which show how the sky looks to the naked eye at different times.

From northern temperate latitudes, including the continental U.S., Canada, and Europe, the ecliptic crosses the southern part of the sky. This means that any bright objects at the *zenith* — the point directly over your head — or in the northern sky cannot be planets. (There are occasional exceptions to this if you are observing from the southernmost parts of the U.S.)

Now that you know how to tell whether you are looking at a star or a planet, you can look around the sky and identify some of the brightest stars. Some people find it easier to identify a few individual bright stars. Others prefer to locate a few favorite constellations or *asterisms* — a few stars, also roughly in the same direction

Fig. 1. The Big Dipper, with an aurora in the sky. Note that the middle star in the handle is double; the fainter star, Alcor, is above the brighter star, Mizar. (Dennis Milon)

from us, that are parts of one or more constellations.

Many people can identify one or two specific constellations or asterisms, even though they don't know any other constellations. (This statement holds true for many professional astronomers.) The most prominent asterism in the sky is the Big Dipper, whose seven stars trace out the shape of a dipper in the northern sky (Fig. 1). The Big Dipper is an asterism rather than a constellation because it makes up only part of the constellation Ursa Major, the Big Bear (Fig. 6, p. 21).

The four stars in the bowl of the Big Dipper make a squarish (actually trapezoidal) shape about 10° across. (Ten degrees is about the width of your fist, if you hold it up at arm's length against the sky.) Curving away from the bowl are the three stars in the handle, which cover another 15°. All the stars in the Big Dipper except the one that connects the handle to the bowl are of about the same brightness, which makes it easy to single out the Dipper in the sky.

Sky observers — including both professional and amateur astronomers — usually express star brightness in magnitudes, the scale of which is described in detail in Chapter 3. The lower the magnitude, the brighter the star. The brightest stars in the sky are magnitude zero (0), or in two cases, magnitudes −0.7 and −1.4, respectively. Figures 3 and 4 on pp. 12–15 show the brightest stars in the sky; the faintest star shown is magnitude 3.5. The naked eye can see stars about 10 times fainter than this, down to those as dim as 6th magnitude under perfect sky conditions.

One difference between the maps or charts in this guide and the real sky is that all the stars in the sky look like points, even though they have different brightnesses. The charts and maps in this guide represent these different brightnesses (magnitudes) as circles of different sizes.

It is often interesting to begin by identifying the brightest star near the zenith. Table 1 (p. 8) lists the 21 brightest stars in the sky. Opposite that table is a display — a Graphic Timetable (Fig. 2) — that shows when the brightest stars visible from midnorthern latitudes are passing their highest points in the sky. On any given date, different stars will be overhead at different times of night; the whole sequence changes with the seasons as the earth orbits the sun. The positions of the stars repeat from year to year.

To use the Graphic Timetable of the Brightest Stars (Fig. 2), run your finger down the side to find your date of observation, then move across the page to find the time of night when you are observing. You will see the names of the brightest stars that are transiting at about that time. An object *transits* when it passes your *meridian* — the imaginary line passing from the point due north on the horizon through the zenith to the point due south on the horizon.

Figure 2 also shows how high the stars are in the sky, in degrees

Table 1. The Brightest Stars in the Sky

Rank	Star	Constellation	Magnitude	r.a. dec. (epoch 2000.0)	
1	Sirius	Canis Major	− 1.46 (dbl)	6ʰ 45ᵐ	− 16° 43′
2	Canopus*	Carina	− 0.72	6ʰ 24ᵐ	− 52° 42′
3	Rigil Kent*	Centaurus	− 0.27 (dbl)	14ʰ 40ᵐ	− 60° 50′
4	Arcturus	Boötes	− 0.04	14ʰ 16ᵐ	+ 19° 11′
5	Vega	Lyra	+ 0.03	18ʰ 37ᵐ	+ 38° 47′
6	Capella	Auriga	+ 0.08	5ʰ 17ᵐ	+ 46° 00′
7	Rigel	Orion	+ 0.12 (dbl)	5ʰ 15ᵐ	− 8° 12′
8	Procyon	Canis Minor	+ 0.38	7ʰ 39ᵐ	+ 5° 14′
9	Achernar*	Eridanus	+ 0.46	1ʰ 38ᵐ	− 57° 14′
10	Betelgeuse	Orion	+ 0.50 (var)	5ʰ 55ᵐ	+ 7° 24′
11	Hadar*	Centaurus	+ 0.61	14ʰ 04ᵐ	− 60° 22′
12	Altair	Aquila	+ 0.77	19ʰ 51ᵐ	+ 8° 52′
13	Aldebaran	Taurus	+ 0.85 (var)	4ʰ 36ᵐ	+ 16° 31′
14	Acrux*	Crux	+ 0.87 (dbl)	12ʰ 27ᵐ	− 63° 05′
15	Antares	Scorpius	+ 0.96 (var)	16ʰ 29ᵐ	− 26° 26′
16	Spica	Virgo	+ 0.98	13ʰ 25ᵐ	− 11° 10′
17	Pollux	Gemini	+ 1.14	7ʰ 45ᵐ	+ 28° 02′
18	Fomalhaut*	Piscis Austrinus	+ 1.16	22ʰ 58ᵐ	− 29° 37′
19	Deneb	Cygnus	+ 1.25	20ʰ 41ᵐ	+ 45° 17′
20	Mimosa*	Crux	+ 1.25	12ʰ 47ᵐ	− 59° 41′
21	Regulus	Leo	+ 1.35	10ʰ 08ᵐ	+ 11° 58′

Notes: (dbl) = double star; combined magnitude of components given.

(var) = variable star; brightest magnitude given.

r.a. = right ascension, in hours and minutes (see p. 412).

dec. = declination, in degrees and minutes (see p. 412).

* = star not visible from midnorthern latitudes. The Graphic Timetable (opposite) shows when the brightest northern stars can be seen above the horizon from midnorthern latitudes; the maximum altitude each star reaches above the northern or southern horizon (for observers at 40° N latitude) is also given, in parentheses.

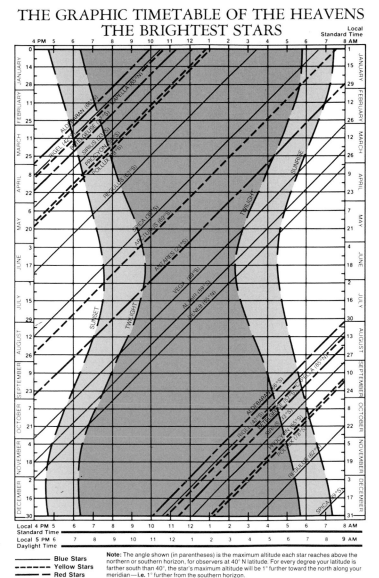

THE GRAPHIC TIMETABLE OF THE HEAVENS
THE BRIGHTEST STARS

Blue Stars ——————
Yellow Stars – – – – –
Red Stars ——————

Note: The angle shown (in parentheses) is the maximum altitude each star reaches above the northern or southern horizon, for observers at 40° N latitude. For every degree your latitude is farther south than 40°, the star's maximum altitude will be 1° further toward the north along your meridian—**i.e.** 1° further from the southern horizon.

Fig. 2. Graphic Timetable of the Brightest Stars (© 1982 Scientia, Inc.)

above the horizon, at their time of transit, for an observer at 40° N latitude. This *altitude* above the horizon is the highest point that each star reaches in the arc it traces across the sky. For example, Sirius, the brightest star in the sky, reaches a maximum of 33° above the southern horizon — slightly more than $\frac{1}{3}$ the altitude of the zenith. Since your fist covers about 10° of sky (when you place your thumb flat on the outside of the fist and hold it at arm's length), you can mark off the altitude above the horizon in 10° segments. You may want to verify first that about nine of your fists indeed cover the 90° from horizon to zenith.

In the region near the ecliptic, a bright object could be a star or a planet. When looking in this part of the sky, do make sure you know which planets are up — above the horizon. (The Graphic Timetables in Chapter 9 provide this information.)

Figures 3 and 4 are pairs of sky maps centered on the north celestial pole and on the south celestial pole, respectively. The *celestial poles* are the imaginary points where the earth's axis, if extended, would meet the celestial sphere. The north and south celestial poles lie above the earth's north and south poles, respectively. As the earth rotates, the sky appears to rotate in the opposite direction around the celestial poles. The sky thus seems to rotate once around the celestial poles every 24 hours. Midway between the celestial poles is the *celestial equator,* which lies on the celestial sphere, above the earth's equator. The celestial equator separates the northern and southern halves of the sky.

Figure 3 shows the northern half of the celestial sphere, and is spread across two pages with some overlap between. This map is centered on the north celestial pole. Below that pole is the Big Dipper. Since the sky appears to rotate around the north or south celestial pole (depending on which hemisphere you are in), whichever pole you can see always remains at a constant height in the sky. (If you are observing from a latitude of 40° N on earth, the north celestial pole will always be tilted up 40° above due north on the horizon; if you are observing from a latitude of 30° N, the pole will always be tilted up 30°, etc.) Observers at midnorthern latitudes will see the Big Dipper appear to revolve around the north celestial pole every 24 hours. For these observers, the Big Dipper is close enough to the north celestial pole that it will never set, and is thus an example of a *circumpolar* asterism.

Figure 4 shows the southern half of the celestial sphere. It includes some stars (near the celestial equator) that midnorthern observers can sometimes see, and some stars that never rise above the horizon at northern latitudes.

The Big Dipper is a particularly handy asterism to know because you can follow lines marked out by its stars and trace them across the sky to other interesting objects. Best known is the line marked by the two stars at the end of the bowl, which are known as *the Pointers.* These two stars point to the North Star, Polaris; to find Polaris, follow a straight line from the Pointers upward from the bowl of the Dipper and try to imagine the line curving slightly as it follows the curve of the sky for about 30°. (This is three fists' width, or about five times the distance between the Pointers, which are separated by $5\frac{1}{2}$°.) Polaris is at the end of an asterism known as the Little Dipper. None of the stars in the Little Dipper is as bright as the five brightest stars of the Big Dipper; the back two stars of the bowl and the two stars between the bowl and Polaris may be hard to see with your naked eye.

Polaris is not an especially bright star, but it is bright enough to be visible ordinarily. It is the brightest star in that region of the sky, so it is not easily confused with other stars. Polaris is within 1° of the true north celestial pole, and is thus of help not only to navigators at sea but also to land-based amateurs navigating around the sky. If you face Polaris, you are facing north. Thus it is best to find Polaris in order to orient yourself before you use any of the charts or maps in this guide.

If you continue along the arc from the Pointers through Polaris, you will come to the Great Square of Pegasus. This pathway and others that you can follow from one constellation to another are marked with dotted lines on Figs. 3 and 4. For example, instead of following the Pointers to Polaris, you can follow the curve of the Big Dipper's handle over about 30° (three fists' width, thumb included) of sky to the bright star Arcturus. If you can follow the same arc for another 30° without hitting the horizon, you will come to the bright star Spica. To remember this, think of "arc to Arcturus," and then "spike to Spica."

If, instead of finding Polaris, you follow the Pointers or the two stars that form the rear of the Big Dipper's bowl in the opposite direction, you will come to the constellation Leo, the Lion, about 35° away. Leo contains the bright star Regulus, which is located at the base of the "sickle" in Fig. 3. You can find other stars and constellations using the pathways marked on Figs. 3 and 4; the angles between some of the stars and constellations are listed below.

Table 2. Angles in the Sky

top stars of bowl of Big Dipper	10°
Pointers of Big Dipper	$5\frac{1}{2}$°
Castor and Pollux (in Gemini)	$4\frac{1}{2}$°
Great Square of Pegasus, width	17°
end stars of Orion's belt	3°
Orion's belt to Betelgeuse or Rigel	9°
end of Orion's belt to Sirius	21°

Note: One fist (thumb included) covers about 10° of sky.

Fig. 3. The brightest stars in the northern half of the sky, with arrows showing some of the pathways that help observers locate and identify them. (Wil Tirion)

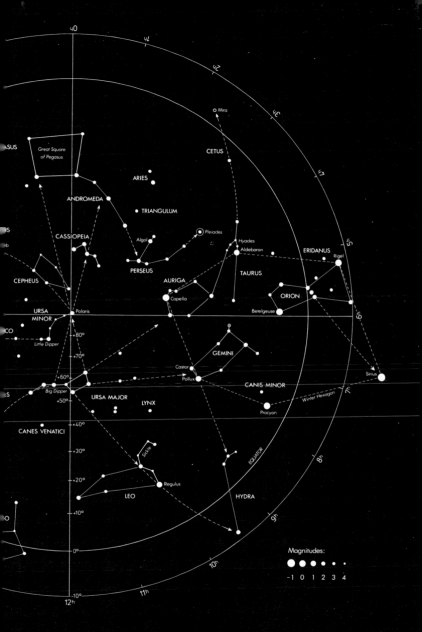

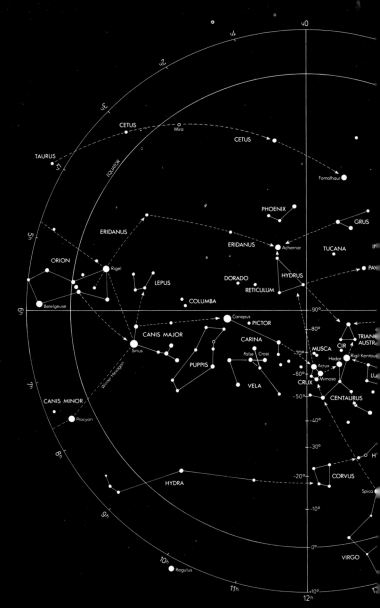

Fig. 4. The brightest stars in the southern half of the sky, with arrows showing some of the pathways that help observers locate and identify them. (Wil Tirion)

14

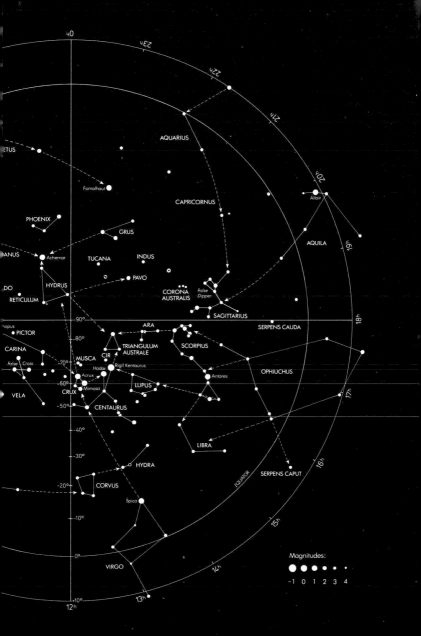

One other asterism is particularly striking and may grab your attention as you scan the sky. This prominent grouping is the straight line marked by three bright stars separated by a total of only 3°; it appears at the extreme right of Fig. 3 and at the extreme left of Fig. 4. These stars mark the belt of Orion, the Hunter, and appear in the evening sky in winter. One of the stars lies on the celestial equator, which explains why Orion is well into the southern part of the sky for an observer at a midnorthern latitude.

About 10° to the north of Orion's belt is the bright reddish star Betelgeuse, and almost 10° to the south of Orion's belt is the bright bluish star Rigel. (Many people pronounce Betelgeuse as "beetle-juice"; others say "beh-tel-jooz." Rigel is pronounced "ry-jel," with the accent on the first syllable.) We shall see in Chapter 4 that the colors of stars reveal their temperatures; Betelgeuse is a cool star and Rigel is a hot one. If you follow the line made by Orion's belt to the east (or left, since you are facing south), you will soon come to the bluish-white star Sirius, the brightest star in the sky.

Those who use this Field Guide extensively will quickly find themselves interested not only in stars and constellations but also in double and variable stars, star clusters, nebulae, and galaxies, all of which are described in Chapter 4. (A set of Graphic Timetables, showing when a few of the most prominent examples of these objects are visible above the horizon also appears in that chapter.) Nebulae — clouds of gas and dust that appear as hazy regions in

Fig. 5. Orion, the Hunter, from the star atlas of Bayer; the first edition was published in 1603. (Smithsonian Institution Libraries)

the sky or as dark masses that we see in silhouette — are fascinating to observe. Farther away in space are galaxies — giant groups of stars, gas, and dust. We live in the Milky Way Galaxy; from our vantage point inside it, the stars, gas, and dust in our galaxy stretch across the sky as a faint band known as the Milky Way. We explain the importance of nebulae and galaxies to the science of astronomy in Chapter 4; they are so beautiful, though, that you don't have to know anything scientific about them to enjoy observing them.

Now that we have had a first look at the sky, let us follow a detailed tour of the sky that we could take at 11 p.m. or midnight in late July or earlier in the evening in August or September. You can use Monthly Sky Map 7 in Chapter 3 to follow this tour at any of the times listed at the lower left of the map. In the next chapter, we set out several less-detailed seasonal tours.

The Late Summer Sky in Detail

Sit and face roughly north. Halfway around to your left will be the Big Dipper; its handle will come in from the left and curve downward, and the bowl will face up toward the right. Follow a line connecting the two Pointers for about 30° (three fists' width) across the sky to the first reasonably bright star, Polaris.

Now continue the same arc to the right and a little further down; you will come to a W in the sky, lying on its left side. The five brightest stars in the W are of about the same brightness. The first two lie in a more or less horizontal line; the third star is up to the right. Look horizontally to the right again to find the fourth star. (A fainter star is in the middle and a little below the line joining the third and fourth stars.) The fifth star is up to the left. These stars outline Cassiopeia (Fig. 7, p. 23).

Cassiopeia is in the Milky Way, which you can see extending upward from Cassiopeia if the sky is dark. The Milky Way appears as a faint band to the naked eye; you can see that it is irregular and contains some dark splotches.

As you look upward along the Milky Way, the first bright star you will come to that is brighter than the stars in Cassiopeia is Deneb in the constellation Cygnus, the Swan. Deneb marks one end of the Northern Cross. If you continue a little further — about three fingers' width (6°) — into the Northern Cross, you will come to another star (whose name is rarely used). Deneb, that next star, and two much fainter stars (further in that direction) form the main axis of the Cross. In the perpendicular direction, at about four fingers' width (8°) to the left and to the right of that second star, are stars marking the crosspiece of the Cross.

You are looking way over your head now; if you look a little further up over the Milky Way and then to your left you will see the brightest star at the top of the sky, which is Vega, in the constellation Lyra, the Lyre. Vega is noticeably brighter than Deneb, though both stars are very bright. Go back to Deneb again and continue along the Milky Way, so that you are now looking way over your head. If you look down to the right a bit, you will find Altair, a star about as bright as Deneb. Deneb, Altair, and Vega make up the Summer Triangle.

At Altair, you are looking way over the top of the sky, so you may as well turn around to face south. Now you are looking more than halfway up in the southern sky. The Milky Way, with its black separation down the middle, is prominent, and Altair is a little to the left of the Milky Way's left edge. It is helpful to find the Summer Triangle again for orientation.

Now turn around again to face the North Star (Polaris), and look back at Cassiopeia. If you continue to the right along the horizontal line marked by the two lower stars of the W, you will come to the constellation Andromeda, which extends upward. If you continue to the right (eastward) past Andromeda, you will come to the Great Square of Pegasus, which is still low in the eastern sky at this time of night, at this time of the year. The Great Square looks like a diamond standing on one of its points (Fig. 8, p. 24); it is about 20° across the diagonal — two fists' width. The Square is especially prominent because there are few stars within it.

Now find the Big Dipper again. Instead of following the Pointers, as you did to find the North Star, follow upward the line defined by the two stars at the rear of the bowl. The first pair of bright stars that you will come to form the end of the bowl of the Little Dipper. Winding around those stars, from the top and around to the left, is an arc of stars that is part of the constellation Draco, the Dragon.

If you look further upward, your view will cross more stars in Draco, and then you will reach Vega again. Just 10° or so (a palm's width) to Vega's left are the stars in Hercules. Further to their left, in the shape of a C, is an arc of stars from the constellation Corona Borealis.

Below Corona Borealis, you will find a bright star that is lower in the west. This is Arcturus, which you can also find by following the arc marked by the handle of the Big Dipper (see p. 11).

We have just completed a network of paths around the sky. Networks like this help clinch identifications of stars and constellations. Earlier in the evening, you could have followed the arc from the Big Dipper through Arcturus, straight on to Spica. This makes a "spike to Spica," though Spica will have set (will be below the horizon) by the time of night described here. But the paths "arc to Arcturus" and "spike to Spica," shown with arrows on Figs. 3 and 4, are valid at any time when these stars are up.

When you face south, you won't see any constellations as prominent as those in the northern sky. The Milky Way continues down more or less perpendicular to the horizon at this time of year. About $\frac{2}{5}$ of the way up from the horizon is Sagittarius. It does not contain any especially bright stars and does not show a readily recognizable pattern, though some people see a teapot there. Sagittarius contains the brightest part of the Milky Way, since the center of our galaxy lies in that direction. You can also find Sagittarius by following the Milky Way from Deneb past Altair and then for an equal distance beyond Altair.

From left to right across the middle of the southern sky goes the ecliptic — the path near which the planets are likely to be found. You can consult the Graphic Timetables of the Heavens in Chapter 9 or check the planetary longitudes in Appendix Table A-8 to see exactly where the planets will be on any given night. The moon also follows a predictable path across the sky that is within a few degrees of the ecliptic. If the moon is out, its brightness will prevent you from seeing very many stars; a full moon is so bright that only the brightest stars in the sky can be seen.

To the right of Sagittarius you will find a reasonably bright star, Antares, at the right edge of the Milky Way. Antares has a reddish tinge. This color is the source of its name, which means "compared with Mars," the red planet. (The Greek name for Mars was "Ares.")

The constellation of Scorpius and a network of stars that wind around Antares make up a fisherman's hook. A telescope will reveal many interesting regions there; some of the most prominent star clusters and nebulae are in Sagittarius and Scorpius.

Now let us follow another network of paths. Facing south, look up to the top of the sky to Vega, and down to Altair. Continue past Altair for about an equal distance to find Capricornus, the Goat. A little bit to the left (east) is Aquarius, with no especially prominent stars. Continuing further to the left we come again to the Great Square of Pegasus.

If you try to count the stars — you may start with those in Cassiopeia — you will soon realize that there are hundreds of them in just that small area of sky. Rather than trying to count all of the stars, though, it is better to learn a few individual stars or groups of stars and get to know them well.

2

A Tour of the Sky

The sun dominates the daytime sky. Sunlight scatters throughout the atmosphere, making the sky blue. This blue sky is brighter than the stars behind it, so we cannot usually see the stars during the daytime. When the sky is clear enough, the moon can often be seen even in the daytime, especially if you know where to look. Chapter 8 discusses the phases of the moon and where to find the moon in the sky.

Shortly before sunset or just after sunrise, the sky becomes dark enough for us to see the brightest planets and stars. Venus is the brightest of these objects and is sometimes bright enough to cast noticeable shadows. If you see an exceedingly bright object in the west at or after sunset — the "evening star" — it is usually Venus. If you see an exceedingly bright object in the east before or at sunrise — the "morning star" — it is also usually Venus. A second bright object in the sky, usually even closer to the sun, may be Mercury.

Jupiter, Mars, and Saturn can also be prominent in the sky, and can be quite far from the sun, so planets visible late at night are from this trio. Mars' reddish tinge is subtle, yet not hard to notice even with the naked eye. Saturn's rings are noticeable with a small telescope but are not visible to the naked eye. Jupiter's moons, belts, and Great Red Spot also require a small telescope to be seen. To tell at a glance which planets are visible in the sky on your night of observation, you can use one of the Graphic Timetables in Chapter 9. The characteristics that will help you distinguish the planets from the bright stars nearby are described in Chapter 1.

Beyond the Solar System

As the sky darkens, the brightest stars become visible. In a city, only a few dozen stars may become visible because the sky remains very bright even at night. But far from the haze of pollution and the competing glimmer of city lights, about 3000 stars may be visible to the naked eye. An equal number of stars are hidden beyond the horizon and most can be viewed by waiting through the night or for a different time of year. The rest of the stars become visible only to observers closer to the equator or in the hemisphere opposite to that from which you are observing. The Atlas Charts and descriptions in Chapter 7 show the entire sky — the stars and constellations that are visible from the southern hemisphere as well as the ones that can be seen from the northern hemisphere. However, since most users of this guide are at midnorthern latitudes, much of the following discussion is designed for them.

The region of the sky near Polaris, the North Star, is visible from midnorthern latitudes all through the year. In this region you can easily see the Big Dipper, the asterism that makes up part of the constellation Ursa Major, the Big Bear. In the autumn evening sky, the bowl of the Big Dipper appears right-side-up, while in the evening sky in the spring, the bowl appears upside-down. Figure 6 shows the Big Bear, including the Big Dipper, from the celestial atlas Johannes Hevelius published in 1690.

On the front side of the bowl of the Dipper, the Pointers point to Polaris, which lies at the end of the handle of the Little Dipper. To find Polaris, follow a line that extends north from the Pointers for about 30°. Thirty degrees is $\frac{1}{3}$ of the distance between the horizon and the zenith. Extend your arm upward from horizontal $\frac{1}{3}$ of the way toward the zenith to get an idea of 30°. (Also, 30° is about

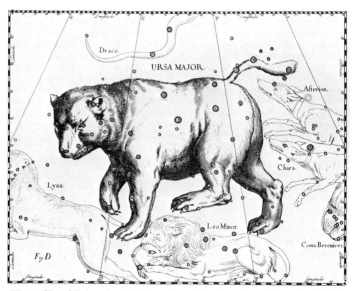

Fig. 6. Ursa Major, the Big Bear, from the star atlas of Hevelius (1690). The handle of the Big Dipper is in the Bear's tail, and the bowl is in its back. The constellation is drawn backwards from the way it appears in the sky because Hevelius drew the celestial sphere as it would appear on a star globe seen from the outside.

three widths of your hand at the end of your outstretched arm.) The altitude of Polaris above the horizon is equal to your latitude on the earth. For example, if you are at latitude 40°, Polaris is 40° above the horizon. If you are at latitude 90°, at the north pole, Polaris is 90° above the horizon, that is, directly overhead.

Polaris remains at this same location in the sky all year. Because its position is stationary, Polaris is visible throughout the year and appears on all the Monthly Sky Maps for the northern hemisphere in Chapter 3. You can use Polaris in all seasons as a reference point for locating other stars. The Big Bear, the Little Bear, and the other circumpolar constellations near Polaris also remain visible above the horizon all year.

Other groups of stars, as we will see below, are visible only during certain seasons of the year. Each night these stars rise above the eastern horizon, travel above and around Polaris, and then set below the western horizon. As the seasons progress, you can see successive groups of constellations making this journey across the sky.

Seasonal Tours

As you read each tour, please follow along with the suitable Monthly Sky Map in Chapter 3. Many of the invisible pathways in the sky we will follow are shown in Figs. 3 and 4 in Chapter 1.

The Autumn Sky (Monthly Map #8)

In the darkening sky on an autumn evening, the Pointers in the bowl of the Big Dipper point upward toward Polaris. From Polaris, follow the Little Dipper, which is upside-down. An American Indian legend holds that the autumn colors spill out of the Little Dipper at this time of year, making the trees turn bright colors.

Continue along the arc from the Pointers to Polaris for an equal distance on the other side of Polaris. You will find a prominent W-shaped constellation, Cassiopeia (Fig. 7). This constellation, like most others, was known by the ancient Greeks and was named after a character in Greek mythology. Cassiopeia was married to Cepheus, the King of Ethiopia, who has his own constellation, which is shaped like a house with a peaked roof, lying west of Cassiopeia.

If we continue upward from Cassiopeia, we find the constellation named after Andromeda, who in Greek mythology was Cassiopeia's daughter. In Andromeda, a faint hazy patch of light is sometimes visible to the naked eye. This light actually comes from the center of the Great Galaxy in Andromeda, the nearest galaxy to our own. (A galaxy is an enormous group of billions of stars, plus dust and gas.) The Andromeda Galaxy (M31) is much farther from us than any of the individual stars we see in the sky, so it marks the farthest that we can see with the naked eye. A telescope is necessary to reveal the spiral shape of this galaxy.

Fig. 7. Cassiopeia, from the star atlas of Bayer (1603). The brightest object shown was an exploding star — a supernova — visible only at the time the chart was drawn. (Smithsonian Institution Libraries)

Now look high in the sky, in a direction south of Andromeda. You will see four stars marking the corners of a square, known as the Great Square of Pegasus (Fig. 8, p. 24). Pegasus was the flying horse in Greek mythology.

If it is very dark out, you can also see the Milky Way passing high overhead, directly through Cassiopeia. The Milky Way — a grouping of stars, dust, and gas in our own galaxy — appears as a hazy band across the sky, with ragged edges, dark patches, and rifts.

Moving east from Cassiopeia along the Milky Way, you will come to the constellation Perseus. In Greek mythology, Perseus was a hero who slew the Medusa, flew off on Pegasus (who is conveniently located nearby in the sky), and then from his winged mount saw Andromeda, whom he saved from the sea monster, Cetus (see p. 136). Binoculars or a telescope will show you a pair of *open clusters,* each a group of many stars, close to each other in Perseus.

On the opposite side of Cassiopeia, beyond Cepheus along the Milky Way, a cross of bright stars is visible directly overhead. This

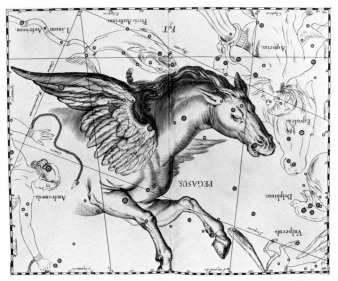

Fig. 8. Pegasus (drawn backwards), from the star atlas of Hevelius (1690). The Great Square is marked by the four largest star images: one (at the upper left) representing the wing; another, where the wing joins the neck; a third, at the top of the leg; and a fourth, now part of Andromeda's head.

prominent grouping of stars, known as the Northern Cross, lies in the constellation Cygnus, the Swan. This swan appears to be flying south, as are the birds at this time of year. Deneb is the star at the Swan's tail. Slightly to the west of the Swan, you will see the bright star Vega in the constellation Lyra, the Lyre. Vega is the third brightest star visible from midnorthern latitudes.

Farther westward, beyond Vega, you will come to the constellation Hercules, named for the Greek hero who performed 12 great labors. With binoculars or a small telescope, we can barely see an object that looks like a hazy mothball. This object is a *globular cluster,* a spherical group of thousands of stars. It is labelled M13, from its place in a catalogue of nonstellar objects — the Messier Catalogue (Table 13, p. 166), compiled in the 18th century by Charles Messier.

The Winter Sky (Monthly Map #11)

At dusk on a winter evening, the Big Dipper is low in the northern sky. The constellations that were easiest to see in the autumn now will appear closer and closer to the western horizon at the same hour on each successive night. By January 1, Cygnus, the Swan, sets in the western sky in the early evening; Cassiopeia is high overhead to our north; Perseus is even higher in the sky.

The Milky Way, the hazy band of gas and dust that gives our galaxy its name, also appears high in the early evening sky at this time of year. Since our home galaxy, the Milky Way Galaxy, is disk-shaped, we see many stars and much dust and gas when we look along the plane of the disk, but few stars and little dust and gas when we look in other directions. The next constellation to the southeast along the Milky Way is Auriga, the Charioteer, with its bright star Capella.

To the south of this part of the Milky Way, you can see a cluster of six or seven stars that are close together in the sky. Though faint, they will catch your eye as your vision sweeps across the sky. These stars are the Pleiades, the Seven Sisters of Greek mythology, the daughters of Atlas. The Pleiades are a cluster of over a hundred stars; the larger the binoculars or wide-field telescope you use, the more you will see. Atlas Chart 10 in Chapter 7 is supplemented by a special chart showing the brightest stars of the Pleiades.

Look further south, and turn this book around so that you are viewing the map of the southern sky on p. 77. (The right edge of the book will now be closest to you.) The most prominent group of stars in the sky in the wintertime is three bright stars in a straight line. They form the belt of Orion, the Hunter (Fig. 9). Extending down from Orion's belt is his sword. On Orion's shoulder you will see the reddish star Betelgeuse, one of the brightest stars visible from midnorthern latitudes. Symmetrically on the other side of Orion's belt, you will find the bluish star Rigel, marking Orion's heel. In Chapter 4, we will see that the colors of stars tell us their temperatures; reddish stars are relatively cool — about 3000°C — and bluish stars are relatively hot — over 10,000°C.

In the area of Orion's sword (Fig. 10) is a hazy region, the Orion Nebula, that is visible with binoculars or small telescopes. A *nebula* is a hazy region of sky that contains clouds of gas or dust (see p. 116). The Orion Nebula (C.Pl. 13) marks the presence of a cloud of gas and dust; in its vicinity new stars are now forming. Many nebulae are really "nurseries" for young stars that have recently been born, and are the site of ongoing stellar birth; we discuss them in more detail in Chapter 4.

Orion seems to be warding off Taurus, the Bull, a constellation you will find by looking beyond Orion's shield. Between the top of the shield and the Pleiades lies a V-shaped group of stars, the Hyades. The reddish star Aldebaran marks the end of one side of the V. The Hyades outline the face of Taurus; the Pleiades ride on the bull's shoulder. The Hyades and the Pleiades are both *open clusters* of stars (also called *galactic clusters*); these clusters are locations where perhaps 100 or more stars are close together in an irregularly shaped group.

At Orion's heel is his dog, Canis Major. Orion's belt points directly to Sirius, the brightest star in the sky. Rising soon after Orion, Sirius appears blue-white and is part of the constellation Canis Major, the Great Dog. Nearby is the yellow-white star Procyon in Canis Minor, the Little Dog. This bright star forms a nearly equilateral triangle with Sirius and Betelgeuse.

The Spring Sky (Monthly Map #2)

As spring approaches, Orion and the V-shaped Hyades move closer to the western horizon each evening, and eventually disappear from our view at sunset. Now you will see a pair of stars, the twins (Castor and Pollux), in the western sky at dusk. Although the two stars are about the same in brightness, Pollux appears slightly reddish, while Castor does not. These stars are in the constellation Gemini, the Twins, which was named after two Roman military gods.

South of the Twins, the bright star Sirius is prominent in the western sky, in the constellation Canis Major, the Great Dog. South of Sirius is the constellation Puppis, the Ship's Stern. This constellation includes a star, zeta Puppis, that is known to be one of the intrinsically brightest and hottest stars in the sky. Zeta

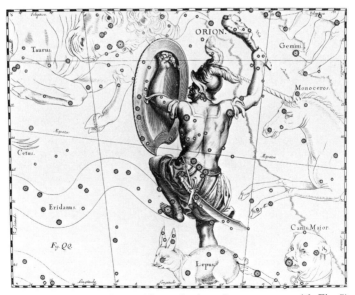

Fig. 9. Orion, the Hunter (drawn backwards — compare with Fig. 5), from the star atlas of Hevelius.

Fig. 10. Orion, photographed with a wide-angle telescope. Note the three stars in Orion's belt. Though Betelgeuse in his shoulder (upper left) and Rigel in his heel (lower right) appear to the eye to have similar brightnesses, Rigel is bluer and so appears brighter on blue-sensitive films such as the kind used here. Orion's sword with its famous nebulae appears directly below the middle star of the belt. (Harvard College Observatory)

Puppis is so far away, though, that it does not appear prominent when we observe this part of the sky. Puppis and several other constellations in this region and farther south were once considered to form a giant constellation of a ship, Argo Navis (Fig. 68, p. 254).

Ursa Major, the Big Bear, appears high in the northern sky on spring evenings, and is tipped so much that any imaginary things in the Big Dipper would spill out. If you follow the Pointers backward, you will see the constellation Leo, the Lion, just to the south of the zenith. To most people, Leo looks more like a backward question mark or sickle than a lion's head. A bright star, Regulus, is at the base of the question mark, at the lion's heart. East of Regulus, a triangle of stars marks the rest of Leo's body. Figure 11 shows how Leo should look. Many people visualize Leo as a lion in the sky with a sickle-shaped head and triangular tail.

On the northern star map, if you follow an arc begun by the stars in the Big Dipper's handle, you will come to a bright reddish star, Arcturus. This star lies in the constellation Boötes, the Herdsman. Farther along the arc from the Big Dipper through Arcturus, you will find another bright star, Spica. This blue-white star rises in the east-southeast in the constellation Virgo, the Virgin.

Look up again toward the northeast, to the lower left of Boötes and Hercules, which rise in the east in the evening. Here in the constellation Lyra you will see Vega, a star that is brighter than Spica.

The Summer Sky (Monthly Map #5)

On summer evenings at sunset, Vega is the brightest star near the zenith. Arcturus, very slightly brighter than Vega, is the reddish star that is also high in the sky to the west. The constellation Cygnus, which includes the prominent stars that make up the Northern Cross, lies east of Vega along the highest part of the Milky Way. Hercules lies about 10° to the west of Vega.

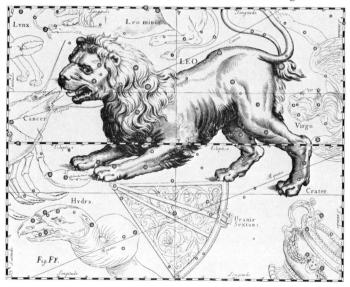

Fig. 11. Leo, the Lion (drawn backwards), from the star atlas of Hevelius.

Fig. 12. The Milky Way, from Cassiopeia (left) to Sagittarius (right). The bright objects above the Milky Way toward the right are stars and star clusters in Scorpius. (Palomar Observatory photo)

Continuing north along the Milky Way from Cygnus, you will find the W-shaped constellation, Cassiopeia. To the southeast of Cassiopeia is Andromeda, near the horizon. Andromeda's Great Galaxy, M31, may be faintly visible to the naked eye. The Great Square of Pegasus is farther south, on the same side of the Milky Way as Andromeda.

The bright star Spica lies toward the southwest, in the constellation Virgo, the Virgin. In the south you will find the bright reddish star Antares in the constellation Scorpius, the Scorpion. ("Antares" means "compared with Ares," the Greek name for Mars, because Antares is also reddish.) The stars of Scorpius wind a great distance across the sky.

To the east of Scorpius, you will see the constellation Sagittarius, the Archer, in which the center of our galaxy lies. The farther south you are, the higher in the sky Sagittarius rises, and the better you can see the beautiful star clouds in the direction of the center of the Milky Way (Fig. 12).

Moving south along the Milky Way, you will come to the star Altair in the constellation Aquila, the Eagle. Altair, Deneb (in the constellation Cygnus), and Vega make up the Summer Triangle.

Every summer around August 12, the Perseid meteor shower occurs. At the height of this shower, as many as one bright meteor per minute may be visible in pre-dawn hours. The meteors streak across the sky like shooting stars or falling stars, though they are only bits of interplanetary dust burning up in the earth's atmosphere. The dates of meteor showers are listed in a table in Chapter 12.

Now that you have followed these seasonal tours, you can use the Monthly Sky Maps in the next chapter to find your way around the sky at any time of year.

3

The Monthly Sky Maps

When we look at the stars at night, we have the feeling that we are underneath a giant bowl. Although the stars are really at many different distances from the earth, which spins underneath them, it is often convenient to picture the sky as the ancients did — as stars attached to a giant sphere rotating overhead. The set of constellations we see appears to rotate $\frac{1}{12}$ of the way around the sky each month, so at the same time of night one month later, we see an additional 30° of constellations in the east and lose sight of 30° of constellations that have set in the west.

It is, of course, very difficult to flatten out a sphere so that it can be reproduced on a flat page. The distortions in the Mercator projection of the earth, in which Greenland is made to look as large as South America merely because it is located much closer to a pole, are familiar to us all.

We have chosen a new way of drawing these star maps to minimize distortions in the regions of the sky that are most often studied. Star maps are made by projecting the positions of stars onto a flat plane as though their shadows were being cast from a given point. We have chosen this point so that the position approximately halfway up the sky to the zenith (45° in altitude) has minimum distortion. Though the constellations near the boundaries of the maps are expanded, the amount of the expansion is not too great, and constellations near the horizon are not compressed. Also, because of the projection used, when you follow a pathway from one constellation to another using certain pairs of stars, the angles between the paths will be the same on the maps as they are in the sky.

How to Use the Maps

These Monthly Sky Maps are designed to be easy to use; no knowledge of celestial coordinates is necessary. Most of the objects plotted on these maps are of magnitude 4.5 or brighter — bright enough to be seen with the naked eye. Seventy-two maps are included. The first 48 are for northern-hemisphere observers; they are specially marked for latitudes between 10° N and 50° N, and can also be used somewhat outside this range.

The second set of 24 maps is for observers in the southern hemisphere; these maps are specially marked for latitudes between the equator and 40° S latitude, and can also be used somewhat outside this range. The words "Sky Map" and the map number are printed in reverse (black on white, not white on black) on these maps, so you can distinguish the southern-hemisphere maps from the northern-hemisphere ones at a glance.

Across the bottom of each map are a series of curved lines that represent the horizon for observers at different latitudes. In the upper middle portion of each map, the zenith (the point directly overhead) for observers at each latitude is marked with a plus sign. The directions east and west are marked at the edges.

We have provided a considerable amount of extra sky past the zenith on each map, to make it easier to orient yourself. You will soon become used to the locations of the northern and southern horizons and the zenith for your latitude. Some people will find it helpful to cut masks out of opaque paper and use them to hide the stars below the horizons for their latitude.

Each set of maps is valid for specific times on specific days, as shown in Table 4 (pp. 34–35). For example, Sky Map 1 — which is valid in our ordinary system of timekeeping for midnight on January 1 — is drawn for *sidereal time* (time by the stars, as described in Chapter 14) of 6^h40^m. Each succeeding map (Sky Map 2, Sky Map 3, and so on) is valid two hours later on the same date or one month later at the same time. Some of the times of night at which the maps are valid are shown to the lower left of the left-hand pages. Arrows near the east and west points show the direction in which the sky rotates through the night.

Each Monthly Sky Map for the northern hemisphere actually consists of two pairs of maps: The first pair has the constellation figures drawn on it; the second pair shows the same stars and omits the outlines of the constellation figures. (For the southern hemisphere, all the maps show constellation outlines.) In each pair of maps, the left-hand page shows what you will see when facing north and the right-hand page shows what you will see when facing south. When facing north, hold the book with its left edge toward you and look down at the map. In ⁺his orientation it will correspond to the sky. When facing south, hold the book with its right edge toward you and look down at the map to have it correspond to the sky.

On the maps, stars of different brightnesses (magnitudes) are shown as dots of different sizes. We describe the magnitude scale of brightness below. The maps in this section include only stars that are bright enough to be visible to the naked eye. In addition, the Milky Way is shown, and a few other objects of special interest. These objects are also shown and described in Chapter 7, where a set of Atlas Charts breaks down the sky into 52 sections. The Atlas Charts (primarily for telescope observers) are more detailed than the Monthly Sky Maps in this chapter; each Monthly Sky Map shows a view of half the sky visible at a given time.

The Magnitude Scale

Astronomers describe the brightness of stars with a scale built on a historical base. In the second century B.C., the Greek astronomer Hipparchus said that the brightest stars were "of the first magnitude," the next brightest group of stars were "of the second magnitude," and so on. The faintest stars visible to the naked eye were "of the sixth magnitude."

This scale was placed on a mathematical basis in the mid-19th century. Measurements showed that a difference of 5 magnitudes corresponded to a factor of about 100 in brightness; the current magnitude scale is defined so that a factor of 100 corresponds to exactly 5 magnitudes. A few stars are even brighter than first magnitude and have been accommodated by having magnitudes of 0 and then negative numbers on the scale. The brightest star in the sky is Sirius, whose magnitude is -1.4. Canopus, the second brightest star, is not visible north of the southern U.S. and has a magnitude of -0.7. Alpha Centauri, the third brightest star (also not visible from midnorthern latitudes), and Arcturus, the fourth brightest star, are slightly brighter than magnitude 0.0. Another dozen stars are fainter than magnitude 0.0 but brighter than magnitude 1.0. Most stars are much dimmer; the number of stars for each whole unit of magnitude increases rapidly as we go to fainter magnitudes.

The magnitude scale is different from most scales we commonly use in that increasing in brightness by one unit on the magnitude scale corresponds to multiplying by a fixed number (about 2.51) on a scale of brightness given in units of energy. For example, if we consider first a star of 3rd magnitude and then ask how much

Table 3. The Magnitude Scale

difference in magnitudes	factor in brightness	
1 mag	2.512	times
2 mag	6.31	times
3 mag	15.85	times
4 mag	39.81	times
5 mag	100	times
6 mag	251	times
7 mag	631	times
8 mag	1585	times
9 mag	3981	times
10 mag	10,000	times
15 mag	1,000,000	times

brighter a star of 2nd magnitude is, we have subtracted 1 on the magnitude scale. The result: the star of 2nd magnitude is about 2.5 times brighter than the star of 3rd magnitude. Thus for each magnitude added or subtracted, stars are about 2.5 times fainter or brighter, respectively. By definition, for each 5 magnitudes added, stars are exactly 100 times fainter. (Thus each magnitude corresponds exactly to the fifth root of 100, or 2.512 . . . , which is about 2.5 times brighter or fainter.) An increase or decrease of two magnitudes corresponds to $(2.512)^2$, or a little over 6 times; three magnitudes corresponds to $(2.512)^3$, or a little over 15 times, etc., as shown in Table 3.

Although Sirius, at magnitude -1.4, is the brightest star, the moon and some of the planets get somewhat brighter in the sky. Venus can be as bright as magnitude -4.4. The full moon is magnitude -12.6, and the sun is magnitude -26.8. Note that the smaller the magnitude number (or the more negative it is), the *brighter* the object is.

Going to fainter magnitudes, 6th magnitude is the faintest that the naked eye can see under the best observing conditions. A medium-sized telescope (with a lens or mirror 6–10 inches, 15–25 cm, in diameter) will allow you to see stars of 10th or 12th magnitude. The best ground-based telescopes can observe to about 24th magnitude. The Space Telescope, scheduled for launch in 1986, should enable us to observe 28th-magnitude objects.

You don't need a telescope to use the Monthly Sky Maps that follow; they are designed for observations with the naked eye or binoculars. If you want to observe regions of the sky in greater detail, you can use the Atlas Charts in Chapter 7, which are designed primarily for telescope observations.

Table 4. Index to Monthly Sky Maps

Standard/D.S.T.	Jan.		Feb.		March		April		May		June	
	1	15	1	15	1	15	1	15	1	15	1	15
6 p.m./7 p.m. 18 h /19 h	10		11		12		1		2		3	
7 p.m./8 p.m. 19 h /20 h		11		12		1		2		3		4
8 p.m./9 p.m. 20 h /21 h	11		12		1		2		3		4	
9 p.m./10 p.m. 21 h /22 h		12		1		2		3		4		5
10 p.m./11 p.m. 22 h /23 h	12		1		2		3		4		5	
11 p.m./midnight 23 h /24 h		1		2		3		4		5		6
midnight/1 a.m. 24 h /1 h	1		2		3		4		5		6	
1 a.m./2 a.m. 1 h /2 h		2		3		4		5		6		7
2 a.m./3 a.m. 2 h /3 h	2		3		4		5		6		7	
3 a.m./4 a.m. 3 h /4 h		3		4		5		6		7		8
4 a.m./5 a.m. 4 h /5 h	3		4		5		6		7		8	
5 a.m./6 a.m. 5 h /6 h		4		5		6		7		8		9
6 a.m./7 a.m. 6 h /7 h	4		5		6		7		8		9	

Note: For the southern hemisphere Sky Maps, add 12 to the above Sky Map numbers.

Table 4 (contd.). Index to Monthly Sky Maps

July		August		Sept.		Oct.		Nov.		Dec.		
1	15	1	15	1	15	1	15	1	15	1	15	Standard/D.S.T.
4		5		6		7		8		9		6 p.m./7 p.m. 18 h /19 h
	5		6		7		8		9		10	7 p.m./8 p.m. 19 h /20 h
5		6		7		8		9		10		8 p.m./9 p.m. 20 h /21 h
	6		7		8		9		10		11	9 p.m./10 p.m. 21 h /22 h
6		7		8		9		10		11		10 p.m./11 p.m. 22 h /23 h
	7		8		9		10		11		12	11 p.m./midnight 23 h /24 h
7		8		9		10		11		12		midnight/1 a.m. 24 h /1 h
	8		9		10		11		12		1	1 a.m. /2 a.m. 1 h /2 h
8		9		10		11		12		1		2 a.m. /3 a.m. 2 h /3 h
	9		10		11		12		1		2	3 a.m. /4 a.m. 3 h /4 h
9		10		11		12		1		2		4 a.m. /5 a.m. 4 h /5 h
	10		11		12		1		2		3	5 a.m. /6 a.m. 5 h /6 h
10		11		12		1		2		3		6 a.m. /7 a.m. 6 h /7 h

Note: For the southern hemisphere Sky Maps, add 12 to the above Sky Map numbers.

SKY MAP 1
Facing North

Northern Latitudes

	TIME	D.S.T.
January 1	24 h	
January 15	23 h	
February 1	22 h	
February 15	21 h	
March 1	20 h	21 h
etc.		

EPOCH 2000.0

36

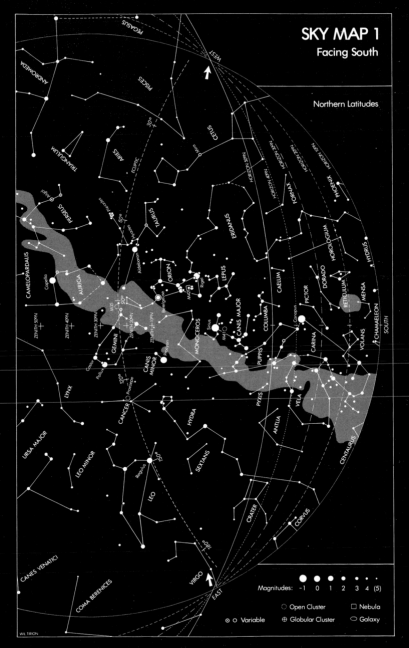

SKY MAP 1
Facing South

Northern Latitudes

Magnitudes: -1 0 1 2 3 4 (5)

◌ Open Cluster   □ Nebula
◉ ○ Variable   ⊕ Globular Cluster   ○ Galaxy

WIL TIRION

37

SKY MAP 1
Facing North

Northern Latitudes

	TIME	D.S.T.
January 1	24 h	
January 15	23 h	
February 1	22 h	
February 15	21 h	
March 1	20 h	21 h
etc..		

EPOCH 2000.0

38

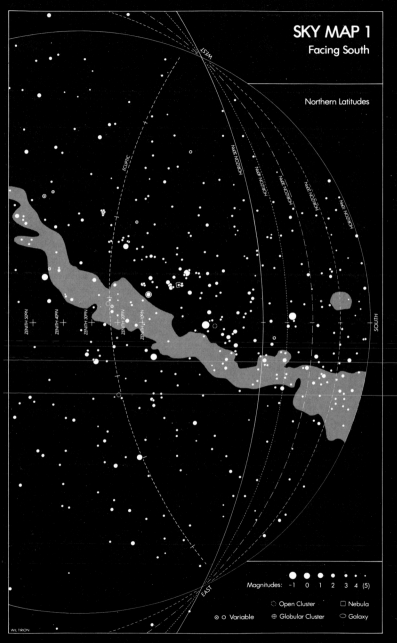

SKY MAP 1
Facing South

Northern Latitudes

ECLIPTIC

WEST

EAST

SOUTH

HORIZON 50°N
HORIZON 40°N
HORIZON 30°N
HORIZON 20°N

ZENITH 50°N
ZENITH 40°N
ZENITH 30°N
ZENITH 20°N
ZENITH 20°N

Magnitudes: −1 0 1 2 3 4 (5)

⊙ ○ Variable

◌ Open Cluster

⊕ Globular Cluster

□ Nebula

○ Galaxy

WIL TIRION

39

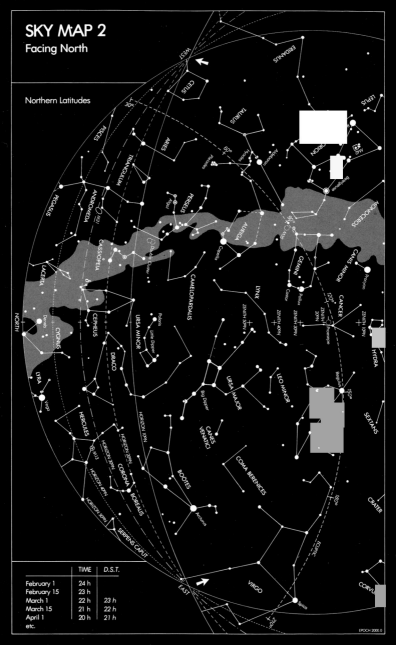

SKY MAP 2
Facing North

Northern Latitudes

	TIME	D.S.T.
February 1	24 h	
February 15	23 h	
March 1	22 h	23 h
March 15	21 h	22 h
April 1	20 h	21 h
etc.		

EPOCH 2000.0

40

SKY MAP 2
Facing South

Northern Latitudes

Magnitudes: -1   3 4 (5)

○ Open Cluster   □ Nebula
◉ ○ Variable   ⊕ Globular Cluster   ○ Galaxy

WIL TIRION

41

SKY MAP 2
Facing North

Northern Latitudes

WEST

NORTH

ZENITH 50°N
ZENITH 40°N
ZENITH 30°N
ZENITH 25°N
ZENITH 19°N

HORIZON 19°N
HORIZON 20°N
HORIZON 30°N
HORIZON 40°N
HORIZON 39°N

ECLIPTIC

EAST

	TIME	D.S.T.
February 1	24 h	
February 15	23 h	
March 1	22 h	23 h
March 15	21 h	22 h
April 1	20 h	21 h
etc.		

EPOCH 2000.0

42

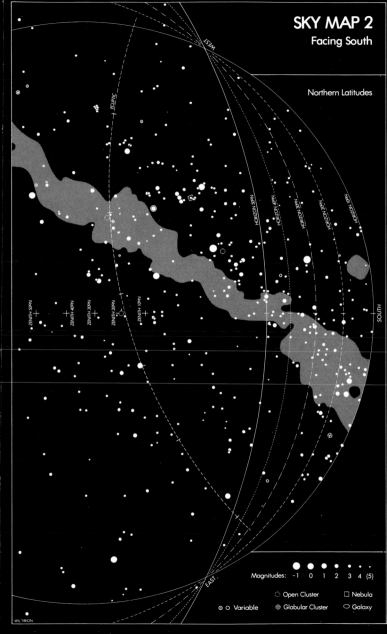

SKY MAP 2

Facing South

Northern Latitudes

Magnitudes: -1 0 1 2 3 4 (5)

◎ ○ Variable   ⊕ Globular Cluster   ○ Galaxy

◌ Open Cluster   □ Nebula

W.I.L. TIRION

43

SKY MAP 3
Facing North

Northern Latitudes

	TIME	D.S.T.
March 1	24 h	01 h
March 15	23 h	24 h
April 1	22 h	23 h
April 15	21 h	22 h
May 1	20 h	21 h
etc.		

EPOCH 2000.0

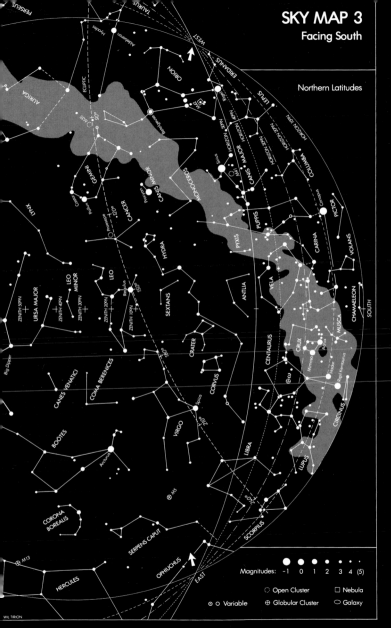

Magnitudes: -1 0 1 2 3 4 (5)

☼ Open Cluster   ▢ Nebula
⊙ ○ Variable   ⊕ Globular Cluster   ○ Galaxy

WIL TIRION

SKY MAP 3
Facing North

Northern Latitudes

WEST

NORTH

EAST

ECLIPTIC

ZENITH 50°N
ZENITH 40°N
ZENITH 30°N
ZENITH 20°N
ZENITH 10°N

HORIZON 50°N
HORIZON 40°N
HORIZON 30°N
HORIZON 20°N
HORIZON 10°N

	TIME	D.S.T.
March 1	24 h	01 h
March 15	23 h	24 h
April 1	22 h	23 h
April 15	21 h	22 h
May 1	20 h	21 h
etc.		

EPOCH 2000.0

46

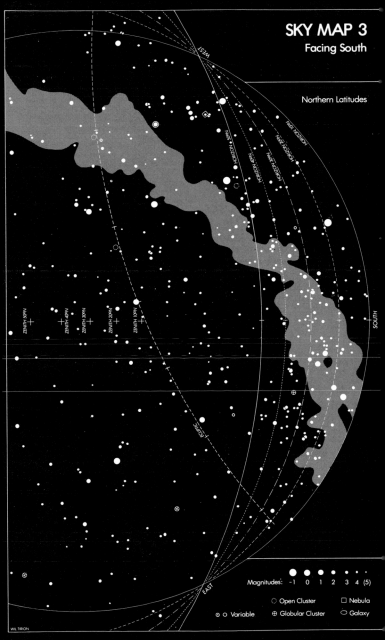

SKY MAP 3
Facing South

Northern Latitudes

WEST

EAST

SOUTH

HORIZON 10°N
HORIZON 20°N
HORIZON 30°N
TRUE HORIZON
HORIZON 50°N

ZENITH 50°N
ZENITH 40°N
ZENITH 30°N
ZENITH 20°N
ZENITH 10°N

ECLIPTIC

WIL TIRION

Magnitudes: -1 0 1 2 3 4 (5)

⦂ Open Cluster   □ Nebula

⦿ ○ Variable   ⊕ Globular Cluster   ○ Galaxy

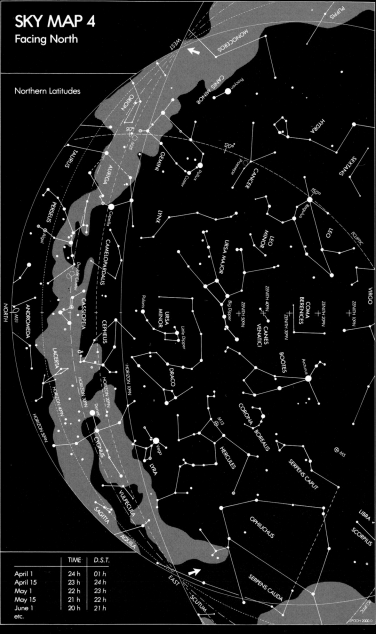

SKY MAP 4
Facing North

Northern Latitudes

	TIME	D.S.T.
April 1	24 h	01 h
April 15	23 h	24 h
May 1	22 h	23 h
May 15	21 h	22 h
June 1	20 h	21 h
etc.		

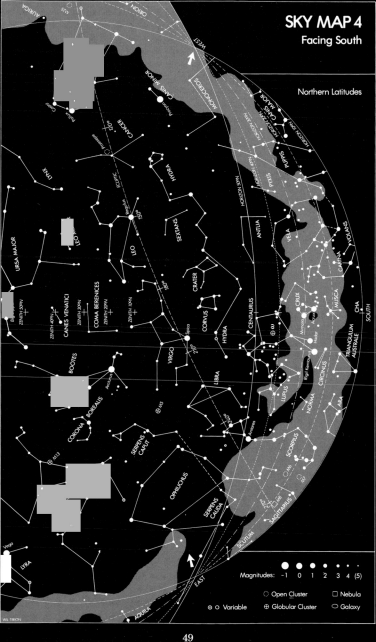

SKY MAP 4

Facing South

Northern Latitudes

Magnitudes: -1 0 1 2 3 4 (5)

◌ ○ Variable   ⊕ Globular Cluster   ○ Galaxy
◌ Open Cluster   □ Nebula

SKY MAP 4
Facing North

Northern Latitudes

WEST

ECLIPTIC

ZENITH 50N
ZENITH 40N
ZENITH 30N
ZENITH 20N
ZENITH 10N

NORTH

HORIZON 50N
HORIZON 40N
HORIZON 30N
HORIZON 20N
HORIZON 10N

EAST

	TIME	D.S.T.
April 1	24 h	01 h
April 15	23 h	24 h
May 1	22 h	23 h
May 15	21 h	22 h
June 1	20 h	21 h
etc.		

EPOCH 2000.0

SKY MAP 4
Facing South

Northern Latitudes

WEST

EAST

SOUTH

ECLIPTIC

HORIZON 50°N
HORIZON 40°N
HORIZON 30°N
HORIZON 20°N

ZENITH 50°N
ZENITH 40°N
ZENITH 30°N
ZENITH 20°N
ZENITH 10°N

Magnitudes: -1 0 1 2 3 4 (5)

⊙ ○ Variable

◌ Open Cluster   □ Nebula
⊕ Globular Cluster   ◯ Galaxy

WIL TIRION

51

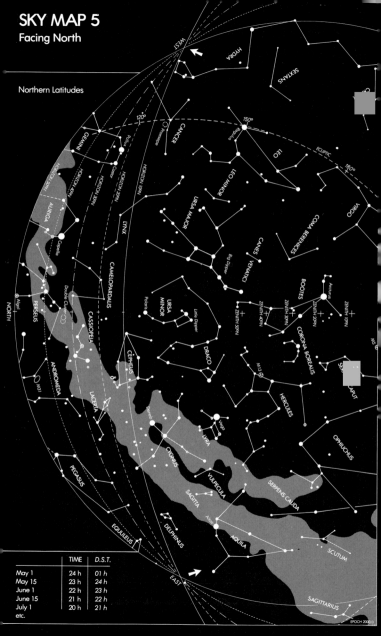

SKY MAP 5
Facing North

Northern Latitudes

	TIME	D.S.T.
May 1	24 h	01 h
May 15	23 h	24 h
June 1	22 h	23 h
June 15	21 h	22 h
July 1	20 h	21 h
etc.		

EPOCH 2000.0

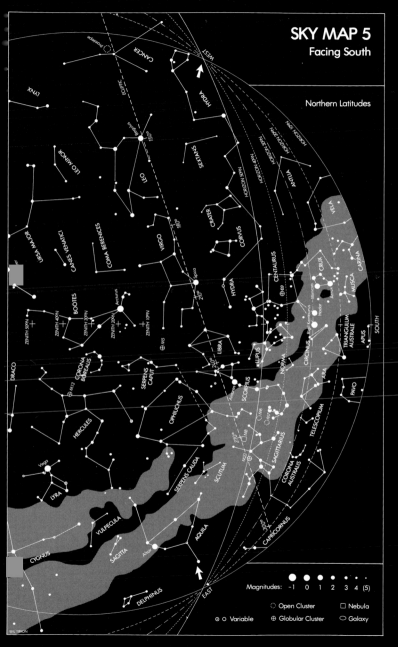

SKY MAP 5
Facing South

Northern Latitudes

Magnitudes: -1 0 1 2 3 4 (5)

◌ Open Cluster   □ Nebula

⊙ ○ Variable   ⊕ Globular Cluster   ○ Galaxy

W.L. BROXON

53

SKY MAP 5
Facing North

Northern Latitudes

WEST

ECLIPTIC

HORIZON NORTH
HORIZON 20°N
HORIZON 30°N
HORIZON 40°N
HORIZON 50°N

ZENITH 50°N
ZENITH 40°N
ZENITH 30°N
ZENITH 20°N
ZENITH NORTH

NORTH

EAST

EPOCH 2000.0

	TIME	D.S.T.
May 1	24 h	01 h
May 15	23 h	24 h
June 1	22 h	23 h
June 15	21 h	22 h
July 1	20 h	21 h
etc.		

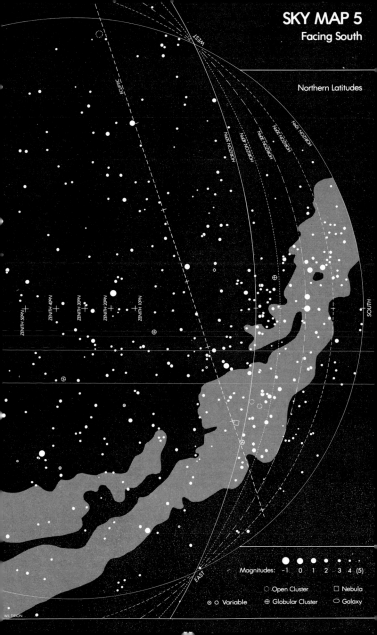

SKY MAP 5
Facing South

Northern Latitudes

ECLIPTIC

TRUE HORIZON
TRUE HORIZON
TRUE HORIZON
TRUE HORIZON

WEST

SOUTH

ZENITH 50°N
ZENITH 40°N
ZENITH 30°N
ZENITH 20°N
ZENITH 10°N

EAST

Magnitudes: -1 0 1 2 3 4 (5)

Open Cluster   □ Nebula
⊕ ○ Variable   ⊕ Globular Cluster   ○ Galaxy

WIL TIRION

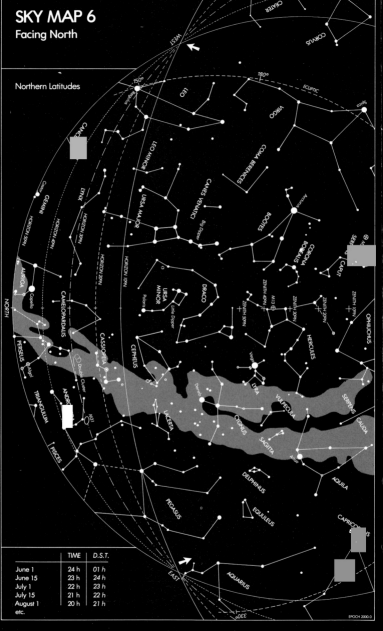

SKY MAP 6
Facing North

Northern Latitudes

	TIME	D.S.T.
June 1	24 h	01 h
June 15	23 h	24 h
July 1	22 h	23 h
July 15	21 h	22 h
August 1	20 h	21 h
etc.		

EPOCH 2000.0

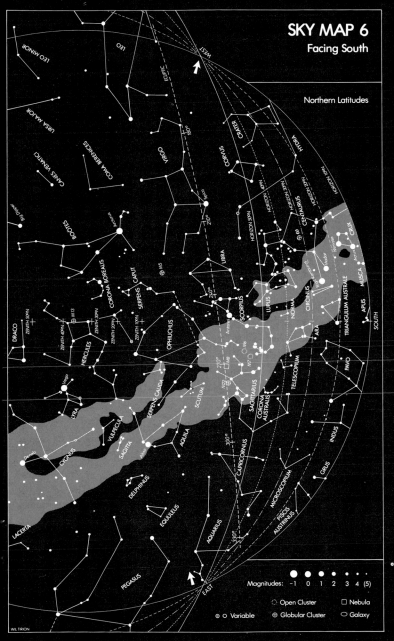

SKY MAP 6
Facing South

Northern Latitudes

Magnitudes: -1 0 1 2 3 4 (5)

☼ Open Cluster   □ Nebula
⊙ ○ Variable   ⊕ Globular Cluster   ○ Galaxy

WIL TIRION

57

SKY MAP 6
Facing North

Northern Latitudes

WEST

ECLIPTIC

HORIZON 20°N
HORIZON 30°N
HORIZON 40°N
HORIZON 50°N
HORIZON 60°N

NORTH

ZENITH 50°N
ZENITH 40°N
ZENITH 30°N
ZENITH 20°N
ZENITH 10°N

EAST

	TIME	D.S.T.
June 1	24 h	01 h
June 15	23 h	24 h
July 1	22 h	23 h
July 15	21 h	22 h
August 1	20 h	21 h
etc.		

EPOCH 2000.0

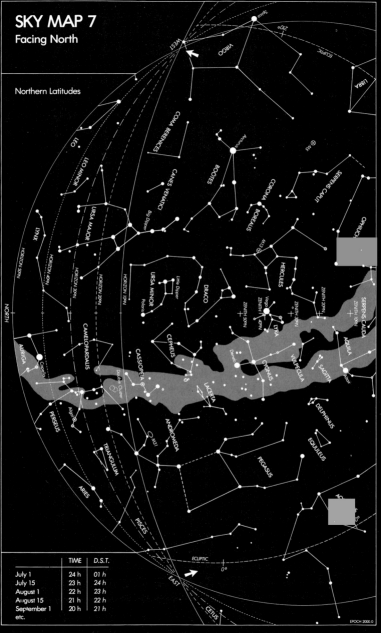

SKY MAP 7
Facing North

Northern Latitudes

	TIME	D.S.T.
July 1	24 h	01 h
July 15	23 h	24 h
August 1	22 h	23 h
August 15	21 h	22 h
September 1	20 h	21 h
etc.		

EPOCH 2000.0

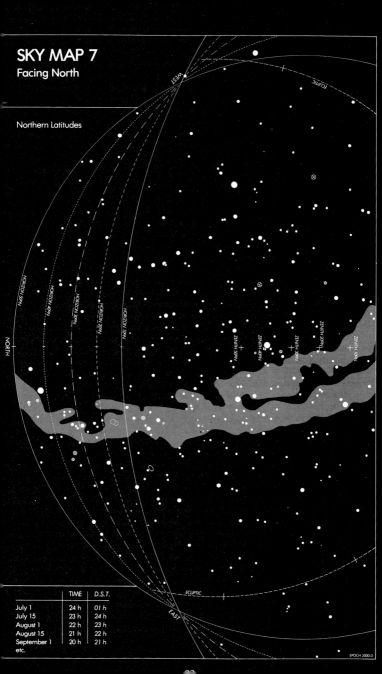

SKY MAP 7
Facing North

Northern Latitudes

	TIME	D.S.T.
July 1	24 h	01 h
July 15	23 h	24 h
August 1	22 h	23 h
August 15	21 h	22 h
September 1	20 h	21 h
etc.		

EPOCH 2000.0

SKY MAP 7
Facing South

Northern Latitudes

WEST

HORIZON 30°N
HORIZON 40°N
HORIZON 50°N

ECLIPTIC

ZENITH 50°N
ZENITH 40°N
ZENITH 30°N
ZENITH 20°N
ZENITH 10°N

SOUTH

EAST

Magnitudes: -1 0 1 2 3 4 (5)

◌ Open Cluster   □ Nebula

⊙ ○ Variable   ⊕ Globular Cluster   ○ Galaxy

WIL TIRION

63

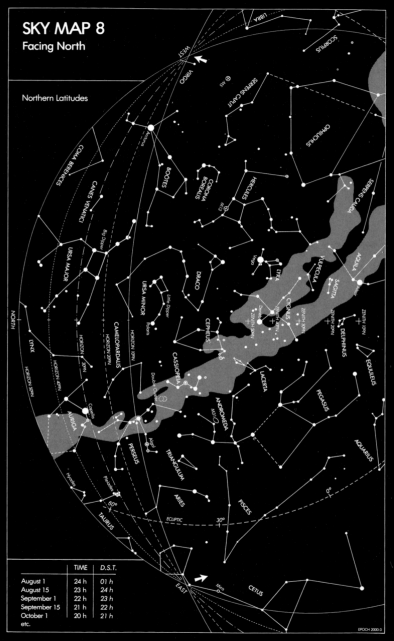

SKY MAP 8
Facing North

Northern Latitudes

	TIME	D.S.T.
August 1	24 h	01 h
August 15	23 h	24 h
September 1	22 h	23 h
September 15	21 h	22 h
October 1	20 h	21 h
etc.		

EPOCH 2000.0

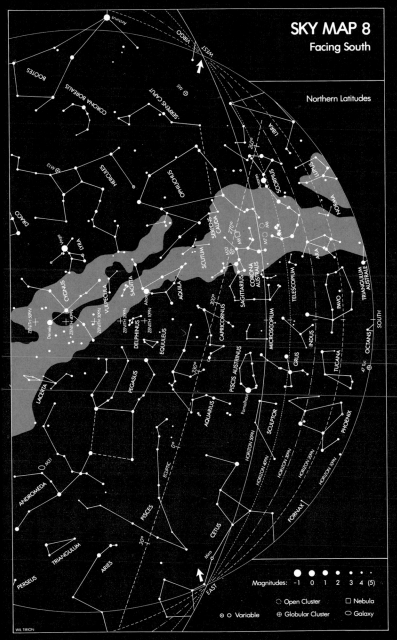

SKY MAP 8
Facing South

Northern Latitudes

Magnitudes: -1 0 1 2 3 4 (5)

◌ Open Cluster   □ Nebula
⊙ ○ Variable   ⊕ Globular Cluster   ○ Galaxy

WIL TIRION

65

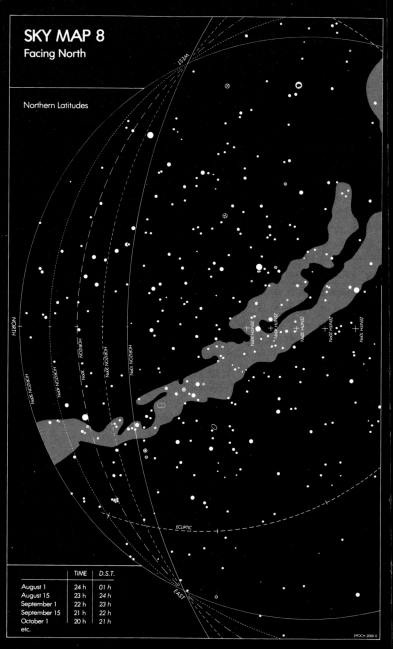

SKY MAP 8
Facing North

Northern Latitudes

NORTH

HORIZON N0°N
HORIZON N30°N
N30°N HORIZON
HORIZON N60°N

WEST

ZENITH N90°N   ZENITH N90°N   ZENITH N90°N   ZENITH N90°N

ECLIPTIC

EAST

	TIME	D.S.T.
August 1	24 h	01 h
August 15	23 h	24 h
September 1	22 h	23 h
September 15	21 h	22 h
October 1	20 h	21 h
etc.		

EPOCH 2000.0

66

SKY MAP 8
Facing South

Northern Latitudes

Magnitudes: -1 0 1 2 3 4 (5)

☼ ○ Variable   ☼ Open Cluster   ⊕ Globular Cluster   ○ Galaxy   □ Nebula

WIL TIRION

SKY MAP 9
Facing North

Northern Latitudes

	TIME	D.S.T.
September 1	24 h	01 h
September 15	23 h	24 h
October	22 h	23 h
October 15	21 h	22 h
November 1	20 h	
etc.		

EPOCH 2000.0

SKY MAP 9
Facing South

Northern Latitudes

Magnitudes: -1 0 1 2 3 4 (5)

☆ Open Cluster   □ Nebula
⊙ ○ Variable   ⊕ Globular Cluster   ○ Galaxy

W.IL TIRION

69

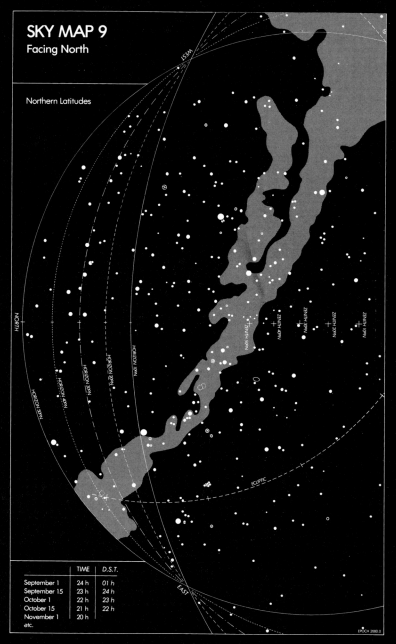

SKY MAP 9
Facing North

Northern Latitudes

	TIME	D.S.T.
September 1	24 h	01 h
September 15	23 h	24 h
October 1	22 h	23 h
October 15	21 h	22 h
November 1	20 h	
etc.		

EPOCH 2000.0

70

SKY MAP 9
Facing South

Northern Latitudes

Magnitudes: -1 0 1 2 3 4 (5)

⊙ ○ Variable   ⊕ Globular Cluster   ○ Galaxy
⊙ Open Cluster   □ Nebula

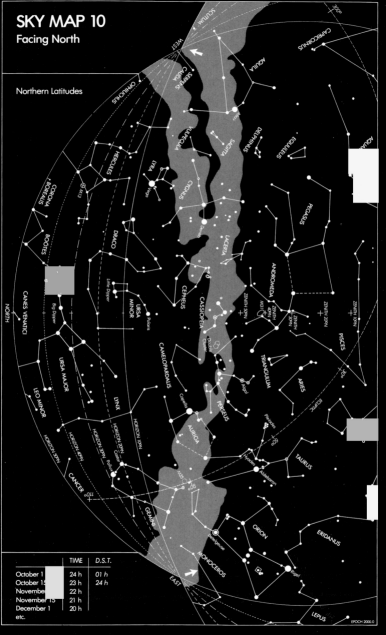

SKY MAP 10
Facing North

Northern Latitudes

	TIME	D.S.T.
October 1	24 h	01 h
October 15	23 h	24 h
November 1	22 h	
November 15	21 h	
December 1	20 h	
etc.		

EPOCH 2000.0

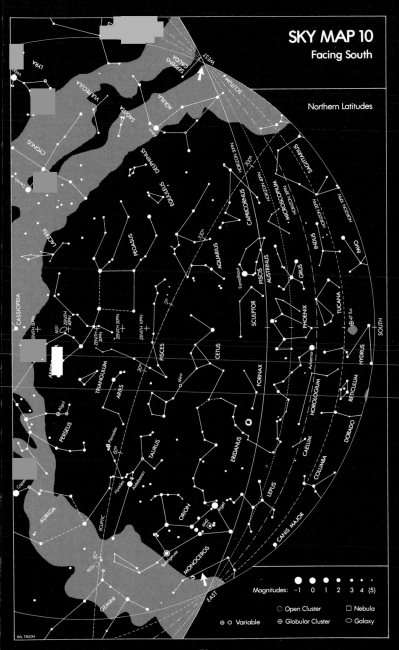

SKY MAP 10
Facing South

Northern Latitudes

Magnitudes: -1 0 1 2 3 4 (5)

◌ Open Cluster   □ Nebula

⊙ ○ Variable   ⊕ Globular Cluster   ○ Galaxy

WIL TIRION

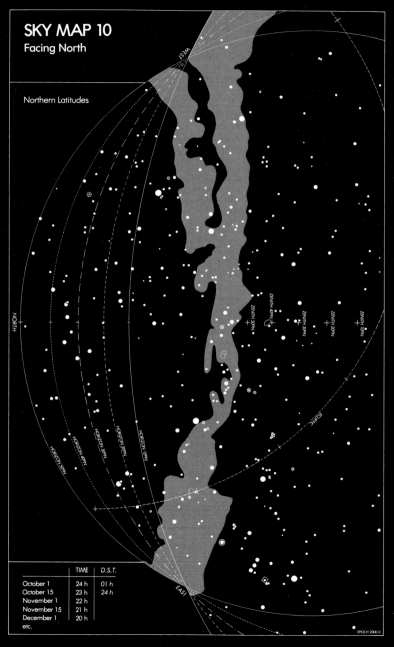

SKY MAP 10
Facing North

Northern Latitudes

	TIME	D.S.T.
October 1	24 h	01 h
October 15	23 h	24 h
November 1	22 h	
November 15	21 h	
December 1	20 h	
etc.		

EPOCH 2000.0

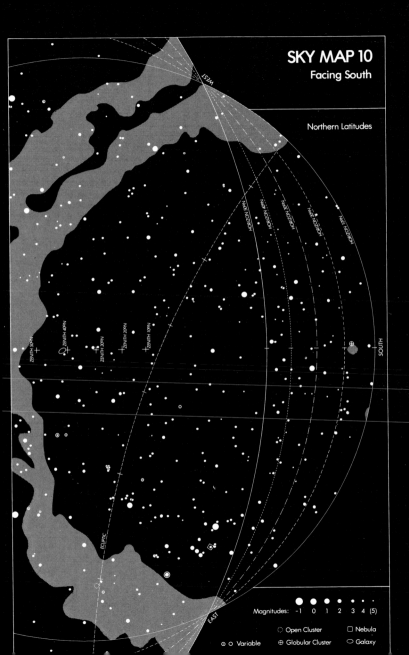

SKY MAP 10
Facing South

Northern Latitudes

Magnitudes: -1 0 1 2 3 4 (5)

○ Open Cluster   □ Nebula
⊙ ○ Variable   ⊕ Globular Cluster   ○ Galaxy

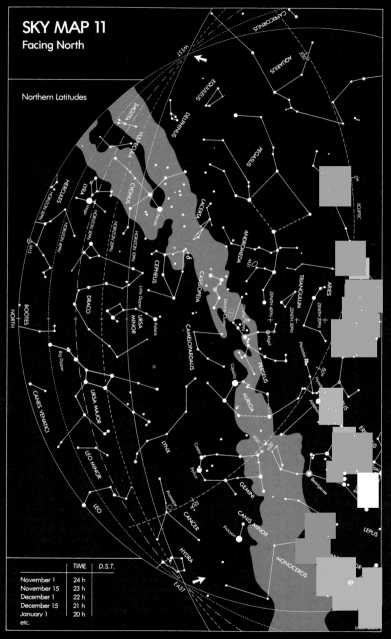

SKY MAP 11
Facing North

Northern Latitudes

	TIME	D.S.T.
November 1	24 h	
November 15	23 h	
December 1	22 h	
December 15	21 h	
January 1	20 h	
etc.		

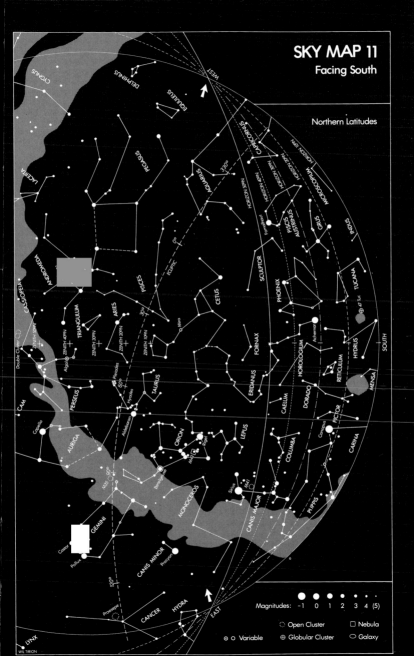

SKY MAP 11
Facing South

Northern Latitudes

Magnitudes: -1 0 1 2 3 4 (5)

○ Open Cluster   □ Nebula
⊙ ○ Variable   ⊕ Globular Cluster   ○ Galaxy

WIL TIRION

SKY MAP 11
Facing North

Northern Latitudes

	TIME	D.S.T.
November 1	24 h	
November 15	23 h	
December 1	22 h	
December 15	21 h	
January 1	20 h	
etc.		

EPOCH 2000.0

78

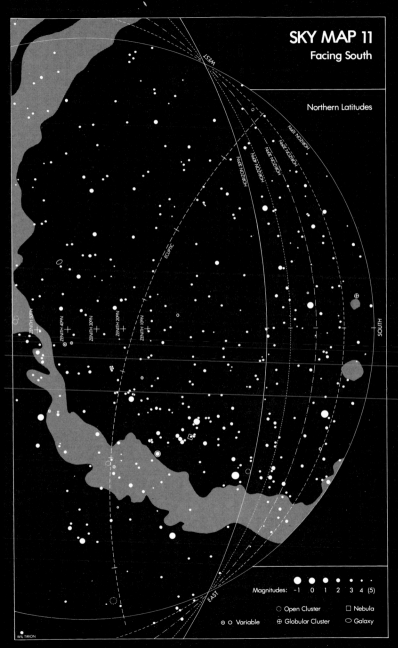

SKY MAP 11
Facing South

Northern Latitudes

Magnitudes: -1 0 1 2 3 4 (5)

◌ Open Cluster   □ Nebula
⊙ ○ Variable   ⊕ Globular Cluster   ○ Galaxy

WIL TIRION

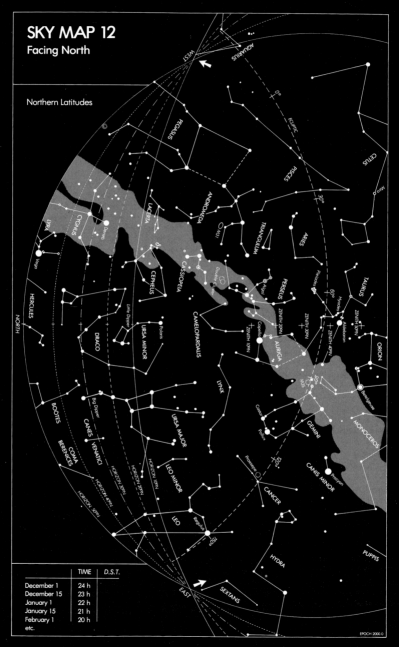

SKY MAP 12
Facing North

Northern Latitudes

	TIME	D.S.T.
December 1	24 h	
December 15	23 h	
January 1	22 h	
January 15	21 h	
February 1	20 h	
etc.		

EPOCH 2000.0

80

SKY MAP 12
Facing South

Northern Latitudes

Magnitudes: -1 0 1 2 3 4 (5)

◌ Open Cluster   □ Nebula

⊙ ○ Variable   ⊕ Globular Cluster   ○ Galaxy

WIL TIRION

81

SKY MAP 12
Facing North

Northern Latitudes

WEST

ECLIPTIC

NORTH

ZENITH 10°N
ZENITH 20°N
ZENITH 30°N
ZENITH 40°N
ZENITH 50°N

HORIZON 50°N
HORIZON 40°N
HORIZON 30°N
HORIZON 20°N
HORIZON 10°N

EAST

	TIME	D.S.T.
December 1	24 h	
December 15	23 h	
January 1	22 h	
January 15	21 h	
February 1	20 h	
etc.		

EPOCH 2000.0

82

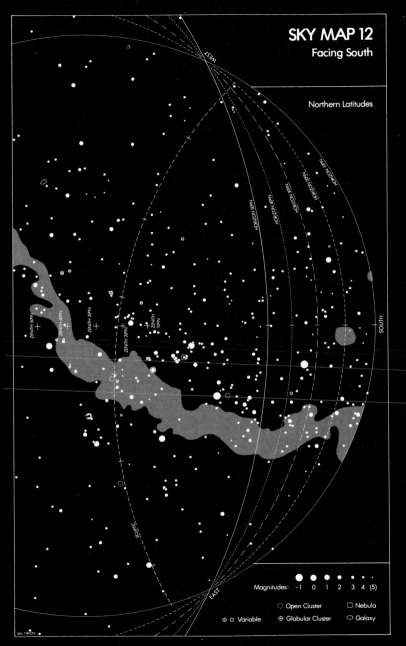

SKY MAP 12
Facing South

Northern Latitudes

HORIZON 50°N
HORIZON 40°N
HORIZON 30°N

ZENITH 50°N
ZENITH 40°N
ZENITH 30°N
ZENITH 20°N
ZENITH 10°N

WEST

SOUTH

EAST

ECLIPTIC

Magnitudes: -1   0   1   2   3   4   (5)

⊙ ○ Variable   ⊕ Globular Cluster   ○ Galaxy

⚬ Open Cluster   □ Nebula

W.L.TIRION

83

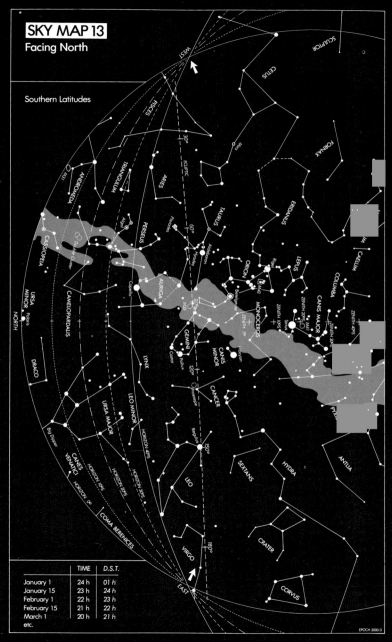

SKY MAP 13
Facing North

Southern Latitudes

	TIME	D.S.T.
January 1	24 h	01 h
January 15	23 h	24 h
February 1	22 h	23 h
February 15	21 h	22 h
March 1	20 h	21 h
etc.		

EPOCH 2000.0

84

SKY MAP 13
Facing South

Southern Latitudes

Magnitudes: -1 0 1 2 3 4 (5)

◌ Open Cluster   □ Nebula
⊚ ○ Variable   ⊕ Globular Cluster   ◯ Galaxy

WIL TIRION ©

85

SKY MAP 14
Facing North

Southern Latitudes

	TIME	D.S.T.
February 1	24 h	01 h
February 15	23 h	24 h
March 1	22 h	23 h
March 15	21 h	22 h
April 1	20 h	
etc.		

EPOCH 2000.0

86

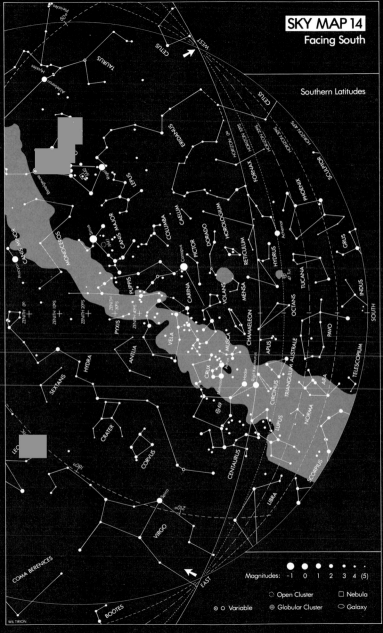

Southern Latitudes

Magnitudes: -1 0 1 2 3 4 (5)

◌ Open Cluster   □ Nebula

⊙ ○ Variable   ⊕ Globular Cluster   ○ Galaxy

WIL TIRION

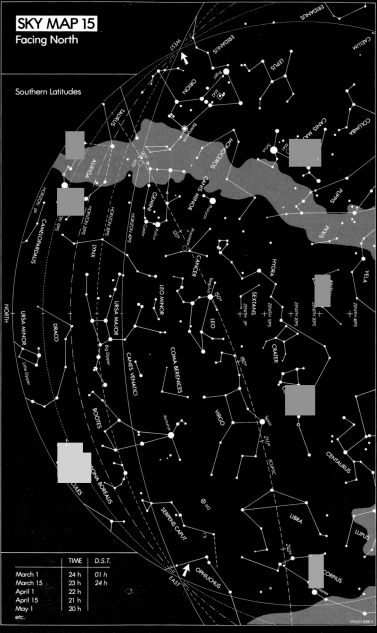

SKY MAP 15
Facing North

Southern Latitudes

	TIME	D.S.T.
March 1	24 h	01 h
March 15	23 h	24 h
April 1	22 h	
April 15	21 h	
May 1	20 h	
etc.		

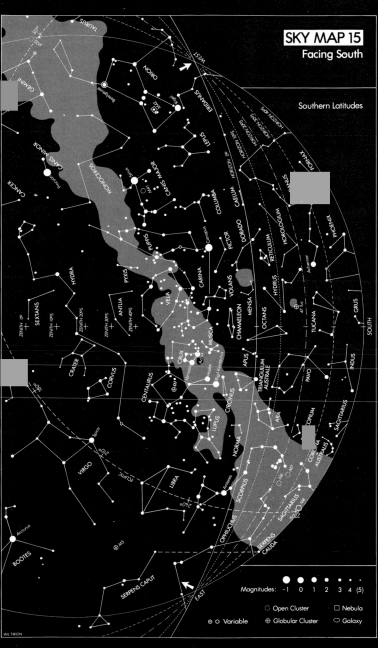

SKY MAP 15
Facing South

Southern Latitudes

Magnitudes: -1 0 1 2 3 4 (5)

☆ Open Cluster   □ Nebula
⊙ O Variable   ⊕ Globular Cluster   ○ Galaxy

WIL TIRION

89

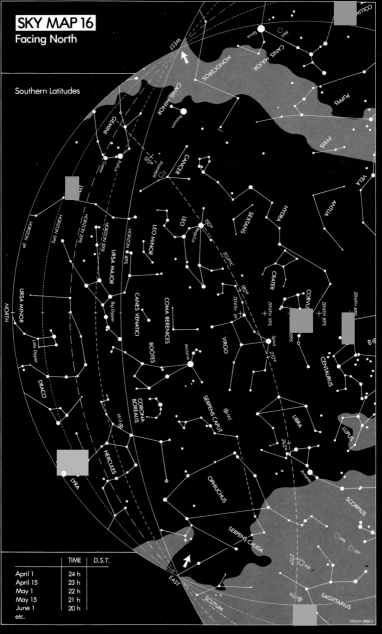

SKY MAP 16
Facing North

Southern Latitudes

	TIME	D.S.T.
April 1	24 h	
April 15	23 h	
May 1	22 h	
May 15	21 h	
June 1	20 h	
etc.		

EPOCH 2000.0

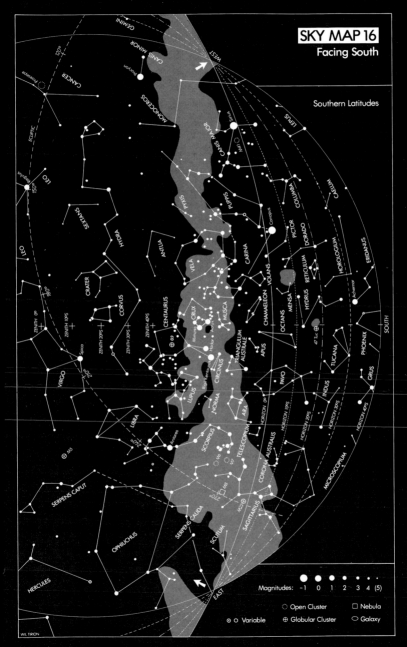

SKY MAP 16
Facing South

Southern Latitudes

Magnitudes: -1 0 1 2 3 4 (5)

◌ Open Cluster   □ Nebula
⊙ ○ Variable   ⊕ Globular Cluster   ○ Galaxy

WIL TIRION

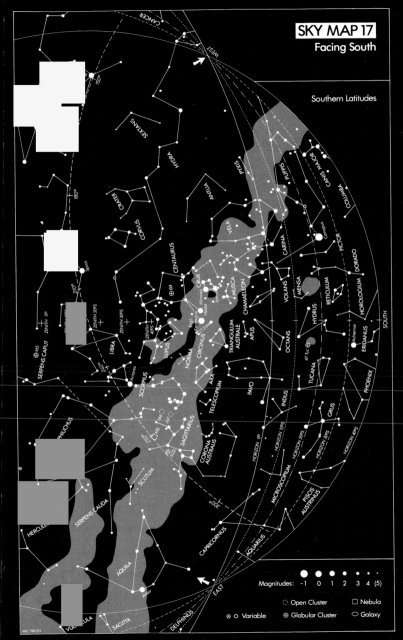

SKY MAP 17
Facing South

Southern Latitudes

Magnitudes: -1 0 1 2 3 4 (5)

○ Open Cluster   □ Nebula
⊙ ○ Variable   ⊕ Globular Cluster   ○ Galaxy

WIL TIRION

93

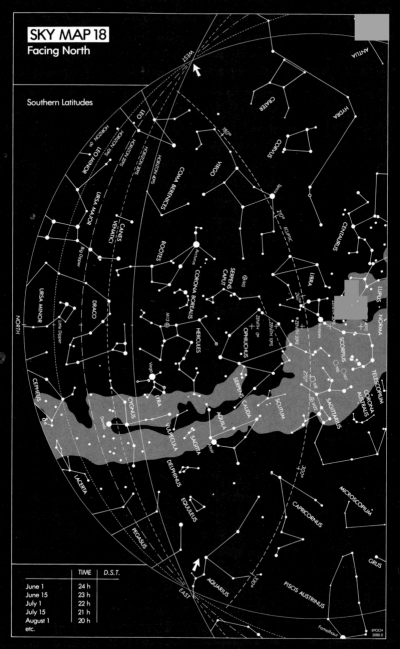

SKY MAP 18
Facing North

Southern Latitudes

	TIME	D.S.T.
June 1	24 h	
June 15	23 h	
July 1	22 h	
July 15	21 h	
August 1	20 h	
etc.		

EPOCH 2000.0

SKY MAP 18
Facing South

Southern Latitudes

Magnitudes: -1 0 1 2 3 4 (5)

○ Open Cluster   □ Nebula
◉ ○ Variable   ⊕ Globular Cluster   ○ Galaxy

WIL TIRION

95

	TIME	D.S.T.
July 1	24 h	
July 15	23 h	
August 1	22 h	
August 15	21 h	
September 1	20 h	21 h
etc.		

EPOCH 2000.0

SKY MAP 19
Facing South

Southern Latitudes

Magnitudes: -1 0 1 2 3 4 (5)

◌ Open Cluster   □ Nebula

⊙ ○ Variable   ⊕ Globular Cluster   ○ Galaxy

W.IL TIRION

SKY MAP 20
Facing North

Southern Latitudes

	TIME	D.S.T.
August 1	24 h	
August 15	23 h	
September 1	22 h	23 h
September 15	21 h	22 h
October 1	20 h	21 h
etc.		

98

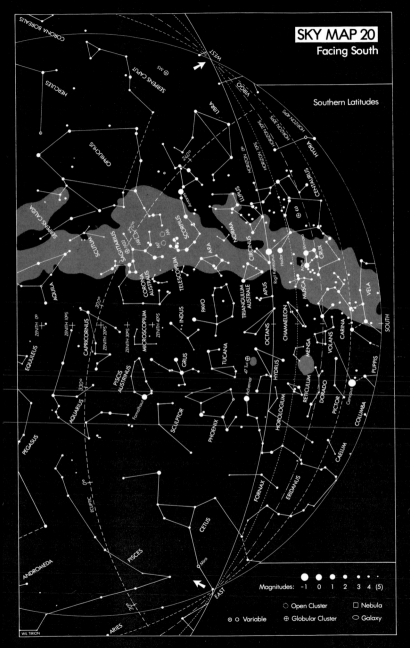

Southern Latitudes

Magnitudes: -1 0 1 2 3 4 (5)

⬡ Open Cluster   ☐ Nebula

⊙ ○ Variable   ⊕ Globular Cluster   ○ Galaxy

WIL TIRION

99

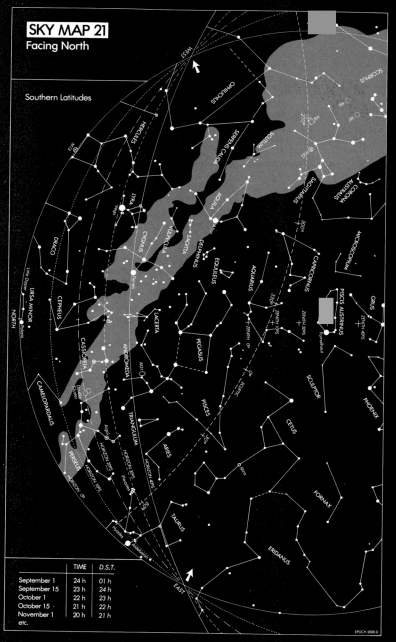

SKY MAP 21
Facing North

Southern Latitudes

	TIME	D.S.T.
September 1	24 h	01 h
September 15	23 h	24 h
October 1	22 h	23 h
October 15	21 h	22 h
November 1	20 h	21 h
etc.		

EPOCH 2000.0

SKY MAP 21
Facing South

Southern Latitudes

Magnitudes: -1 0 1 2 3 4 (5)

☼ ○ Variable   ⊕ Globular Cluster   □ Nebula   ○ Galaxy
☼ Open Cluster

WIL TIRION

101

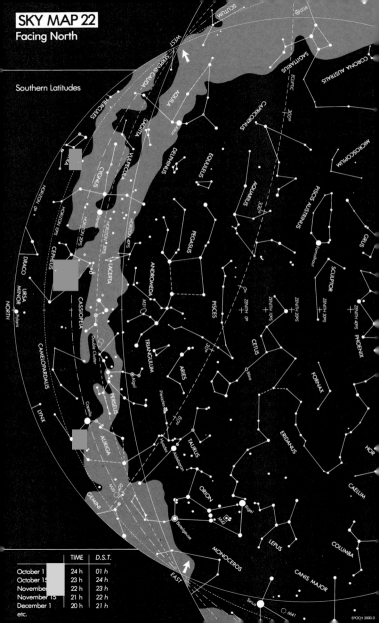

SKY MAP 22
Facing North

Southern Latitudes

	TIME	D.S.T.
October 1	24 h	01 h
October 15	23 h	24 h
November	22 h	23 h
November 15	21 h	22 h
December 1	20 h	21 h
etc.		

EPOCH 2000.0

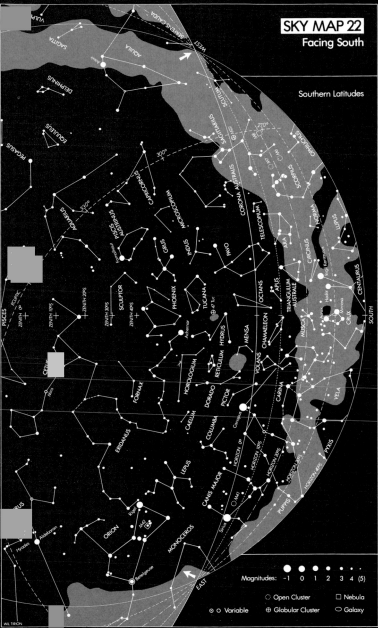

SKY MAP 22
Facing South

Southern Latitudes

Magnitudes: -1 0 1 2 3 4 (5)

○ Open Cluster
⊕ Globular Cluster

□ Nebula
○ Galaxy

⊙ ○ Variable

WEST

EAST

SAGITTA

AQUILA

DELPHINUS

EQUULEUS

PEGASUS

PISCES

ECLIPTIC

CETUS

Mira

ERIDANUS

FORNAX

CAELUM

LEPUS

ORION

Rigel

M42

Betelgeuse

Hyades   Aldebaran

TAURUS

SERPENS CAUDA

SCUTUM

OPHIUCHUS

SERPENS CAUDA

SCORPIUS

SAGITTARIUS

CORONA AUSTRALIS

CAPRICORNUS

MICROSCOPIUM

AQUARIUS

PISCIS AUSTRINUS

Fomalhaut

GRUS

INDUS

PAVO

TELESCOPIUM

ARA

NORMA

LUPUS

CIRCINUS

TRIANGULUM AUSTRALE

CENTAURUS

MUSCA

CRUX

SOUTH

PHOENIX

TUCANA

47 Tuc

HYDRUS

OCTANS

APUS

CHAMAELEON

MENSA

VOLANS

CARINA

VELA

PYXIS

PUPPIS

SCULPTOR

Achernar

RETICULUM

HOROLOGIUM

DORADO

PICTOR

Canopus

COLUMBA

CANIS MAJOR

Sirius

M41

MONOCEROS

ZENITH 0°
ZENITH 10°S
ZENITH 20°S
ZENITH 30°S
ZENITH 40°S

HORIZON 0°
HORIZON 10°S
HORIZON 20°S
HORIZON 30°S
HORIZON 40°S

WIL TIRION

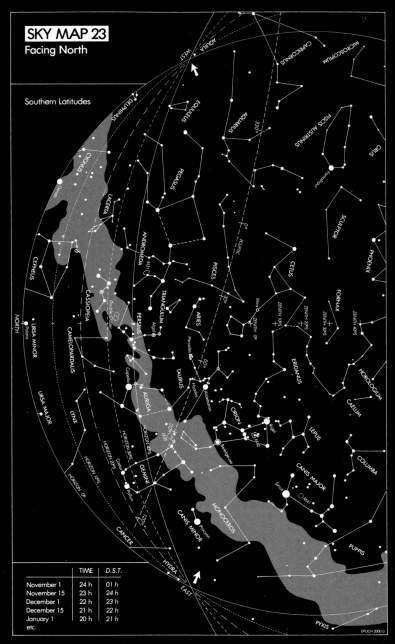

Southern Latitudes

WEST

AQUILA

DELPHINUS

EQUULEUS

CAPRICORNUS

MICROSCOPIUM

AQUARIUS

PISCIS AUSTRINUS

GRUS

PEGASUS

330°

ECLIPTIC

0°

CETUS

SCULPTOR

PHOENIX

CYGNUS

Deneb

LACERTA

ANDROMEDA

M31

PISCES

300°

FORNAX

ZENITH 40°S

HOROLOGIUM

CEPHEUS

CASSIOPEIA

Double Cluster

PERSEUS

TRIANGULUM

Algol

ARIES

Pleiades

Mira

30°

ZENITH 10°S

ZENITH 0°

ZENITH 20°S

ZENITH 30°S

ERIDANUS

NORTH

URSA MINOR

Polaris

CAMELOPARDALIS

Capella

AURIGA

TAURUS

Hyades

Aldebaran

60°

ORION

Rigel

CAELIUM

LEPUS

URSA MAJOR

LYNX

HORIZON 40°S

HORIZON 30°S

HORIZON 20°S

Castor

GEMINI

Pollux

M42

Betelgeuse

M42

COLUMBA

HORIZON 10°S

CANIS MAJOR

Sirius

M41

HORIZON 0°

CANCER

CANIS MINOR

Procyon

MONOCEROS

120°

HYDRA

PUPPIS

EAST

PYXIS

	TIME	D.S.T.
November 1	24 h	01 h
November 15	23 h	24 h
December 1	22 h	23 h
December 15	21 h	22 h
January 1	20 h	21 h
etc.		

EPOCH 2000.0

Southern Latitudes

Magnitudes: -1 0 1 2 3 4 (5)

○ Open Cluster   □ Nebula
⊙ ○ Variable   ⊕ Globular Cluster   ○ Galaxy

WIL TIRION

105

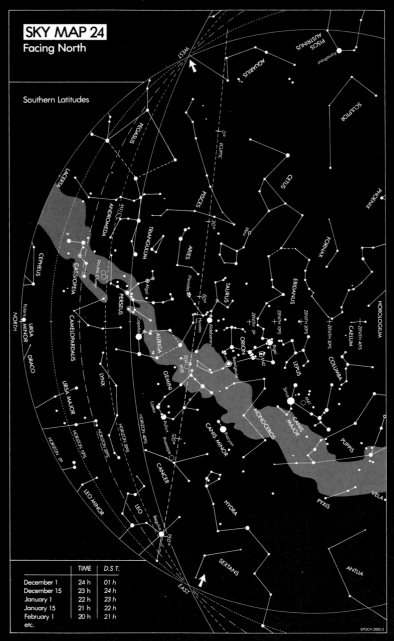

SKY MAP 24
Facing North

Southern Latitudes

	TIME	D.S T.
December 1	24 h	01 h
December 15	23 h	24 h
January 1	22 h	23 h
January 15	21 h	22 h
February 1	20 h	21 h
etc.		

EPOCH 2000.0

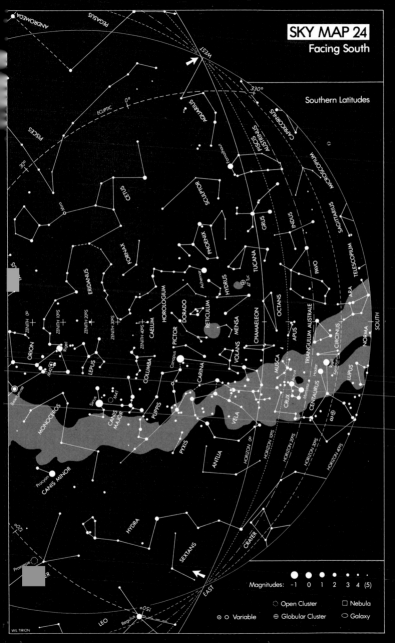

Magnitudes: -1 0 1 2 3 4 (5)

◌ ○ Open Cluster   ◻ Nebula

⊙ ○ Variable   ⊕ Globular Cluster   ○ Galaxy

W.L. TIRION

4

Stars, Nebulae, and Galaxies

The universe contains many types of objects, and most of them are shown on the Atlas Charts in Chapter 7. In this chapter, we will give a brief description of the different types of objects whose positions in the sky are fixed to the celestial sphere, a fictitious bowl that appears to rotate around the celestial poles. At the end of the chapter, we provide lists and graphic displays of a selection of the most interesting objects in the sky. For your convenience, we have chosen the objects so that some of them will be visible at each time of year.

STARS

You have already seen the brightest stars on the Monthly Sky Maps in Chapter 3; they are also listed in Table A-2 in the Appendix. Astronomers measure the brightness of stars on the magnitude scale, which is described in Chapter 3 (p. 32).

Both the Monthly Sky Maps and the detailed Atlas Charts in Chapter 7 use circles of different sizes to represent the magnitudes of the stars. Remember that the higher the magnitude, the fainter the star. Stars as faint as 6th magnitude can be seen by the naked eye under perfect conditions, but depending on your eyesight and the brightness of the night sky, the visual limit is usually much lower. It would not be uncommon if your sky were bright enough to prevent you from seeing stars any fainter than 3rd magnitude. Indeed, in cities, the sky itself may be so bright from streetlights and other sources of "light pollution" as to prevent you from seeing any stars at all!

Star Temperatures

All stars are balls of gas, so hot that they are glowing. Stars have different temperatures, ranging from about 2,100°C (about 3,800°F) up to about 50,000°C (about 90,000°F). Just as an iron poker glows more and more as it is heated, starting with red hot and then becoming white hot as it grows even hotter, a star's color also indicates its temperature. The coolest stars are reddish balls of glowing gas; the hottest stars glow blue-white. These colors can even be seen with the naked eye. In the constellation Orion, for example, it is not hard to see that the star Betelgeuse, in Orion's shoulder, looks much redder and is therefore much cooler than the star Rigel, in Orion's heel.

Astronomers always give star temperatures using the kelvin scale, which uses Celsius degrees but begins at absolute zero, the coldest temperature possible. Absolute zero is $-273°C = -459.4°F$.

Spectral Types

Depending on its temperature, each star has its own *spectral type,* which is determined by analyzing its spectrum — the band of colors into which the light from the star can be broken down. The rainbow displays the visible part of the spectrum: red, orange, yellow, green, blue, indigo, and violet; a mnemonic for the order of colors is the name of an imaginary fellow, ROY G. BIV.

When a narrow band of light from the sun or another star is spread out into its spectrum, with blue, say, at the left and red at the right, the spectrum turns out to be crossed by dark lines, each indicating the absence of some particular color. These *spectral lines* are the signatures of different chemical elements at different temperatures. Each element has unique sets of spectral lines for various temperature ranges, so the spectrum of a star reveals the chemical composition of the star's atmosphere. These spectral lines are dark because they are absorbing some of the energy from deeper layers of the sun or the star; they are thus called *absorption lines.* All stars have spectra that are continuous bands of radiation crossed by absorption lines.

When hot gas shines by itself, it may give off *emission lines,* specific colors that are brighter than the adjacent colors. A few stars have some emission lines in addition to their absorption lines; the presence of emission lines is noteworthy because it usually shows that there is hot gas surrounding the star. Although emission lines are rare in the spectra of stars, the hot gas in most nebulae is so tenuous that these nebulae have emission-line spectra. They are, therefore, called *emission nebulae,* and appear reddish because the red emission line from heated hydrogen gas is the strongest contributor to the visible part of their spectra.

In this book, we discuss mainly the visible part of the spectrum because our discussions are meant for optical observers. But professional astronomers are increasingly studying other parts of the spectrum. The whole spectrum is a set of waves of different wavelengths. Gamma rays, x-rays, and ultraviolet have wavelengths shorter than those of visible light. Infrared and radio waves have wavelengths longer than those of visible light. Of those other parts of the spectrum, only some radio studies can be carried out by amateur observers; along with a few parts of the infrared, radio waves are the only part of the spectrum besides the visible part that comes through the earth's atmosphere.

In the original determination of spectral types of stars, capital letters were assigned in alphabetical order to indicate the strength of hydrogen's spectral lines relative to the spectral lines from other elements, with A representing the greatest strength. But astronomers later realized that hydrogen radiation is strongest for stars of intermediate temperature rather than for the hottest or coolest stars. So the modern list in order of temperature is out of alphabetical order. From the hottest to the coolest, the order is: O B A F G K M, remembered by generations from the mnemonic "Oh, be a fine girl, kiss me." The actual temperatures and other properties corresponding to the spectral types are given in Appendix Table A-3. Appendix Table A-2 gives spectral types and other data for the 287 brightest stars in the sky.

Double and Variable Stars

Most stars in the sky that appear to be single are actually made up of two or more objects revolving around each other. Sirius, for example, is really a *double* or *binary star;* the bright, bluish-white star we see with the naked eye has a much fainter companion (known as Sirius B) revolving around it. Double stars (which are called "double" even when three or more stars are present) can be interesting to observe, especially when the stars in a pair or multiple system present a beautiful contrast of colors in binoculars or a telescope (see Table 5, p. 127). In some double-star systems, the stars periodically block each other as they orbit, making the total brightness we see vary. These particular binary star systems are examples of *eclipsing binaries.*

Some stars vary in brightness all by themselves. Sometimes a star actually periodically changes in size and therefore brightness. These stars are called *variable stars.*

Since double stars and variable stars are particularly interesting for casual star observers and amateur astronomers to observe, we devote a whole chapter (Chapter 6) to them. Two short tables of the double and variable stars that are easiest to observe appear at the end of this chapter; the times when they are visible above the horizon are shown in a Graphic Timetable (Fig. 24). You will find more charts that will help you observe interesting doubles and variables in Chapter 6.

Distances to the Stars

All the stars are so far away that we measure the distance in terms of how long it takes their light to reach us. Light travels very rapidly — 300,000 km/sec (186,000 miles/sec). This is 1,080,000,000 km/hr, 10 million times faster than a car. Light can travel seven times the distance around the earth in a single second, can travel from the moon to the earth in slightly more than a second, and can travel from the sun in eight minutes. Yet light takes over four years to reach us from the nearest star, Proxima Centauri. We call the distance that light travels in a year *one light-year,* and say that

Proxima Centauri is thus four light-years away. Only a few dozen stars are within 20 light-years of our sun.

To compare stars, it is useful and important to know how bright they would be if they were all at the same distance from us, so that only their intrinsic differences in brightness — how bright they actually are — and not merely their positions affect the result. The standard distance that astronomers use for this purpose is about 32.6 light-years.

Measuring distances and angles

Astronomers measure the distance to the nearest stars by a method of triangulation called *trigonometric parallax:* the nearest stars appear to move slightly with respect to the more distant stars. To see a similar effect, hold your thumb up with your arm outstretched, and look at it first with one eye closed and then, without moving your head, with the other eye closed. Notice how your thumb seems to move slightly across the background. Now do the same thing with your arm bent so that your thumb is closer to your eye — your thumb will appear to move more. Similarly, the more a star appears to move as the earth orbits the sun, the closer it is.

Astronomers figure out how far away from the earth and the sun they would have to be before the distance between the earth and the sun covered an angle of only 1 arc second (where 1 arc second is $\frac{1}{60}$ of 1 arc minute, 1 arc minute is $\frac{1}{60}$ of one degree (1°), and 1° is $\frac{1}{360}$ of the way around the sky). Thus 1 arc second is a tiny angle, $\frac{1}{3600}$ of a degree of arc. The distance at which the angle between the earth and sun is only 1 arc second is called 1 *parsec* (from the words "parallax," for the method of measuring distance, and "second," for the actual angle). The value of 1 parsec works out to be 3.262... light-years. The standard distance astronomers use in order to determine the absolute magnitudes of stars is 10 parsecs.

The brightness a star would have at this distance of about 32.6 light-years from us is called its *absolute magnitude.* The absolute magnitude is a measure of the intrinsic brightness of the star.

At the turn of the century, Ejnar Hertzsprung and Henry Norris Russell plotted many stars on the same graph, removing the effect of distance on brightness by either plotting a group of stars that were all at the same distance or by using the absolute magnitudes. When they plotted the temperature on one axis and the magnitude on the other, they found that the points representing most stars fell on the graph in a narrow band — the *main sequence.* The brighter stars are hotter for the most part, and the fainter stars are cooler.

A few stars are brighter than main-sequence stars of the same temperature. These exceptional stars are thus known as *giants,* and a few even more exceptional stars are brighter still, and are known as *supergiants.* Giants and supergiants turn out to be especially large in addition to being especially bright. A few stars are fainter than main-sequence stars of the same temperature. These exceptionally faint stars are known as *white dwarfs*. They turn out to be extraordinarily small for stars — only the size of the planet earth.

The Life Histories of Stars

Much of the astronomical research over the last few decades has been devoted to studying the life cycles of stars. Stars begin as collapsing clouds of gas in interstellar space, and become stars when nuclear fusion begins in their interiors, fusing hydrogen into helium. At that point, stars join the *main sequence*. During this period of their lives, energy from the fusion creates a pressure that pushes outward, balancing the inward force of gravity. Depending on how hot and bright a star is, it spends more or less time as a main-sequence star. The hottest stars, though more massive than the sun, use their fuel at such a furious pace that they have relatively short lifetimes — perhaps only 100,000 years. Stars like the sun live on the main sequence for about 10 billion years; the sun is now about halfway through that lifetime. Stars cooler and fainter than the sun can live much longer, 50 billion years or more.

In the type of nuclear fusion that goes on inside the sun and most other stars, four hydrogen nuclei — the central parts of hydrogen atoms — fuse to become one helium nucleus. The resulting helium has slightly less mass than the four hydrogens that went into forming it. The difference is converted into energy in the amount $E = mc^2$, a fact and formula discovered by Albert Einstein in 1905.

When a star begins to die, its fate depends on how much mass it has. Let us first consider stars with about as much mass as the sun, ranging from the least massive stars (with 0.07 — 7% of — the mass of the sun) to stars with about twice as much mass as the sun. As the stars use up all the hydrogen in their interiors, and so can no longer fuse hydrogen into helium, they can no longer fight off gravity and their interiors begin to contract. During the contraction, energy is released and their outer layers are pushed out. These outer layers become large and cool, and the stars become *red giants*.

Eventually, the outer layers of some of the red giants become disconnected and float outward, carrying perhaps 20% of the star's mass out into space. The expanding shell of gas is known as a *planetary nebula* (see p. 116). Planetary nebulae are marked with a special symbol on the Atlas Charts in Chapter 7. The escaped nebula bares the inner layers of the star, which appear blue because they are so hot, hotter even than the hottest normal stars. The cool, colorful outer layers and the bluish central star of a planetary nebula show in Color Plates 17, 20, 27, 48, and 50.

After only 50,000 years or so — a twinkling of an eye in the lifetime of a star — the planetary nebula drifts away, and the central star cools. Eventually the central star contracts until it is about the size of the earth. If a mass less than 1.4 times the mass of the sun is left, the star remains at this stage indefinitely. It is then a *white dwarf*. White dwarfs are faint because they are so small. Thus we see that objects like red giants, planetary nebulae, and white dwarfs, originally discovered as independent classes of objects, are really different stages in the lifetimes of ordinary stars.

Sometimes a white dwarf is part of a double-star system, and its gravity attracts material from the other star in the system. When this material hits the surface of a white dwarf, it may begin nuclear fusion briefly, and the system brightens considerably. We see it as a *nova*. Novae were originally thought of as new stars, but are now known to be stars that suddenly become brighter or visible because of this effect. A few famous novae are marked on the charts.

A star that contains more than about twice as much mass as the sun comes to a more spectacular end. After it becomes a red giant, it grows bigger yet and becomes a supergiant. At that point it may explode completely, becoming a *supernova* (C.Pl. 26). A supernova may rival in brightness the whole galaxy it is in. It was only in the 1920s that it was realized that supernovae and novae are completely different phenomena. A supernova completely devastates the star, forming heavy elements because of the high temperature and then spewing these elements out into space. The heavier elements in the human body (all those heavier than iron) were formed in supernovae that exploded before our solar system formed. When our sun and planets formed, they incorporated these heavy elements. Also, it seems increasingly likely that our solar system formed when a nearby supernova made a cloud of gas and dust start to collapse. So supernovae were important to the existence of humanity.

In a few cases, the remnants of a supernova can still be seen in space, as in the Crab Nebula (C.Pl. 1). Still more supernovae are detected from the radio waves they emit, which are studied with radio telescopes on earth or with x-ray telescopes in orbit. Though we know, from studying other galaxies, that supernovae go off every 30 years or so in each galaxy, no supernovae have been visible in our galaxy since 1604. Astronomers hope to observe such a spectacular event, but have no way of predicting it.

During a supernova explosion, the center of the star can be extremely compressed and can collapse until it is quite small. Stars that have 1.4–4 times the mass of the sun collapse until they are about 20 km (12 miles) across. At that point, the neutrons in the star cannot be pushed closer together. Most of the star is then a gas containing neutrons only, or perhaps the still smaller particles called quarks that make up neutrons. The star is now a *neutron star*.

Neutron stars are too small and faint to be seen directly in visible light. But some neutron stars give off beams of radio radiation. As the star rotates very rapidly — once every second or so — in some cases this beam of radiation sweeps by the earth, the way a lighthouse beacon apparently gives pulses of light as it turns. We detect pulses of radio waves from the neutron star, and so call the object a *pulsar*. Even though we can't see pulsars, we have marked a few prominent pulsars on the Atlas Charts in Chapter 7. Over 300 pulsars are now known.

Other neutron stars exist in binary systems (double stars). The gravity of the neutron star sometimes pulls mass off the other star. This mass falls on the neutron star with such force that it heats up enough to give off x-rays. X-ray telescopes in orbit have detected signs of a number of these *binary x-ray sources*.

In a few cases, a mass greater than about four times that of the sun remains after a supernova explosion. Then nothing can stop the collapse of the star. Because of its gravity, it becomes more and more compressed. In cases where the gravity is so high, scientists must use the theory of gravity developed by Einstein in 1917 to explain the situation. Einstein's theory is called "the general theory of relativity." The theory predicts that even light will be bent so much by the strong gravity that the light will no longer be able to escape. Neither could any object escape. The star has become a *black hole*.

Black holes are even fainter than neutron stars, so we must look for signs of their presence other than their optical appearance. They are too small to hope to see them block out light from behind. Our best chance of detecting a black hole is if it is located in a binary system. Then the black hole should be attracting material from its companion star. The matter will swirl around the black hole before it enters, and heat up until it gives off x-rays. So x-ray sources are candidates we must study in order to see if other evidence tells us that they are black holes. We conclude that they are black holes when they have too much mass to wind up as neutron stars. We measure the mass of the invisible star in a binary by studying the motions of its visible companion star. The effect is similar to deducing that an invisible dancer is present by watching the movements of the companion. The most likely black-hole candidate in our galaxy is in the constellation Cygnus; we have marked its location on Atlas Chart 19 in Chapter 7.

STAR CLUSTERS

Sometimes stars form in groups. With binoculars or a small telescope, we might see a dozen or so stars in an irregular grouping in a certain area, though the cluster may actually contain hundreds of stars. The stars are contained within a region about 30 light-years across. Usually, using a larger telescope or making a longer photographic exposure will reveal more stars. The Pleiades (Fig. 13) and the Hyades in Taurus (Fig. 53, p. 190) are among the most famous examples of this type of cluster, which is often called an *open cluster*. Such groupings are also called *galactic clusters,* because they lie in the plane of our galaxy. Open clusters are relatively young on a cosmic scale; the Pleiades formed only 100 million years ago. Long-exposure photos of this cluster (Fig. 13, C.Pl. 15) show that some of the gas and dust from which the stars formed is still present nearby.

Fig. 13. (*top*) The Pleiades in Taurus. With the naked eye, most people can see six stars in this open cluster; binoculars or telescopes reveal dozens of stars. (Royal Observatory, Edinburgh) (*bottom*) A longer exposure of the Pleiades reveals that dust around the stars reflects light toward us. (Lick Observatory photo)

Fig. 14. The globular cluster M13 in Hercules. (Palomar Observatory photo)

In other places in the sky, thousands or hundreds of thousands of stars of a common origin may be located within 300 light-years or so, forming a huge spherical ball. These groupings are called *globular clusters*. In the northern sky, the globular cluster M13 (Fig. 14, C.Pl. 4) in the constellation Hercules is the easiest to see. A globular cluster may look like a hazy mothball to the naked eye or when viewed through a small telescope; larger telescopes are necessary to see individual stars in this type of cluster.

Scientists have found that all the globular clusters are very old — perhaps 10 billion years old. Open clusters, on the other hand, have a wide range of ages. Some may even be forming now. A number of interesting open and globular clusters that can be seen at various times of year are listed in Table 7 at the end of this chapter; the times when the clusters are visible are plotted on a Graphic Timetable (Fig. 25).

NEBULAE

Nebulae — clouds of gas and dust that appear hazy to our view — are some of the most beautiful objects to observe in space. The word nebulae (singular, *nebula*) comes from the Greek word for "cloud." A few particularly interesting nebulae are listed in Table 8 at the end of this chapter; the times when they can be observed are shown on a Graphic Timetable (Fig. 26).

Some nebulae represent shells of gas thrown off by dying stars. Planetary nebulae (p. 112) and the remnants of supernovae (p. 113) are examples. Planetary nebulae — often called planetaries, for short — got their name because they often appear in small telescopes as small greenish disks; their telescopic images resemble those of the planets Uranus and Neptune, but they have no other similarity to planets. Planetary nebulae often appear greenish (C.Pl. 48) because hot oxygen in a certain condition emits a lot of green radiation. Two planetary nebulae are included in Table 8 and Fig. 26; others are listed in Appendix Table A-5.

Other nebulae represent gas and dust still surrounding young stars. The dust reflects the starlight to us, making a *reflection nebula*. The nebulae surrounding the stars in the Pleiades (C.Pl. 15) are reflection nebulae.

Some nebulae glow; we call them *emission nebulae*. Other nebulae are dark, and absorb radiation from behind. We call them *dark nebulae* or *absorption nebulae* (C.Pl. 45).

The Horsehead Nebula in Orion (C.Pl. 12) is an example of both emission and absorption nebulae. The reddish cloud of gas is an emission nebula, consisting of glowing hydrogen gas; the "horsehead" (seen in silhouette against the reddish glow) is an absorption nebula. The color of an emission nebula does not show up except in long-exposure photographs; this type of nebula emits a reddish glow because ionized hydrogen radiates strongly at one specific wavelength (color) in the red. (Hydrogen is ionized whenever it is hot.) Notice that on the left side of Fig. 15 (and below the horsehead in C.Pl. 12) we see fewer stars; stars are still present behind the nebulae but many are hidden by the absorbing gas and dust. In this nebula, some of the absorbing material intrudes on the glowing gas and forms the shape of a horse's head in silhouette.

The North America Nebula (C.Pl. 53) in the constellation Cygnus is another example of an emission nebula with an absorption nebula that defines the boundaries we see. In this case, the part of the emission nebula we see is roughly the same shape as the continent of North America. Notice how few stars are visible in what would be the Gulf of Mexico because of the dark absorption nebula there.

Some nebulae contain small dark nebulae in which stars may be

Fig. 15. The Horsehead Nebula (NGC 2024) in Orion. North is at left. (The Kitt Peak National Observatory)

forming. The nebula M16 in Serpens (C.Pl. 5) is a good example. Since radio waves and infrared rays penetrate the nebulae better than visible light does, scientists interested in the birth of stars are now studying nebulae not only in visible light but also in other parts of the spectrum.

The Great Nebula in Orion (C.Pl. 13) turns out to be a particularly exciting place. The emission nebula we see is a blister sticking out toward us from a large cloud of dust and gas in which stars are actually forming now. Though we can't see through the emission nebula with visible light, infrared and radio waves penetrate the dark cloud behind it. The absorbing gas and dust in that cloud prevent radiation from the emission nebula from entering and breaking up the gas clouds from which stars will soon form.

Fig. 16. This drawing of the entire sky shows 7,000 stars plus the Milky Way. (Lund Observatory, Sweden)

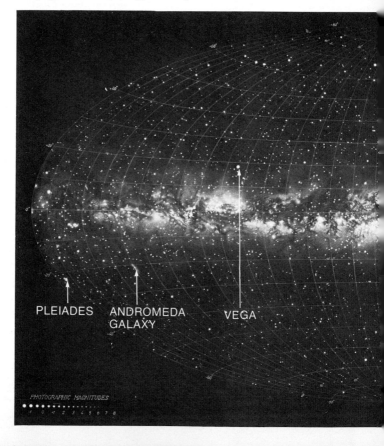

PLEIADES   ANDROMEDA   VEGA
           GALAXY

PHOTOGRAPHIC MAGNITUDES

The Milky Way Galaxy

A *galaxy* is an island of matter in space — a giant collection of gas, dust, and millions or billions of stars. The galaxy we live in, which includes about a trillion (a thousand billion) stars, is called the *Milky Way Galaxy*. Most of the Milky Way Galaxy is in the shape of a disk; the earth is approximately halfway out from the center. Since the earth is located inside the Milky Way Galaxy, various parts of the galaxy are always visible in our sky. The disk of the Milky Way Galaxy is thin. When we look toward or away from the center or in any other direction in the plane of that disk, we see many stars and much gas and dust; in those directions we see a band of white across the sky, mottled with some dark material. People have long called the band of light that appears to cross the sky by the name the *Milky Way* because of its appearance (Fig. 16); it is the Milky Way from which our galaxy draws its name.

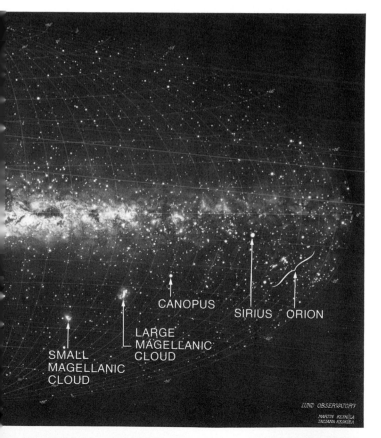

CANOPUS

SIRIUS   ORION

LARGE
MAGELLANIC
CLOUD

SMALL
MAGELLANIC
CLOUD

LUND OBSERVATORY

MARTIN KESKÜLA
TATJANA KESKÜLA

If we look in directions other than that of the plane of our galaxy, our line of sight quickly passes out of the galaxy, and we do not see many stars or much gas or dust; in those directions we see only some of the nearer stars against a dark background. Imagine living deep in a groove of a phonograph record: when you look toward the center or toward the outside edge of the record, you look through a lot of vinyl; when you look upward or downward, though, you quickly see out of the record.

Most of the open clusters lie in the disk of our galaxy, so we see them in or near the Milky Way. Most of the nebulae lie in the disk as well. The globular clusters, though, form a large spherical halo centered at the center of our galaxy. The halo extends far above and below the center of our galaxy; thus these globular clusters do not, for the most part, lie in or near the Milky Way. Indeed, it was the realization around 1920 that the globular clusters seemed to be forming a spherical halo around something that led Harlow Shapley to deduce that whatever was at the center of that halo was the center of our galaxy. This was the proof that the sun does not lie in the center of the galaxy or of the universe.

When we look away from the Milky Way, we are able to look out of our galaxy and see more distant objects, such as other galaxies. (Galaxies and nebulae are known as *deep-sky objects*.) Still, when we look at photographs of these distant objects, the individual points of light we see are usually stars in the foreground, stars that are relatively close to us — in our galaxy, rather than in the distant galaxies.

GALAXIES

The Milky Way is only one of millions of galaxies in the universe. Most galaxies outside our own are too faint and distant to be seen with the naked eye or binoculars, but it is fascinating to study their shapes in a telescope. A few interesting examples of galaxies are listed in Table 8; the times of the year when they are visible are shown on a Graphic Timetable (Fig. 26).

Classes of Galaxies

A basic scheme for classifying galaxies was worked out in the 1920s by Edwin Hubble. More recent and elaborate methods, such as those of Gerard de Vaucouleurs, have also been developed. The Milky Way Galaxy is an example of a *spiral galaxy* — a galaxy in which several "arms" seem to unwind from a central region. Hubble classified the spiral galaxies with an *S* (for *spiral*) followed by an *a, b,* or *c,* depending on the increasing looseness of the arms of the galaxy. Thus a galaxy with tightly wound arms is classified as an Sa; one with less tightly wound arms, like our galaxy or the Andromeda Galaxy (C.Pl. 9), is an Sb; and one with loosely wound arms is an Sc (C.Pl. 10).

We cannot move around to different sides of a galaxy to see it from different perspectives. A telescope reveals so many galaxies, though, that we can see examples of each class from different angles — broadside (face-on), somewhat tilted, or edge-on (Figs. 17, 18, and 19). From studying all these galaxies we know, for example, that an Sa galaxy has a smaller bulge at its center than an Sc galaxy does.

The youngest stars and most of the gas and dust are located in the arms of a spiral galaxy. Since the hottest stars burn out quickly, whenever we see a hot star (which we can recognize by its blue color), it must be relatively young. Color photographs of galaxies (such as C.Pl. 10) show that the central regions are relatively yellow, indicating that older stars are dominant there, while the arms are relatively blue and therefore contain relatively young stars.

Some spiral galaxies have a central bar from which the arms unwind, and are classified with a *B* (for *bar*): SBa, SBb, and SBc (see Fig. 75 opposite Atlas Chart 34 in Chapter 7).

Most galaxies do not have spiral arms, but have an elliptical shape instead. They are classified with an *E* (for *elliptical*), with a number from 0–7, showing how far out-of-round they appear. E0 galaxies appear spherical (Fig. 20), while E7 galaxies appear quite

Fig. 17. (*left*) A broadside view of NGC 4622 in Centaurus, a type Sb spiral galaxy similar to our own. (The Cerro Tololo Inter-American Observatory)
Fig. 18. (*right*) M81 (NGC 3031) in Ursa Major, a type Sb spiral galaxy seen at an angle of 59° from face-on. (The Kitt Peak National Observatory)

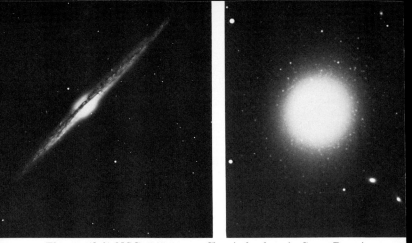

Fig. 19. (*left*) NGC 4565, a type Sb spiral galaxy in Coma Berenices. (Palomar Observatory photo)

Fig. 20. (*right*) M87 (NGC 4486) in Virgo, a peculiar elliptical galaxy at the core of the Virgo Cluster of galaxies. The faint dots around it are its globular clusters. Short exposures show a jet of gas being expelled from the galaxy's core; violent activity must be going on there, perhaps because a giant black hole is present. (The Kitt Peak National Observatory)

oblong. Elliptical galaxies are quite old, contain only old stars, and have no gas and dust. They come in a wide range of sizes, from huge to relatively small on a galactic scale.

In between the most elliptical of the elliptical galaxies and the spiral galaxies with the most tightly wound arms is a transitional class of galaxy: S0. An S0 galaxy has a disk but no arms. It is currently thought that the shapes of galaxies result from the different conditions under which they formed, and that galaxies do not evolve from one type to another.

Some galaxies of each type have a noticeable peculiarity in addition to one of the basic shapes described above. A galaxy might have a jet of gas sticking out (C.Pl. 24), for example, or be wrapped with extra dust (C.Pl. 42). These galaxies are labelled (*pec*), for *peculiar*, as in E7 (pec).

Other galaxies have no special shape, and are known as *Irregular* (*Irr*). Some irregular galaxies seem to have a trace of spiral structure, and might really be the equivalent of Sd spirals (that is, spirals with very loose arms). Others are truly irregular. The nearest type-Irr galaxies to us are the two small satellite galaxies of the Milky Way Galaxy. Since they were first seen by Magellan's crew as they sailed far enough south, these satellite galaxies are known as the Magellanic Clouds. They are best visible from southern latitudes (see Fig. 94 opposite Atlas Chart 52 in Chapter 7), and cannot be seen from the continental United States. In a telescope, the Large Magellanic Cloud (LMC) seems to show traces of spiral structure (C.Pl. 34), while the Small Magellanic Cloud (SMC) seems completely irregular (Fig. 21).

All galaxies except the very closest ones are moving away from us. We can tell this by studying their spectra, which are shifted toward the red end of the spectrum. A *redshift* — a lengthening of the wavelength of any wave — results from motion away from you, just as the length of a sound wave increases as a motorcycle passes you and starts to recede from you. Edwin Hubble discovered that the farther away a galaxy is, the faster it is receding. We measure how fast galaxies are receding by studying the redshifts in their spectra. This method works no matter how far away the galaxy is, as long as we can take a long enough exposure to measure a redshift. Once we measure the redshift, we use *Hubble's Law,* which links redshift with distance, to find out how far away galaxies are. This is the only method we have for finding the distances to the farthest objects in the universe.

The enormous distances to galaxies give us an indication of the immense scale of the universe. The nearest galaxy that northern-hemisphere observers can see is the Great Galaxy in Andromeda, sometimes simply called the Andromeda Galaxy (C.Pl. 9). The center of this galaxy is dimly visible to the naked eye, but it is so far away that its light has been travelling for 2.2 million years — at a speed of more than 186,000 miles per second— to reach us. When we observe the Andromeda Galaxy, we are looking 2.2 million years back in time.

All galaxies apparently come in groups. Our Milky Way Galaxy, the Andromeda Galaxy, and M33 (C.Pl. 10) are the three spirals in our Local Group of some two dozen galaxies, which also includes a couple of giant elliptical galaxies and many dwarf elliptical and irregular galaxies.

Fig. 21. The Small Magellanic Cloud, with the globular cluster 47 Tucanae (NGC 104) at right and the globular cluster NGC 362 above (to the north). (Harvard College Observatory)

The Local Group is an outlying member of a large *cluster of galaxies* that covers much of the constellation Virgo and is thus known as the Virgo Cluster (Fig. 22). The bulk of the galaxies in the Virgo Cluster are about 60 million light-years away. Other prominent clusters are in Coma Berenices, about 300 million light-years away, and in Hercules, about 500 million light-years away. (These values are computed assuming that Hubble's constant — the rate at which velocity increases with distance — is 75 km/sec/megaparsec = 23 km/sec per million light-years.) Clusters of galaxies have recently been studied particularly well from x-ray telescope observations, which have revealed the presence of gas in between the galaxies.

Many clusters of galaxies are linked together as *superclusters.* In between the superclusters are giant voids where no galaxies are found.

Quasars

It came as a surprise in 1960 when some of the sources of celestial radio waves appeared to be pointlike — "quasi-stellar" — instead of looking like galaxies. These "quasi-stellar radio sources" — *quasars,* for short — were discovered three years later to be extremely far away, when Maarten Schmidt discovered that the spectrum of one of them showed an extreme redshift. A class of radio-quiet quasars — quasi-stellar objects with huge redshifts — has since been discovered. We now know of over a thousand quasars.

The redshift can be expressed as the fraction or percentage by which the wavelengths of light are shifted. For speeds much slower than the speed of light, this fraction is the same as the fraction that the recession speed of the quasar is of the speed of light. For example, a redshift (called z) of 0.2 means that the wavelengths are shifted toward the red by 0.2 times (20% of) the original wavelength. It is also true that the galaxy or quasar emitting such light is receding from us at 0.2 times the speed of light, or 60,000 km/sec. A quasar with redshift of 1.0 has its wavelengths shifted by 1.0 times (100% of) the original wavelength; each wavelength is doubled. But Einstein's formulas from the special theory of relativity must be used to calculate how fast the quasars are receding; their velocities are always less than the speed of light.

The most distant quasar is redshifted by $z = 3.78$. Few quasars with redshifts greater than 3 (300%) have been found, even though instruments exist that would be sensitive enough to do so. The farther out the quasar is, the farther back in time the light reaching us was emitted. That few quasars of the largest redshifts are found indicates that we are seeing back to the instant at which the first quasars were formed — somewhat over 10 billion years ago.

Since quasars are so far away yet still send us some visible light and relatively strong radio radiation, the quasars must be astonishingly bright. Astronomers now agree that giant black holes are

Fig. 22. Part of the Virgo Cluster of galaxies, including the elliptical galaxies M84 (right center) and M86 (far right). NGC 4438 is the distorted galaxy at upper left. The positions of these galaxies are shown on Chart 27A in Chapter 7. (The Kitt Peak National Observatory)

probably present in the centers of quasars. These black holes would contain millions of times the mass of the sun. As gas is sucked into the black holes, it heats up and gives off energy. Quasars may be a stage in the evolution of galaxies, for there is evidence that some quasars have structure around them not unlike that of galaxies. Since we see most quasars quite far away, we are seeing them far back in time, and we can conclude that the epoch at which quasars were brightest took place long ago.

Only one quasar — 3C 273, in the constellation Virgo (Fig. 23) — is bright enough to be accessible with amateur telescopes, but we have plotted a handful of quasars on the Atlas Charts in Chapter 7 for illustrative purposes.

Cosmology

Hubble's observations that the most distant galaxies are receding from us faster than closer galaxies can be explained if the universe were expanding in a way similar to the way a giant loaf of raisin bread rises. If you picture yourself as sitting on a raisin, all the other raisins will recede from you as the bread rises. Since there is more dough between you and the more distant raisins, the dough will expand more and the distant raisins will recede more rapidly than closer ones. Similarly, the universe is expanding. We have the same view no matter which raisin or galaxy we are on, so the fact that all raisins and galaxies seem to be receding doesn't say that we are at the center of the universe. Indeed, the universe has no center. (We would have to picture a loaf of raisin bread extending infinitely in all directions to get a more accurate analogy.)

Fig. 23. The quasar 3C 273, in Virgo, is magnitude 13.6. (See also Chart 27A.) The long jet of gas coming from it is a sign of violent activity there. (Palomar Observatory photo)

Since the universe is uniformly expanding now, we can ask what happened in the past. As we go back in time, the universe must have been more compressed, until it was at quite a high density 15 billion or so years ago. Most models of the origin of the universe say that there then was a *big bang* that started the expansion. A new model — the *inflationary universe* — holds that the early universe grew larger rapidly for a short time before it settled down to its current rate of expansion. The big bang itself may not have occurred; the first matter could have formed as a chance fluctuation in the nothingness of space.

The temperatures were so high in the first microsecond of the universe that not even the chemical elements had formed. But as the universe cooled, first basic atomic particles, such as protons (hydrogen nuclei), and then heavier atoms formed. Still, the universe was opaque until it cooled down to 100,000°C. When it reached that temperature, protons and electrons combined to make hydrogen atoms and the universe became transparent. The hot radiation that permeated the universe at that time has cooled as the universe has continued to expand, and is now detectable as a faint glow in sensitive radio telescopes. The glow has a temperature of only 3 kelvins (3°C above absolute zero), and is our best evidence that the early universe was hot and dense.

Astronomers ask what will happen to the universe in the future. The evidence is not all in. One possibility is that the universe is *open* — it will continue to expand forever. Another possibility is that the universe will eventually stop its expansion and begin to contract. We know that this cannot happen for at least 50 billion years more — at least three times longer than the current age of the universe — because we can observe the rate at which the universe is now expanding. Still, if the universe does contract in the long run, then we have a *closed universe* that will wind up in a *big crunch.* The inflationary model indicates that the universe will expand forever, but at a decreasing rate.

Observations from new telescopes on the ground, from Space Telescope in orbit, from the Very Large Array of radio telescopes stretching over a Y-shaped pattern with a 13-mile (21-km) radius in New Mexico, and from other instruments are expected to help us forecast the future of the universe.

Table 5. Selected Double Stars

Name	Constellation	Magnitudes	Color	Separation of Components (in arc sec)
gamma Andromedae	Andromeda (Chart 9)	2.3 & 5.1	orange & blue	10″
beta Monocerotis	Monoceros (Chart 24)	5.2, 5.6, 4.7	blue	2.8″ & 7.4″
iota Cancri	Cancer (Chart 13)	4.2 & 6.6	yellow & blue	31″
gamma Virginis	Virgo (Chart 27)	3.6 & 3.6	yellow	4″
beta Scorpii	Scorpius (Chart 29)	2.9 & 5.1		14″
beta Cygni (Albireo)	Cygnus (Chart 18)	3.2 & 5.4	yellow & green	35″

Note: The times when these doubles are visible are shown in the Graphic Timetable (Fig. 24) on p. 129. The positions of these and other double stars are given in Table 10 (pp. 148–149) and are plotted on the Atlas Charts in Chapter 7.

Table 6. Selected Variable Stars

Name	Constellation	Type (*see p. 151*)	Magnitude (Range)	Period (Days)
Mira (omicron Ceti)	Cetus (Chart 22)	long-period	3.4–9.2	332
Algol (beta Persei)	Perseus (Chart 10)	eclipsing binary	2.2–3.5	2.9
delta Cephei	Cepheus (Chart 8)	Cepheid variable	3.6–4.3	5.4
zeta Geminorum	Gemini (Chart 12)	Cepheid variable	3.7–4.3	10.2
beta Lyrae	Lyra (Chart 18)	eclipsing binary	3.4–4.3	12.9
RR Lyrae	Lyra (Chart 18)	RR Lyrae variable	7.0–8.0	0.6

Note: The times when these variables are visible are shown in the Graphic Timetable (Fig. 24) on p. 129. The positions of these and other variable stars are given in Table 11 (p. 155) and Table 12 (p. 157); they are also plotted on the Atlas Charts in Chapter 7. Special finding charts for Mira, Algol, and beta Lyrae appear at the end of Chapter 6 (see pp. 159–160).

Table 7. Open and Globular Clusters

Name	Constel- lation	Type	Mag.	Diam.	r.a. dec. (2000.0)
M103	Cassiopeia	open	7	6′	01ʰ33ᵐ +60°41′
h and χ Persei	Perseus	open	*	45′, 45′	02ʰ23ᵐ +57°06′
M45 (Pleiades)	Taurus	open	3	120′	03ʰ48ᵐ +24°06′
M79	Lepus	globular	7	8′	05ʰ24ᵐ −24°31′
M35	Gemini	open	6	29′	06ʰ09ᵐ +24°20′
M44 (Praesepe)	Cancer	open	4	90′	08ʰ40ᵐ +19°59′
M3	Canes Venatici	globular	6	19°	13ʰ42ᵐ +28°23′
M5	Serpens	globular	6	20′	15ʰ18ᵐ +02°05′
M13	Hercules	globular	6	23′	16ʰ41ᵐ +36°27′
M92	Hercules	globular	6	12′	17ʰ17ᵐ +43°09′
M23	Sagittarius	open	6	27′	17ʰ57ᵐ −19°01′
M24	Sagittarius	open	6	†	18ʰ17ᵐ −18°29′
M11	Scutum	open	7	12′	18ʰ50ᵐ −06°16′
M15	Pegasus	globular	6	12′	21ʰ30ᵐ +12°10′
M39	Cygnus	open	6	32′	21ʰ31ᵐ +48°26′

*The double cluster; a 1° field is necessary to include both.
† M24 is the star cloud in the direction of the galactic center.

Table 8. Nebulae and Galaxies

Name	Type	Constel- lation	Diam.	ra. dec. (2000.0)
M31	spiral	Andromeda	2° × 30′	00ʰ42ᵐ +41°16°
M1 (Crab)	supernova	Taurus	6′ × 4′	05ʰ35ᵐ +22°01′
M42 (Orion)	emission	Orion	60′	05ʰ35ᵐ −05°23′
M81	spiral	Ursa Major	16′ × 10′	09ʰ56ᵐ +69°03′
M49	elliptical	Virgo	2′	12ʰ29ᵐ +07°59′
M51 (Whirlpool)	spiral	Canes Venatici	12′ × 6′	13ʰ30ᵐ +47°11′
M20 (Trifid)	emission	Sagittarius	29′	18ʰ03ᵐ −23°02′
M57 (Ring)	planetary	Lyra	1′	18ʰ54ᵐ +35°01′
M27 (Dumbbell)	planetary	Vulpecula	8′ × 4′	20ʰ00ᵐ +22°43′

Notes: Diameters are expressed in degrees (°) and arc minutes (′). Abbreviations; spiral = spiral galaxy; supernova = supernova remnant; emission = emission nebula; elliptical = elliptical galaxy; planetary = planetary nebula.

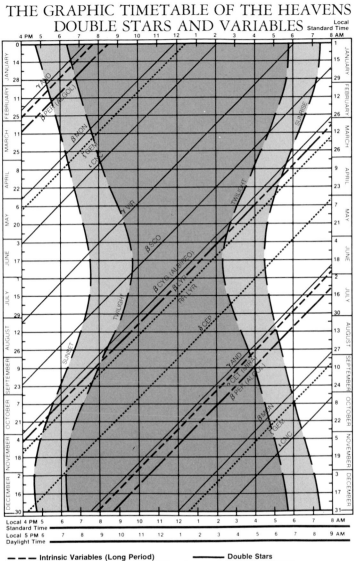

THE GRAPHIC TIMETABLE OF THE HEAVENS
DOUBLE STARS AND VARIABLES

- – – Intrinsic Variables (Long Period)   ——— Double Stars
- ·········· Intrinsic Variables (Cepheids & RR Lyrae)   —— · — Eclipsing Variables

Fig. 24. Graphic Timetable of Double and Variable Stars. (© 1982 Scientia, Inc.)

THE GRAPHIC TIMETABLE OF THE HEAVENS
OPEN AND GLOBULAR CLUSTERS

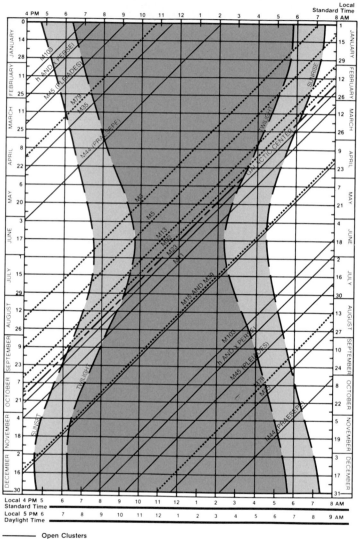

——— Open Clusters
·········· Globular Clusters

Fig. 25. Graphic Timetable of Star Clusters. (© 1982 Scientia, Inc.)

THE GRAPHIC TIMETABLE OF THE HEAVENS
GALAXIES AND NEBULAE

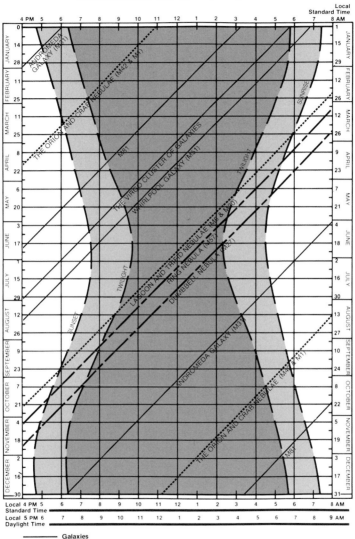

Fig. 26. Graphic Timetable of Galaxies and Nebulae. (© 1982 Scientia, Inc.)

5

The Constellations

Today we know that the stars in any given constellation do not necessarily have any physical relationship to one another. Some stars within a constellation may be relatively close to earth, while others may be relatively far away. All we know is that the stars are in roughly the same direction as seen from earth. Yet, just as it was convenient for the earliest observers of ancient civilizations to divide up the heavens into constellations or groups of seemingly related stars, it remains convenient for us to associate each star with one and only one constellation.

History of the Constellations

Exactly when and where the first system of constellations was devised is not known. Cuneiform texts and artifacts from the civilizations of the Euphrates Valley suggest that the lion, the bull, and the scorpion were already associated with constellations by 4000 B.C. Many scholars have been intrigued by the fact that there is some similarity between the names given to constellations by civilizations separated by vast distances. Perhaps a very ancient common tradition for naming a few groups of stars will ultimately be found. For the most part, however, the constellations of different civilizations appear to have developed independently of one another.

Of the 88 constellations listed by the International Astronomical Union in the definitive compilation of 1930, more than half were known to the ancients. Early records of the Greek constellations are found in the poetry of Homer, from about the 9th century B.C., and of Aratus, from about the 3rd century B.C. Sometime between the lives of these two poets, probably in the mid- to late 5th century B.C., the *ecliptic* — the path the sun appears to follow across the celestial sphere in the course of a year — was identified in Babylon and perhaps in Greece as well. The Babylonians divided the ecliptic into the 12 parts of the *zodiac* — the band of constellations through which the sun, moon, and planets move (as seen from earth) in the course of the year.

During the 2nd century A.D., the Egyptian astronomer Ptolemy catalogued information about 1022 stars, grouped into 48 constellations. Not surprisingly, Ptolemy's catalogue comprises only stars visible from the latitude of Alexandria, where he lived and wrote.

The *Almagest,* Ptolemy's chief work, remained the last word on the constellations until the 16th century, when European voyages of discovery took navigators into southern latitudes. The first star atlas, published by Johann Bayer in 1603, included 12 new constellations visible in southern-hemisphere skies. In 1624, the German astronomer Jakob Bartsch added three new constellations in the spaces that existed between previously named constellations. Bartsch also listed as a separate constellation the grouping we know as Crux, the Cross, whose four chief stars Ptolemy had noted as part of the constellation Centaur. (The name Crux also reveals an attempt to depaganize the heavens that was typical of this period.) Earlier in the 17th century, Tycho Brahe likewise elevated to the status of constellation the asterism Coma Berenices, Berenice's Hair, which the ancients had thought of as part of Leo or Virgo.

Seven more constellations visible from midnorthern latitudes were added by the German astronomer Johannes Hevelius in 1687. The visit of Nicolas Louis de La Caille to the Cape of Good Hope in 1750 resulted in 14 additional southern constellations. There have been other attempts since that time to invent new constellations, but official acceptance has been withheld. However, since the mid-1800s it has become traditional to omit Ptolemy's largest constellation, Argo Navis (Argo the Ship), and to list in its stead three constellations representing the ship's keel (Carina), stern (Puppis), and sails (Vela), in addition to the compass (Pyxis) invented by La Caille.

The current list of constellations is the one adopted by the International Astronomical Union in 1928, and codified in a list two years later. The IAU defined a constellation as one of 88 regions into which they divided the entire sky; each area of the sky belongs to one and only one of these regions. The boundaries zigzag so that the lines that separate constellations do not wreak havoc with ancient figures. A few stars that had been previously thought of as part of another constellation wound up in a new one — for example, one of the four stars of the Great Square of Pegasus is now officially in Andromeda. But on the whole, the IAU division provided a great simplification.

The lines dividing constellations were drawn along lines of right ascension and declination for the year 1875.0. (Astronomers use decimals to indicate parts of a year; 1875.0 means the beginning of the year 1875.) But because of *precession* (the drifting of the direction of the earth's axis among the stars), the lines between constellations have drifted slightly; the Atlas Charts in Chapter 7 show that the lines dividing constellations are no longer aligned so neatly with lines marking the coordinate scales.

Precession has also changed the dates when the sun appears to drift through each constellation of the zodiac, so that the sun is not actually in the sign listed in the newspapers' daily horoscopes. The sun actually passes through 13, not 12, constellations in the course of a year. Further, all or part of 24 constellations are actually in the zodiac, if we define it as the region within about 8° of the ecliptic — the band in which we find the first eight planets. (If we include Pluto, unknown to the ancients, the band would be wider still.)

The list below divides the constellations into three groups: the constellations of the traditional zodiac, listed in order around the sky; other constellations catalogued by Ptolemy, arranged here alphabetically; and constellations added since 1600, also listed alphabetically. Since in many cases more than one mythological story is associated with the same constellation, the list does not aim at providing full coverage of all the myths. For each constellation, we list the numbers of the Atlas Charts on which the constellation is found in Chapter 7. A list of standard abbreviations for the constellations and the genitive form of their names used to make star names (*e.g.* alpha Centauri, in Centaurus) appears in Appendix Table A-1.

Beautiful drawings of some of the constellations from the old star atlases of Bayer and Hevelius appear throughout this *Field Guide*.

Constellations of the Zodiac

Aries, the Ram. The golden fleece of this ram was the prize ultimately carried off by Jason, leader of the Argonauts. (Charts 10, 22, and 23)

Taurus, the Bull. Zeus disguised himself as a snow-white bull in order to attract Europa, Princess of Phoenicia. Drawn to the animal by its beauty, she climbed onto its back. Zeus then swam with his passenger to Crete, where he revealed his identity to her and won her. (Charts 10, 11, 23, and 24)

Gemini, the Twins. These are the two devoted twins and, some say, half-brothers. Gemini is the Roman word for the Dioscuri, who in Greek mythology were sons of Leda, the wife of Tyndarus, the king of Sparta. Though Castor was the son of Tyndarus, Pollux (Polydeuces in Greek) may have been the son of Zeus, which would have made him immortal. After Castor's death, Pollux was overwhelmed with grief, and wanted to share his immortality with his twin. Finally Zeus reunited them by placing them together in the heavens. (Charts 12, 24, and 25)

Cancer, the Crab. When Hercules was struggling with Hydra (see p. 139), Juno sent this crab to attack him. The crab did not succeed in its mission, and instead was crushed. But Juno rewarded the crab for its attempt by placing it among the stars. (Charts 13, 25, and 26)

Leo, the Lion. The thick, tough skin of this, the fiercest lion in the world, became the trademark of Hercules (see p. 139) after the hero succeeded in strangling the lion to death, thus successfully completing his first Labor. (Charts 13, 14, 26, and 27)

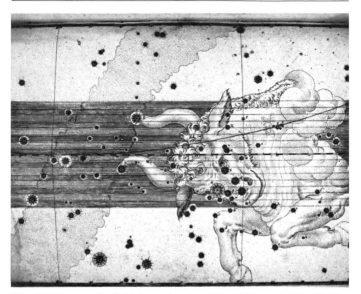

Fig. 27. Taurus, the Bull, from the star atlas of Bayer (1603). (Smithsonian Institution Libraries)

Virgo, the Virgin. Virgo has been identified with a variety of heroines. The association with Ceres, goddess of the harvest, helps explain the name of the brightest star in Virgo, Spica, which means an ear of wheat or corn. (Charts 27, 28, 39, and 40)

Libra, the Scales. This constellation was probably invented at the time the zodiac was established. The Arabic names of the two brightest stars in Libra mean the southern and the northern claw, confirming that at one time Libra was a part of Scorpius. The constellation was later associated with the scales held by Astraea, goddess of justice. (Charts 28, 29, and 40)

Scorpius, the Scorpion. According to one account, Apollo, who was concerned about maintaining his sister Artemis' virginity, sent the scorpion to kill Orion (see p. 140). Both Orion and Scorpius were placed as far from one another in the heavens as possible, so as to avoid further trouble between them, though Scorpius still pursues Orion around the celestial sphere. (Charts 29 and 41)

Sagittarius, the Archer. This constellation has been associated with the centaur Chiron, who was known for his marksmanship as well as for his expertise in medicine and music. Others say that Chiron was too civilized a model for Sagittarius; they say that Chiron is associated with the constellation

Sagittarius because he identified it in the sky to help guide the Argonauts on their voyage to Thessaly. (Charts 30, 31, 41–43)

Capricornus, the Sea Goat. The last three constellations of the zodiac are in the region of the sky called the Sea. From early antiquity this constellation has been depicted with the head and body of a goat and with the tail of a fish. According to one interpretation, the goat, an expert climber, represents the sun's climb from its lowest position in the sky, which is in this constellation. The fish's tail may represent the seasonal rains. (Charts 31, 32, and 43)

Aquarius, the Water Carrier. Early artifacts depict this constellation as a man or boy pouring water from a bucket, also representing the rainy season. (Charts 21, 31, 32, and 44)

Pisces, the Fish. Venus and Cupid escaped from the monster Typhon (see under Piscis Austrinus, p. 141) by disguising themselves as these fish and jumping into the Euphrates River. (Charts 9, 21, and 22)

Other Constellations Catalogued by Ptolemy

Andromeda. Several constellations are associated with the story of Andromeda, the daughter of Cassiopeia and Cepheus. When Cassiopeia boasted that her beauty exceeded that of the sea nymphs, the slighted beauties asked Neptune, god of the sea, to punish the braggart. Neptune sent the sea monster Cetus to ravage the kingdom of Cepheus. When Cepheus consulted an oracle for advice, he was informed that only the sacrifice of Andromeda to Cetus, the sea monster, would appease the gods. Andromeda was duly chained to a rock by the sea but her fate was averted by the arrival of Perseus (see p. 140), who turned Cetus into stone by flashing the face of Medusa before the monster. (Charts 9, 10, and 20)

Aquila, the Eagle. This bird was rewarded with a place in the heavens for having brought from earth the handsome Ganymede to serve as cupbearer of the gods. (Charts 30 and 31)

Ara, the Altar. The Olympian gods raised this altar to commemorate their victory over the Titans. (Charts 41 and 51)

Argo Navis, the Ship Argo. This ship carried the Argonauts from Thessaly to Colchis in search of the Golden Fleece. No longer considered a single constellation, this grouping has been subdivided into four constellations: Carina, Puppis, Pyxis, and Vela (see pp. 142 and 145).

Auriga, the Charioteer. No story definitively explains the figure that this constellation's stars supposedly define — a charioteer (minus chariot and horse), holding the reins in his right hand, a goat on his left shoulder, and two small kids in his left arm. One myth associates Auriga with Erecthonius, the lame son of Vulcan and Minerva. Erecthonius invented the chariot, which not only enabled him to get around but also won him a place in the heavens. (Charts 3, 11, and 12)

Boötes, the Herdsman. No single myth is definitively associated with this constellation. According to one story, Boötes was rewarded with a place in the heavens for his invention of the plow. (Charts 15, 16, 28, and 29)

Canis Major, the Greater Dog. This constellation has been associated with several mythical dogs, including the hound of Actaeon, but it has also been known as the Dog of Orion. Orion loved to hunt wild animals, such as Lepus, the Hare (see p. 139); Canis Major, at Orion's heel in the sky, seems ready to pounce on the Hare. (Charts 24, 25, and 36)

Canis Minor, the Lesser Dog. Among the dogs with which this constellation has been associated is Helen's favorite, which allowed Paris to abduct her without resistance. (Chart 25)

Cassiopeia. See also Andromeda (above). When Cassiopeia objected to the wedding of Perseus and Andromeda, Perseus exhibited the head of Medusa. As a result, his enemies, including Cassiopeia, were turned into stone. Neptune placed Cassiopeia in the heavens, but in order to humiliate her, he arranged it so that at certain times of the year she would appear to be hanging upside down. (Charts 1, 2, and 9)

Centaurus, the Centaur. This constellation is often identified with Chiron (see under Sagittarius). When Hercules accidentally wounded Chiron with one of his poisoned arrows, Chiron suffered greatly, but, as an immortal, could not find release in death. He resolved his problem by offering himself as a substitute to Prometheus, a Titan who was suffering because he had stolen fire from heaven for the benefit of mankind. The gods had punished Prometheus by chaining him to a rock, where each day a vulture devoured his liver, which was restored each night. At the request of Hercules, Zeus agreed to release Prometheus if a substitute could be found for him. Chiron offered his immortality to Prometheus and went to Tartarus, where Prometheus had been imprisoned by Zeus. Zeus placed Chiron among the stars. See also Corona Australis (below). (Charts 38–40, 49, and 50)

Cepheus, a king of Ethiopia. See Andromeda (above). (Charts 1, 2, and 8)

Cetus, the Whale. A sea monster; see Andromeda. (Charts 21–23, 33 and 34)

Corona Australis, the Southern Crown. Also called Corona Austrina. The Southern Crown is said to represent a crown of laurel or olive, placed on victors in games and on those who perform great service to their peers. According to one story, this constellation symbolizes a laurel wreath ready to be placed on the brow of Chiron in acknowledgment of his service to Prometheus. See Centaurus (above). (Chart 42)

Corona Borealis, the Northern Crown. This constellation is generally associated with Ariadne, daughter of Minos, king of

Crete. Each year, as a tribute to Crete, 14 young people from Athens were fed to the Minotaur, the monster that Minos kept in a labyrinth. When Theseus arrived from Athens as one of the prospective victims, Ariadne fell in love with him. She offered to help him escape the Minotaur if he promised to take her back to Athens as his bride. Theseus agreed. Ariadne kept her part of the bargain, and after Theseus killed the Minotaur, they sailed off. They stopped on the island of Naxos, where Theseus and the Athenians abandoned the sleeping Ariadne. When she awoke, clamoring for vengeance, she found Dionysus, god of the vine, who married her immediately. As a wedding gift, Dionysus gave her a crown studded with gems. When Ariadne died, Dionysus set the crown among the stars. (Charts 16 and 17)

Corvus, the Raven. According to some, this is the raven that Apollo sent to guard his beloved, Coronis, during his absences. During one of the god's absences, Coronis fell in love with another, and was unfaithful to Apollo. When the raven told Apollo what had happened, Apollo rewarded the bird by placing it in the sky. (Charts 27 and 39)

Crater, the Cup. This constellation, which is supposed to resemble a goblet, has been associated with various gods and heroes, including Apollo, Bacchus, Hercules, and Achilles. (Charts 27 and 38)

Cygnus, the Swan. According to one legend, this constellation is related to the story of Phaethon, a mortal who learned that his father was Helius, the sun god. Helius rashly promised to let Phaethon drive the chariot of the sun across the sky. Phaethon soon lost control, and his reckless driving threatened to destroy the earth with the sun's heat. Zeus intervened by hurling a thunderbolt at Phaethon, who fell into the Eridanus River (see below). Phaethon's devoted friend, Cygnus, in his grief dived into the water in search of the body. Apollo took pity on Cygnus, changed him into a swan, and placed the swan in the sky. (Charts 7, 8, 18, and 19)

Delphinus, the Dolphin. In one story, the dolphin successfully convinced the sea goddess Amphitrite to marry Poseidon, from whom she had been fleeing. As a reward, Poseidon placed Delphinus among the stars. (Charts 19 and 31)

Draco, the Dragon. Among the monsters with whom this constellation has been associated is the dragon slain by Cadmus, brother of Europa, on the site of the city that was later to become Thebes. After Cadmus accomplished this feat, Athena instructed him to plant some of the dragon's teeth. Armed men grew up from the soil; when Cadmus threw stones into their midst, they began to fight among themselves. All but five died, and the survivors helped Cadmus build the city of Thebes. (Charts 4–8)

Equuleus, the Little Horse. This group of stars is associated with Celeris, the brother of Pegasus (see p. 140). Mercury gave Celeris to the hero Castor. (Chart 32)

Eridanus, the River. The waters of the River Eridanus are said to steam perpetually as a result of Phaethon's fiery fall (see under Cygnus, above). (Charts 22–24, 34, 35, and 46)

Hercules. The most famous of all the Greek heroes, Hercules was worshiped throughout the Mediterranean world. Best known for his Twelve Labors, he performed many incredible deeds throughout his life. His death provided an equally dramatic end to his career. After accidentally killing a young man, Hercules decided to go into exile with his wife, Deianira. When they came to a river, Hercules swam on ahead, leaving Deianira in the charge of Nessus, a centaur who offered to carry her across the water on his back. When Nessus tried to rape Deianira, Hercules shot the centaur with a poisoned arrow. Before Nessus died, he suggested to Deianira that she keep his blood as a love charm, should she ever need to revive Hercules' interest in her. Some time later, learning of her husband's interest in another woman, Deianira rubbed the centaur's dried blood on one of Hercules' robes. The hero's body burned as soon as he put the robe on, and when he tried to remove the robe, his skin came off with it. When Deianira learned of the effects of the robe, she hanged herself. Hercules built and mounted a funeral pyre, which only the compassionate Philoctetes had the courage to light. Instantly, a flash of lightning was seen. The pyre burned out completely, leaving no bones behind. It was assumed that Hercules' body had been carried to Olympus. (Charts 17, 18, 29, and 30)

Hydra, the Water Snake. Successfully eliminating this many-headed monster was Hercules' Second Labor — a difficult task because whenever one head was cut off, two new ones grew in its place. Hercules solved this problem by having his nephew, Iolaus, burn the stump of each head as soon as Hercules cut them off, thus preventing new heads from sprouting. Because Iolaus assisted him in this labor, Hercules was required to substitute another one in its place. Do not confuse this constellation with Hydrus, a southern constellation (p. 143). (Charts 25, 26, 37–39, and 40)

Lepus, the Hare. According to some, the Hare was placed in the heavens to be near its hunter, Orion. (See Canis Major, p. 137.) (Charts 24, 35, and 36)

Lupus, the Wolf. One story associates this constellation with the impious Lycaon, who doubted Zeus' claim to divinity. In order to test Zeus, Lycaon served the king of gods the flesh of a child. To punish Lycaon for this impious act, Zeus transformed him into a wolf. (Charts 40 and 50)

Lyra, the Lyre. This is the instrument given by Apollo to Orpheus, the most famous poet and musician in Greek legend. When his wife Eurydice died, Orpheus was told he could bring her back from the underworld on condition that he not look back at her until she was in the light of the sun. Orpheus led her out, playing on the lyre, but when he reached the light of the sun, he looked back. Because Eurydice had not yet reached the sunlight, he lost her forever. He remained inconsolable and refused the advances of all the women who tried to win his love. One day a group of women whom he had scorned attacked him; drowning out his music, they tore him apart, and threw his head and lyre into the Hebrus River. Apollo intervened, however; Orpheus' head was placed in a cave, his limbs were buried at the foot of Mount Parnassus, and his lyre was placed among the stars. (Chart 18)

Ophiuchus, the Serpent-bearer. This group is generally identified with Asclepius, the first physician and surgeon, who accompanied the Argonauts. He was so successful in curing the ill and wounded that Pluto began worrying about the declining immigration into the underworld. Pluto convinced Zeus to strike Asclepius with a thunderbolt and put him among the constellations. The snake twined around a staff remains the symbol of medicine today, perhaps because of the association of the snake's periodic sloughing off of its skin with the renewal of life. (Charts 29, 30, and 41)

Orion, the Hunter. When this giant hunter met Artemis, goddess of the hunt and of the moon, her brother Apollo feared for her virginity. Apollo sent Scorpius, the Scorpion (see p. 135), to attack Orion, who leaped into the sea to escape. Apollo then tricked his unsuspecting sister into shooting at a dark spot on the waves that in actuality was Orion. The goddess tried to have Asclepius revive Orion, her hunting companion, but the physician had been killed by Zeus' thunderbolt. (See under Ophiuchus, above.) Artemis then placed Orion in the heavens, where he would be pursued eternally by the scorpion. (Charts 11, 12, 23, and 24)

Pegasus. This is the winged horse that sprang up from the blood of Medusa after Perseus killed her (see below). Pegasus was tamed by Bellerophon, whose successes against monsters and enemies had gone to his head. When Bellerophon decided to ride Pegasus up to Olympus, the gods were offended, so Zeus sent a gadfly to sting Pegasus. Pegasus reared in pain, causing Bellerophon to topple off. Bellerophon was lamed and blinded, but Pegasus continued to climb Olympus, and earned a place among the stars as a constellation. (Charts 9, 19–21, and 32)

Perseus. Armed with a polished shield given to him by Athena, Perseus killed Medusa, the only one of the Gorgons who was mortal. The three Gorgons were winged monsters so horrible to behold that all who looked upon them were turned into stone. Athena told Perseus to use the shield as a mirror. He could thus avoid looking directly at the Gorgons. After he cut off Medusa's head, the winged horse Pegasus sprang from her blood. The head of Medusa enabled Perseus to overcome many enemies, as well as to kill the monster Cetus (see under Andromeda, p. 136). (Charts 2, 3, 9–11)

Piscis Austrinus, the Southern Fish. Also known as Piscis Australis, this constellation is linked, by some, to the story of the monster Typhon. After Zeus and the Olympians overthrew the

Fig. 28. Perseus (drawn backwards), from the star atlas of Hevelius (1690).

Titans, sons of Gaea, Gaea gave birth to another son, Typhon. She incited Typhon to attack the Olympians, who assumed various animal forms to evade him — Venus, for example, assumed the form of a fish. (See also under Pisces, p. 136.) (Charts 43 and 44)

Sagitta, the Arrow. This constellation has been associated with several different arrows, including the one that Apollo used to slay the three one-eyed giants called the Cyclopes. (Charts 18, 19, and 31)

Serpens, the Serpent. This constellation is associated with Ophiuchus (see above). (Serpens Caput, the head, is on Charts 16 and 29; Serpens Cauda, the tail, is on Chart 30)

Triangulum, the Triangle. This minor constellation has been associated with different geographical locations. Not surprisingly, because of its similarity in shape to the Greek letter Δ (delta), it was sometimes called Delta, and thus was associated with Egypt and the Nile, whose delta provided fertile land. It was also associated with the island of Sicily, whose three promontories give it a triangular shape. (Charts 9 and 10)

Ursa Major, the Great Bear. Zeus fell in love with Callisto, daughter of Lycaon (see under Lupus, p. 139). Together they had a son, Arcas. Callisto was changed into a bear by one of the gods —

some say by Artemis, who was angry with Callisto, her formerly chaste companion; some say by Zeus, who was anxious to protect Callisto from the jealousy of his wife, Hera; and some say by Hera. (Charts 4–6, 13–15)

Ursa Minor, the Lesser Bear. When Arcas reached manhood, he was about to shoot a bear, unaware that it was his mother, Callisto. To protect Callisto, Zeus changed Arcas into a bear as well, and carried them both off by their tails to the heavens, where they became constellations. Annoyed at this honor, Hera took her revenge by convincing Poseidon not to allow the bears to bathe in the sea. For this reason, Ursa Major and Ursa Minor are circumpolar constellations, never sinking below the horizon. (Charts 2 and 6)

Constellations Added Since 1600

Note that the names of constellations reflect the times in which they were named; we find here many names of machines in addition to allusions to classical myths.

Antlia, the Pump. La Caille called this constellation the Machine Pneumatique, and in Germany it is called the Luft Pumpe, or air pump. (Charts 37 and 38)

Apus, the Bird of Paradise. The bird for which this constellation was named was originally found in Papua New Guinea. (Charts 50 and 51)

Caelum, the Chisel. La Caille formed this constellation from stars between Columba and Eridanus. (Chart 35)

Camelopardalis, the Giraffe. Bartsch first outlined this constellation, claiming that it represented the camel that brought Rebecca to Isaac. The constellation lies in the large space between Perseus, Auriga, and the Bears. (Charts 2 and 3)

Canes Venatici, the Hunting Dogs. Hevelius formed this constellation out of stars between Ursa Major and Boötes, to represent the two greyhounds held in leash by Boötes (see p. 137). (Chart 15)

Carina, the Keel [of Argo]. See Argo Navis, p. 136. (Charts 47–49)

Chamaeleon, the Chameleon. This small constellation lies below Carina, separated from the south pole by Octans. (Charts 48 and 49)

Circinus, the Drawing Compasses. This constellation, south of Lupus, was added by La Caille. (Chart 50)

Columba, the Dove. Petrus Plancius, a 16th-century Dutch theologian and mapmaker, formed this constellation south of Lepus, to represent the dove that Noah sent out from the ark. (Charts 35 and 36)

Coma Berenices, Berenice's Hair. Long recognized as an asterism, Coma Berenices was first listed as a separate constellation in 1602 by Tycho Brahe. Berenice was the wife of Ptolemy Euergetes of Egypt in the middle of the 3rd century B.C. She vowed to sacrifice her hair to Venus if her husband returned safely from war. After his safe return she honored her vow, but her hair disappeared overnight. A Greek astronomer pointed to the constellation, claiming it to be the missing hair, which Venus must have placed in the sky. Coma Berenices contains the north galactic pole (the north pole with respect to the orientation of our Milky Way Galaxy). (Charts 15, 27, and 28)

Crux, the Cross. The ancient Greeks considered the four chief stars of Crux as part of the Centaur, which surrounds Crux on three sides. There is no central star to the cross, so it looks more like a kite to the eye. (Chart 49)

Dorado, the Swordfish. The Large Magellanic Cloud lies within this constellation. (Chart 47)

Fornax, the Furnace. La Caille formed this group from stars within the southern bend of the River Eridanus. (Chart 34)

Grus, the Crane. A southern constellation. (*Grus americana* is the scientific name for the whooping crane.) (Charts 43–45, and 52)

Horologium, the Clock. One of La Caille's constellations. (Charts 34 and 46)

Hydrus, the Water Snake. Not to be confused with the Ptolemaic constellation Hydra, Hydrus lies between the Large and Small Magellanic Clouds. (Chart 46)

Indus, the Indian. This constellation is supposed to represent a native American, with arrows in both hands. (Charts 43 and 52)

Lacerta, the Lizard. Hevelius formed this constellation from stars lying between Cygnus and Andromeda. The shape supposedly was determined by the available space between the older constellations. (Charts 1, 8, and 20)

Leo Minor, the Lesser Lion. Hevelius formed this constellation from stars lying between the zodiacal constellations Leo and Ursa Major. The name reflects his belief that this star grouping was similar in nature to the two other groups. (Charts 13 and 14)

Lynx. Hevelius chose this name, explaining that only those with the eyes of a lynx could discern this star group. (Charts 3, 4, 12, and 13)

Mensa, the Table. La Caille named this constellation Mons Mensae, after Table Mountain, south of Cape Town, South Africa, where he did much of his work. (Chart 47)

Microscopium, the Microscope. La Caille formed this constellation from the stars south of Capricornus and west of Piscis Austrinus. (Chart 43)

Monoceros, the Unicorn. The Unicorn is generally attributed to Bartsch, but some claim that it is older. (Charts 24 and 25)

Musca, the Fly. In early catalogues, this group, south of the Southern Cross and northeast of Chamaeleon, was sometimes also called the Bee. (Chart 49)

Fig. 29. Monoceros, the Unicorn (drawn backwards), from the star atlas of Hevelius (1690).

Norma, the Square. One of La Caille's constellations, Norma lies immediately to the north of the Southern Triangle, Triangulum Australe (see below). (Charts 40 and 50)

Octans, the Octant. La Caille named this constellation to recognize John Hadley's invention of the octant in 1730. It includes the south celestial pole. (Charts 45–52)

Pavo, the Peacock. According to myth, Argos, the builder of the ship Argo, was changed into a peacock by Hera after she brought his ship to the heavens. So it was fitting that one of the 12 new constellations in the southern skies should be named after the peacock, thus reuniting Argos and his ship. (Charts 51 and 52)

Phoenix. Another one of the 12 new constellations in the southern skies, named after the mythical bird. The phoenix was said to live 500 or 600 years in the Arabian wilderness, to burn itself on a funeral pyre, and to rise from its ashes to live again. In ancient China, Egypt, India, and Persia, the phoenix was an astronomical symbol representing natural cycles. (Charts 33, 45, and 46)

Pictor, the Painter. La Caille named this grouping Equuleus Pictoris, or Painter's Easel. (Equuleus also means "small horse.") The name has been shortened to Pictor. The constellation lies south of Columba. (Charts 35 and 47)

Puppis, the Stern (of Argo). See Argo Navis. (Charts 25, 36–37)

Pyxis, the Compass (of Argo). See Argo Navis. (Chart 37)

Reticulum, the Reticle. This constellation was named by La Caille in honor of his reticle, the network of fine lines he placed in the focus of his telescope in order to make his observations in the southern hemisphere. The constellation, however, had been drawn earlier by Isaak Habrecht of Strasbourg. (Chart 46)

Sculptor. La Caille named this grouping l'Atelier du Sculpteur, the Sculptor's Studio, but the name has been shortened. The constellation lies between Cetus and Phoenix and contains the south galactic pole. (Charts 33 and 44)

Scutum, the Shield. Hevelius formed this constellation from several stars in the Milky Way, between the tail of Serpens and the head of Sagittarius. The constellation is supposed to represent the coat of arms of John Sobieski, king of Poland, in honor of his successful resistance against the Turkish attack on Vienna in 1683. (Chart 30)

Sextans, the Sextant. Hevelius formed this constellation, between Leo and Hydra, to recognize the importance of the sextant to his measurements of the stars. (Chart 26)

Telescopium, the Telescope. La Caille's original formation of this constellation, between Ara and Sagittarius, encroached upon several older constellations. For example, the telescope's stand was in Sagittarius. Later cataloguers redrew the outlines to avoid this kind of overlapping. (Charts 42, 51, and 52)

Triangulum Australe, the Southern Triangle. Much more prominent than its northern counterpart, Triangulum (p. 141), this group lies south of Norma. (Chart 50)

Tucana, the Toucan. Another one of the southern constellations named for exotic birds, this constellation is home to the Small Magellanic Cloud. (Chart 45)

Vela, the Sails (of Argo). See Argo Navis. (Charts 37, 38, 48, and 49)

Volans, the Flying (Fish). Shortened from its original name, Piscis Volans. (Chart 48)

Vulpecula, the Fox. A constellation added by Hevelius, Vulpecula is home to the Dumbbell Nebula (C.Pl. 8). Shortened from Vulpecula cum Ansere, the Fox with the Goose. (Charts 18 and 19)

6

Double and Variable Stars

DOUBLE STARS

When we look at the sky, most of the stars we see seem to be single points of light. But more detailed investigation has revealed that most of the stars in our galaxy are actually in *double star* systems — systems containing two or more stars. The periods with which the components of double star systems orbit each other range from hours up through centuries.

In some cases, the stars are distant enough from each other and the star system is close enough to us that we can actually see that the star system contains more than one object. A system of this type is called a *visual binary*. In other cases, even though we may not be able to actually see more than one star, we can tell that the system is a double star because the component stars sometimes block each other as they orbit, making the total brightness of the system vary. This type of variable is an *eclipsing binary star* (Fig. 31). Sometimes the fact that a star is double can be told only by examination of its spectrum, the breakup of its light into its component colors. The spectrum might show distinct contributions from each of the member stars or might show signs of the stars' motions. A double star that has been identified as such through variations of its spectrum is called a *spectroscopic binary*. Still another possibility is that we can see one star appear to wobble in the sky slightly from side to side over the years, as a result of the gravity of an otherwise invisible companion. This type of double star is called an *astrometric binary*, since it is the delicate measurements of the science of astrometry ("star measuring") that reveals it to us as a double.

Sometimes we see something that looks like a double star, but which is really two stars that happen to be almost in the same line of sight, even though they are at quite different distances from us and therefore are not close to each other. These stars, which are not physically linked, are called *optical doubles*, but are not considered true double stars.

Double stars can be recognized on the Atlas Charts in the next chapter by the horizontal lines drawn through the dots that indicate their brightnesses. Many interesting doubles are listed in Table 10 (pp. 148–149). A few double stars, including a selection of doubles that are visible at different times around the year, are listed in Table 5 (p. 127); the times when they are highest in the sky are shown in Fig. 24 on p. 129. Table 10 includes information

(text continues on p. 150)

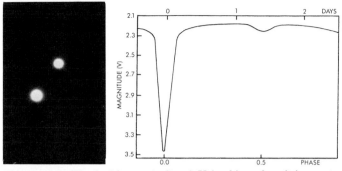

Fig. 30. (*left*) The double star 61 Cygni. Using binoculars, it is easy to see that this is a double. Its components, of magnitudes 5.2 and 6.0, are separated by 30 arc seconds. The components of this system orbit each other every 650 years. The system is only 11 light-years from earth. (Lick Observatory photo)

Fig. 31. (*right*) The light curve of Algol (beta Persei), an eclipsing binary. Algol is popular with variable-star observers because of its wide range in brightness over a conveniently short period. (Anthony D. Mallama)

Table 9. Minima of Algol

1984 Jan. 2, 5:21; Feb. 2, 18:23; Mar. 2, 10:36; Apr. 2, 23:38; May 1, 15:49; Jun. 2, 4:47; Jul. 3, 17:43; Aug. 1, 9:49; Sep. 1, 22:43; Oct. 3, 11:38; Nov. 1, 3:45; Dec. 2, 16:43.

1985 Jan. 3, 5:44; Feb. 3, 18:46; Mar. 1, 14:09; Apr. 2, 3:11; May 3, 16:11; Jun. 1, 8:20; Jul. 2, 21:17; Aug. 3, 10:11; Sep. 1, 2:17; Oct. 2, 15:11; Nov. 3, 4:07; Dec. 1, 20:17.

1986 Jan. 2, 9:17; Feb. 2, 22:19; Mar. 3, 14:32; Apr. 1, 6:44; May 2, 19:44; Jun. 3, 8:42; Jul. 2, 0:50; Aug. 2, 13:45; Sep. 3, 2:39; Oct. 1, 18:44; Nov. 2, 7:41; Dec. 3, 20:39.

1987 Jan. 1, 12:50; Feb. 2, 1:52; Mar. 2, 18:05; Apr. 3, 7:06; May 1, 23:18; Jun. 2, 12:16; Jul. 1, 4:23; Aug. 1, 17:18; Sep. 2, 6:12; Oct. 3, 9:07; Nov. 1, 11:14; Dec. 3, 0:12.

1988 Jan. 3, 13:12; Feb. 1, 5:25; Mar. 3, 18:27; Apr. 1, 10:40; May 2, 23:40; Jun. 3, 12:38; Jul. 2, 4:45; Aug. 2, 17:40; Sep. 3, 6:34; Oct. 1, 22:40; Nov. 2, 11:36; Dec. 1, 3:45.

1989 Jan. 1, 16:45; Feb. 2, 5:47; Mar. 2, 22:00; Apr. 3, 11:02; May 2, 3:13; Jun. 2, 16:11; Jul. 1, 8:18; Aug. 1, 21:13; Sep. 2, 10:07; Oct. 1, 2:13; Nov. 1, 15:09; Dec. 3, 4:07.

Note: The first minimum of each month is given; add 2 days 20 hours 49 minutes successively to find other minima in a given month.
(Calculations by Roger W. Sinnott, *Sky & Telescope*)

Table 10. Double Stars

ADS	Name	r.a. (2000.0) h m	dec. (2000.0) ° ′	Mag- nitudes		PA °	Sep. ″
558	55 Psc	0 39.9	+21 26	5.5	8.7	194	6.6
561	α (alpha) Cas	0 40.5	+56 32	2.2	8.9		63
671	η (eta) Cas	0 49.1	+57 49	3.5	7.5	312	12.4
940	42 φ (phi) And	1 09.5	+47 15	4.6	5.5	133	0.5
996	ζ (zeta) Psc	1 13.7	+07 35	5.6	6.6	63	23.0
1129	ψ (psi) Cas	1 25.9	+68 08	4.7	9.6		25
1507	γ (gamma) Ari	1 53.5	+19 18	4.6	4.7	0	7.8
1563	λ (lambda) Ari	1 57.9	+23 36	4.8	7.3		37.5
1615	α (alpha) Psc	2 02.1	+02 46	4.2	5.2	273	1.6
1630	γ¹ (gamma¹) And	2 03.9	+42 20	2.2	5.1	63	9.8
1630	γ² (gamma²) And	2 03.9	+42 20	5.5	6.3	106	0.5
1697	ι (iota) Tri	2 12.4	+30 18	5.3	6.9	71	3.9
1860	ι (iota) Cas	2 29.1	+67 24	4.6	6.9	232	2.5
1477	α (alpha) UMi	2 31.8	+89 16	2.0	8.9		18
2080	γ (gamma) Cet	2 43.3	+03 14	3.6	6.2	297	2.8
2157	η (eta) Per	2 50.7	+55 54	3.8	8.5		28
—	θ¹,² (theta¹,²) Eri	2 58.3	−40 18	3.2	4.3		7.4
2850	32 Eri	3 54.3	−02 57	4.7	6.2	347	6.7f
2888	ε (epsilon) Per	3 57.9	+40 01	2.9	8.0	9	9.0
3137	φ (phi) Tau	4 20.4	+27 21	5.1	8.7		50
3321	α (alpha) Tau	4 35.9	+16 31	0.9	10.7		120
3823	β (beta) Ori	5 14.5	−08 12	0.2	6.7	206	9.2
4179	λ (lambda) Ori	5 35.1	+09 56	3.6	5.5	44	4.3
4186	θ¹ (theta¹) Ori	5 35.3	−05 23	5.1	6.7		
4241	σ (sigma) Ori	5 38.8	+02 36	3.8	6.6	84	12.9
4566	θ (theta) Aur	5 59.7	+37 13	2.6	7.1	320	3.0
5107	β (beta) Mon	6 28.8	−07 02	4.6	5.1	132	7.2
5400	12 Lyn	6 46.2	+59 27	5.4	6.0	74	1.7
5423	α (alpha) CMa	6 45.2	−16 43	−1.5	8.5	5	4.5
5654	ε (epsilon) CMa	6 58.6	−28 58	1.5	7.8	160	7.4f
5983	δ (delta) Gem	7 20.1	+21 59	3.6	8.2	223	5.9
6175	α (alpha) Gem	7 34.6	+31 53	1.9	2.9	73	3.0
6321	κ (kappa) Gem	7 44.5	+24 24	3.6	9.4	240	7.1
6988	ι¹ (iota¹) Cnc	8 46.7	+28 46	4.0	6.6	307	30.4
7203	σ² (sigma²) UMa	9 10.4	+67 08	4.9	8.2	357	3.6
7402	23 UMa	9 31.5	+63 04	3.7	9.2	269	22.8
7724	γ (gamma) Leo	10 20.0	+19 51	2.2	3.5	124	4.4
7979	54 Leo	10 55.6	+24 45	4.5	6.4	110	6.6
8119	ξ (xi) UMa	11 18.2	+31 32	4.3	4.8	60	1.3
8148	ι (iota) Leo	11 23.9	+10 32	4.0	6.7	132	1.4

Notes: The first column (ADS) is the number in R. G. Aitken's *New General Catalogue of Double Stars* (1932). Often, the primary component of a double-star system is called A and the secondary is called B, with C being a third star in the system. Under Magnitudes, v = variable. Under Separations (Sep.), f = fixed (unchanging). (Based on a compilation by Roger W. Sinnott; courtesy of Sky Publishing Corp.)

Table 10 (contd.). Double Stars

ADS	Name	r.a. (2000.0) h	m	dec. (2000.0) °	′	Mag- nitudes		PA °	Sep. ″
8489	2 CVn	12	16.1	+40	40	5.9	9.0	260	11.5
8531	17 Vir	12	22.5	+05	18	6.5	8.6	337	20.6
—	α (alpha) Cru	12	26.6	−63	06	1.6	2.1	114	4.7
8630	γ (gamma) Vir	12	41.7	−01	27	3.5	3.5	287	3.0
8706	α (alpha) CVn	12	56.0	+38	19	2.9	5.5	229	19.4
8891	ζ (zeta) UMa	13	23.9	+54	56	2.3	4.0	151	14.4
—	α (alpha) Cen	14	39.6	−60	50	0.0	1.2	214	19.7
9338	π (pi) Boo	14	40.7	+16	25	4.9	5.8	108	5.7
9343	ζ (zeta) Boo	14	41.2	+13	44	4.5	4.6	303	1.0
9372	ε (epsilon) Boo	14	45.0	+27	05	2.5	5.0	339	2.8
9375	54 Hya	14	46.0	−25	27	5.2	7.2	126	8.8
9413	ξ (xi) Boo	14	51.4	+19	06	4.7	6.9	326	7.0
9494	44 Boo	15	03.8	+47	39	5.3	6.0	44	1.6
9617	η (eta) CrB	15	23.2	+30	17	5.6	5.9	27	1.0
9701	δ (delta) Ser	15	34.8	+10	32	4.1	5.2	179	3.9
9737	ζ (zeta) CrB	15	39.4	+36	38	5.1	6.0	305	6.3f
9909	ξ (xi) Sco AB	16	04.4	−11	22	4.9	4.9	44	0.7
9913	β (beta) Sco	16	05.4	−19	48	2.7	4.9	21	13.6
10074	α (alpha) Sco	16	29.4	−26	26	0.9v	5.5	276	2.4
10087	λ (lambda) Oph	16	30.9	+01	59	4.2	5.2	22	1.5
10157	ζ (zeta) Her	16	41.3	+31	36	2.9	5.5	83	1.6
10345	μ (mu) Dra AB	17	05.3	+54	28	5.7	5.7	25	1.9
10417	36 Oph	17	15.4	−26	36	5.1	5.1	151	4.8
10418	α (alpha) Her	17	14.7	+14	23	3.2	5.4	107	4.7
10424	δ (delta) Her	17	15.0	+24	50	3.1	8.7	243	9
10526	ρ (rho) Her	17	23.7	+37	09	4.6	5.5	316	4.0
10993	95 Her	18	01.5	+21	36	5.1	5.2	258	6.5
11005	τ (tau) Oph	18	03.1	−08	11	5.2	5.9	280	1.8
11046	70 Oph	18	05.5	+02	30	4.2	6.0	220	1.5
11336	39 Dra	18	24.0	+58	48	5.1	7.8	352	3.7
11635	ε¹ (epsilon¹) Lyr	18	44.3	+39	40	5.0	6.1	353	2.6
11635	ε² (epsilon²) Lyr	18	44.4	+39	37	5.2	5.5	80	2.4
12540	β (beta) Cyg	19	30.7	+27	58	3.2	5.4	54	34.4f
12880	δ (delta) Cyg	19	45.0	+45	08	2.9	6.3	225	2.2
13007	ε (epsilon) Dra	19	48.2	+70	16	3.9	7.0	12	3.3
13632	α¹ (alpha¹) Cap	20	17.6	−12	31	4.3	9.0	221	45.5
13645	α² (alpha²) Cap	20	18.1	−12	33	3.6	10.0	172	6.6
—	β (beta) Cap	20	21.0	−14	47	3.1	6.2		
14279	γ (gamma) Del	20	46.7	+16	08	4.3	5.2	268	9.8
14296	λ (lambda) Cyg	20	47.4	+36	30	4.9	6.1	11	0.9
14636	61 Cyg	21	06.9	+38	45	5.2	6.0	148	29.7
15032	β (beta) Cep	21	28.6	+70	34	3.2	7.8	250	13.7
15270	μ (mu) Cyg	21	44.1	+28	45	4.8	6.1	307	1.5
15971	ζ (zeta) Aqr	22	28.8	−00	01	4.3	4.5	207	1.9
17140	σ (sigma) Cas	23	59.0	+55	45	5.0	7.2	326	3.1

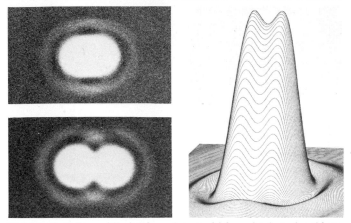

Fig. 32. (*left*) Dawes' limit, the point at which we begin to be able to distinguish that two objects are present, falls between parts A (top left) and B (bottom left). (Chris Jones, Union College)

Fig. 33. (*right*) Dawes' limit corresponds to the brightness distribution at right. (David E. Stoltzmann, Honeywell Systems and Research Center)

on the *position angle* — the angle measured eastward (counterclockwise), with the brighter star at the center of the "clock," for convenience, from the direction north around to the fainter star. The position angle and the apparent separation (in angular units — minutes or seconds of arc) between components of double systems change over time.

Some double star systems are particularly beautiful to observe with a telescope, even a small one, because the component stars are of different colors. Whenever a contrast of colors is seen in the members of a double, it results from the different temperatures of the individual stars.

One eclipsing binary system, Algol, is particularly easy to observe. It is also known as beta Persei because it is the second brightest star in Perseus. The name "Algol," meaning "the demon star," comes from the Arabic words *ras-al-ghul,* meaning the head of the demon. Figure 31 shows Algol's variations in brightness over a 2.9-day period. Algol is such a popular object to observe that we have listed some of the times of its minima in Table 9 (p. 147). A finding chart for Algol (Fig. 40) appears at the end of this chapter. On this and on the other finding charts there, decimal points are omitted from the magnitudes listed, since the decimal points might be confused with stars.

An English amateur astronomer in the last century, William Dawes, worked out a rule of thumb to give the telescope aperture (the diameter of the lens or mirror) necessary to separate (resolve) double stars into their components, that is, to tell that two stars are present in a system (Figs. 32 and 33). His rule is that an aperture of I inches should make barely detectable the presence of a pair of 6th-magnitude stars separated by $4.6/I$ seconds of arc; in the metric system, an aperture of C centimeters should resolve the pair of stars separated by $11.7/C$ seconds of arc. Dawes' rule assumes good sky conditions for viewing, and that the stars are not too different in magnitude.

VARIABLE STARS

In addition to eclipsing binaries — stars that vary because they eclipse each other periodically — there are many stars that vary in brightness all by themselves. All such stars are known as *variable stars.*

The first variable star to be discovered in each constellation has usually been given the letter R, followed by the genitive (possessive) form in Latin of the constellation's name (see Appendix Table A-1). For example, R Andromedae, the first variable star discovered in Andromeda, is a long-period variable. The second variable star in each constellation was given the name S; S Andromedae was apparently a bright star that newly appeared in Andromeda in 1885, though we now know that it was a supernova. The next variable was T, and so on up to Z. Then the lettering scheme started over again with RR and continued to RZ, then SS (not SR) up to SZ, TT up to TZ, and so on up to YY, YZ, and ZZ. Following are AA to AZ, BB to BZ, and so on up to QZ, omitting the letter J (which might be confused with I). This lettering scheme provides names for up to 334 variable stars in each constellation. Additional variables in each constellation are numbered following a V, for variable: V1500 Cygni. Stars that have Bayer designations (Greek letters) have retained them, even when they are variables.

Following a star's brightness over a period of time is called "finding its light curve." A star's light curve is a graph of its brightness as it varies over time. The light curve of SS Cygni (Fig. 34), for example, shows that this variable star changes irregularly in brightness, increasing every 50 days or so from its normal level of 12th magnitude up to 8th magnitude. A finding chart for SS Cygni (Fig. 39) appears at the end of this chapter. SS Cygni is an example of a *dwarf nova.* U Geminorum and AY Lyrae are other prominent members of the class; in fact, dwarf novae are often called *U Gem stars.*

Sometimes stars are variable as a result of an actual change in a star's size and therefore in its brightness. One type of variable, known as a *long-period variable,* takes weeks to complete a cycle of variation. This type of variable is also called a *Mira variable,*

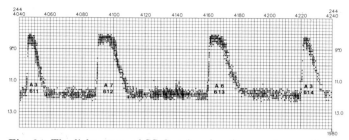

Fig. 34. The light curve of SS Cygni, a dwarf nova whose outbursts occur about every 50 days. (AAVSO)

after the prototype star Mira (omicron Ceti) in the constellation Cetus, the Whale. Sixteen cycles of variation for Mira are shown at the top of Fig. 35; a finding chart for this star (Fig. 41) also appears at the end of this chapter.

Mira was the very first variable star (except for novae and super-novae) to be discovered. It was noticed as a variable with the naked eye in 1596, even before the invention of the telescope. Since it changed from 9th magnitude, which is invisible to the naked eye, to 3rd magnitude, this star seemed to appear anew in the sky from time to time. It was named "Mira," the Latin word for "wonder-ful," because of its wonderful changes in brightness. R Boötis, which ranges from 7th to 13th magnitude (Fig. 35), is another ex-ample of this class.

RR Lyrae variables, named after the prototype star RR Lyrae in the constellation Lyra (Atlas Chart 18), on the other hand, have very short, regular periods of less than one day (Fig. 36). Since RR Lyrae stars are found mostly in globular clusters, they are also known as *cluster variables.* The fact that all variables of this type have essentially the same absolute magnitude — about 0.5 — was important for measuring the distance to the globular clusters in which RR Lyrae stars are located. This, in turn, led to Harlow Shapley's discovery in the 1920s that the sun is not at the center of our galaxy.

Cepheid variables, named after the star δ (delta) Cephei, have a distinct pattern to their variations (Fig. 37), with periods (cycles) that can range from about one day up to a few days. RR Lyrae stars and Cepheids are two types of variables that have been very important in helping astronomers determine the scale of distances in the universe. All RR Lyrae stars have about the same intrinsic brightness, so any differences in apparent brightness from one RR Lyrae star to the next result only from differences in their dis-tances from us. Each Cepheid variable, on the other hand, can have a different intrinsic brightness, but it is fairly easy for astron-

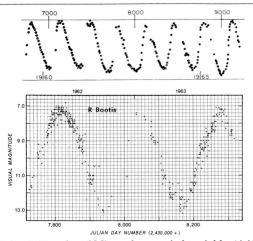

Fig. 35. (*top*) 16 cycles of Mira, a long-period variable. (AAVSO) (*bottom*) The light curve of R Boötis, a Mira variable with a period of 223 days. (AAVSO)

omers to calculate the distance to a Cepheid once they know its period of variation. (The period of variation can be measured easily at a telescope; then the intrinsic brightness can be calculated from the star's period. From a comparison of this intrinsic brightness with the star's apparent brightness, astronomers can calculate the distance to the star.) Cepheid variables are giant stars, so large that they can be detected in some of the nearby galaxies; they provide the major method we have of measuring distances to these galaxies. All measurements to more distant galaxies depend, in turn, on the measurements of distances to nearby galaxies.

A star like *R Coronae Borealis* (abbreviated R CrB) varies in brightness because it occasionally gives off a cloud of opaque material — essentially soot — that masks the surface that we ordinarily see. R CrB thus fades from 6th magnitude to 11th magnitude or even fainter (see the light curve facing Atlas Chart 16 in Chapter 7, and the finding chart for this star at the end of this chapter). RY Sagittarii, which covers about the same range in magnitude, is another example of the class.

Flare stars, such as UV Ceti or YZ Canis Minoris, occasionally flare up within minutes by one to six magnitudes.

RV Tauri stars, such as R Scuti and V Vulpeculae, vary with a period of one to five months. During each period these stars have one deep minimum and one shallow minimum; at a deep minimum, the brightness of these yellow supergiants may drop by as much as three magnitudes.

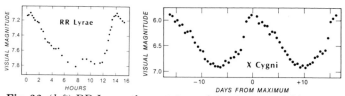

Fig. 36. (*left*) RR Lyrae, the prototype of the cluster variables, has a period of 13.6 hours. It rises rapidly to maximum brightness, and declines more slowly. This light curve is based on 330 visual observations by Erich Leiner, a German amateur astronomer. (*Sky & Telescope*)
Fig. 37. (*right*) The light curve of X Cygni, a Cepheid variable with a 17-day period. (AAVSO)

For several periods, *semiregular variables* have periods lasting months or years; then they are irregular in their variations for a while.

Variable double stars such as *RS Canum Venaticorum* (abbreviated RS CVn and informally pronounced "RS Can Ven") are unusual eclipsing binaries that are especially interesting to study for those who have photoelectric equipment available. (Photoelectric equipment that can measure magnitudes accurately to 0.01 magnitude is needed.) RS CVn is a binary in which both stars are somewhat more massive than our sun, but generally one member is smaller and hotter and the other is a larger and cooler subgiant. The stars orbit each other very closely every 4.8 days. Every time the hotter star is eclipsed by the subgiant, the light drops by about 1 magnitude. One of the brightest RS CVn-type binaries is λ (lambda) Andromedae, which varies between visual magnitudes 3.70 and 4.05 every 54 days. Other RS CVn stars have periods between 0.6 and 80 days. The light variations arise as one star, with large starspots (comparable to the sunspots on our sun) darkening up to 40% of one hemisphere, rotates on its axis. Enormous bursts of radio waves coincide with the violent flare-ups of the star's extremely active outer atmosphere.

Variable stars are marked on the Atlas Charts in Chapter 7 with both inner and outer circles whenever possible. The inner circles give their faintest magnitude and the outer circles give their brightest magnitude. A number of variables are listed in Tables 11 and 12; several also appeared in Table 6 and Fig. 24 in Chapter 4. The stars listed in Tables 6 and 11 have shorter periods, and are thus easier to observe as variables, than the other stars listed in Table 12.

Novae — newly visible stars — can be considered as variables in one sense; they are marked with special symbols on the Atlas Charts, however. A nova brightens when gas from a large companion star falls on the white-dwarf member of a binary system, triggering nuclear fusion on the white dwarf's surface. The supernovae of 1572 and 1604 are marked on the charts with the nova symbol, since the distinction between novae and supernovae (p. 113) was not known at the time they were seen.

Do not confuse *pulsars* — neutron stars that give off regular pulses of radio waves and which are not ordinarily detectable optically — with variable stars, stars that vary in brightness in visible light.

Variable-star observers often give dates in terms of *Julian days,* a system of calendar-keeping in which the days are numbered consecutively from January 1, 4713 B.C. The Julian day system eliminates the need to calculate the number of days in a month or the effects of leap years. It also eliminates the need to worry about changes from the Julian to the Gregorian or other calendars. Also, Julian days begin at noon, so the Julian date does not change in the course of a night's observing. Portions of a day are customarily indicated by decimals in this system, instead of by hours, minutes, and seconds. For example, noon on January 1, 1985 is the beginning of Julian Day 2,446,067.0. Every tenth Julian Day is listed in Appendix Table A-8.

The American Association of Variable Star Observers, 187 Concord Ave., Cambridge, Mass. 02138, is an international organization of amateur astronomers devoted to the study and cataloguing of variable stars. They have a central registry of all variable star observations, and calculate many light curves. They are always glad to have new members and to receive observations of variable stars from amateur astronomers. The AAVSO can provide finding charts for hundreds of variable stars. They provided suggestions for the variables included in Table 12 as a set of first objects for amateurs who want to start making variable-star observations and reports.

Table 11. Short-period Variable Stars

Name	Constel- lation	Type	r.a. (2000.0) h m s	dec. (2000.0) °	Mag- nitude (Range)	Period (Days)
Algol (beta Persei)	Perseus	E	03 08 10	+40 57.4	2.2–3.5	2.9
zeta Geminorum	Gemini	C	07 04 07	+20 34.2	3.7–4.3	10.2
beta Lyrae	Lyra	E	18 50 05	+33 21.8	3.4–4.3	12.9
RR Lyrae	Lyra	RR	19 25 28	+42 47.8	7.3–8.1	0.6
delta Cephei	Cepheus	C	22 29 10	+58 24.9	3.6–4.3	5.4

Notes: C = Cepheid variable; E = eclipsing binary; RR = RR Lyrae variable (cluster variable — see p. 152). Finding charts for beta Lyrae and beta Persei (Algol) appear at the end of this chapter (Fig. 40).

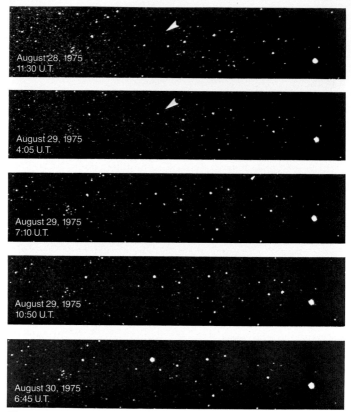

Fig. 38. The brightening of Nova Cygni 1975 was recorded in this series of 25-min. exposures taken by an amateur astronomer as part of a search for meteors. The bright star Deneb is at right. (© 1975 Ben Mayer)

The AAVSO is a major link between professional astronomers and amateurs who make visual observation of variable stars. The light curves and special observing projects set up in collaboration with the AAVSO and the International Amateur-Professional Photoelectric Photometry association (IAPPP) often provide important scientific data. For example, objects observed by professional astronomers in the x-ray region of the spectrum from satellites aloft often correspond to optical objects whose variations in brightness can be followed by amateurs on the ground. The address for the IAPPP is Fairborn Observatory, 1247 Folk Rd., Fairborn, Ohio 45324.

Table 12. Long-period Variable Stars

Name	r.a. (2000.0) h m s	dec. ° ′	Type	Magnitude (Range)	Period (Days)
T Cas	00 23 15	+55 47.6	M	[7.9–11.9]	445
o Cet	02 19 21	−02 58.7	M	[3.4–9.2]	332
R Tri	02 37 01	+34 16.1	M	[6.2–11.7]	266
R Hor	02 53 51	−49 53.3	M	[6.0–13.0]	404
X Cam	04 45 43	+75 06.1	M	[8.1–12.6]	144
R Pic	04 46 10	−49 14.8	SR	6.7–10.0	164
L₂ Pup	07 13 32	−14 38.4	SR	2.6–6.2	140
S CMi	07 32 43	+08 19.1	M	[7.5–12.6]	333
U Gem	07 55 06	+22 00.6	UG	8.2–14.9	103
R Car	09 32 15	−62 47.3	M	[4.6–9.6]	309
R Leo	09 47 33	+11 25.7	M	[5.8–10.0]	312
L Car	09 45 15	−62 30.4	C	3.4–4.1	35
S Car	10 09 22	−61 32.9	M	[5.7–8.5]	150
R UMa	10 44 38	+68 46.5	M	[7.5–13.0]	302
T UMa	12 36 23	+59 29.2	M	[7.7–12.9]	257
S UMa	12 43 59	+61 05.5	M	[7.8–11.7]	226
T Cen	13 41 46	−33 35.8	SR	5.5–9.0	91
R CVn	13 48 57	+39 32.6	M	[7.7–11.9]	329
R Cen	14 16 34	−59 54.8	M	5.9–10.7	546
S Boo	14 22 53	+53 48.6	M	[8.4–13.3]	271
V Boo	14 29 45	+38 51.7	SR	7.0–12.0	258
R Cre	15 48 34	+28 09.2	RCB	5.8–14.8	†
R Sex	15 50 42	+15 08.0	M	[6.9–13.4]	356
U Her	16 25 48	+18 53.6	M	[7.5–12.5]	406
R Dra	16 32 39	+66 45.2	M	[7.6–12.4]	246
R Oph	17 07 46	−16 05.6	M	[7.6–13.3]	303
T Her	18 09 06	+31 01.3	M	[8.0–12.8]	165
R Sct	18 47 28	−05 42.3	RV	4.5–8.2	†
R Cyg	19 36 49	+50 12.0	M	[7.5–13.9]	426
R Vul	21 04 23	+23 49.3	M	[8.1–12.6]	136
T Cep	21 09 32	+68 29.4	M	6.0–10.3]	388
SS Cyg	21 42 42	+43 35.0	UG	8.2–12.4	50
R Peg	23 06 39	+10 32.6	M	[7.8–13.2]	378
V Cas	23 11 39	+59 41.9	M	7.5–12.2	229

Notes: C = Cepheid variable; M = Mira — long-period variable; RCB = R Coronae Borealis variable; RV = RV Tauri variable; SR = semiregular variable; UG = U Geminorum variable (SS Cygni). Brackets indicate an average range of magnitude; † = irregular, no period.

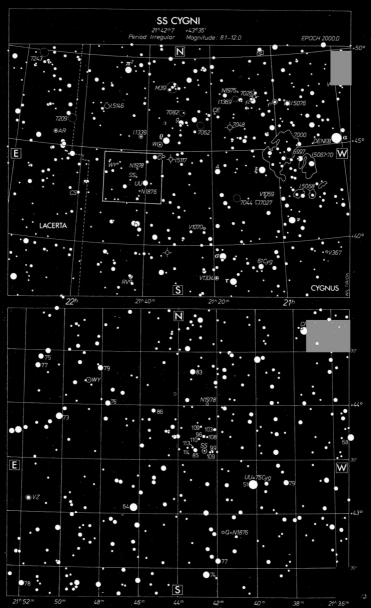

Fig. 39. SS Cygni. (Wil Tirion) Note: Decimal points are omitted from magnitudes on this chart, to avoid confusion with points representing stars.

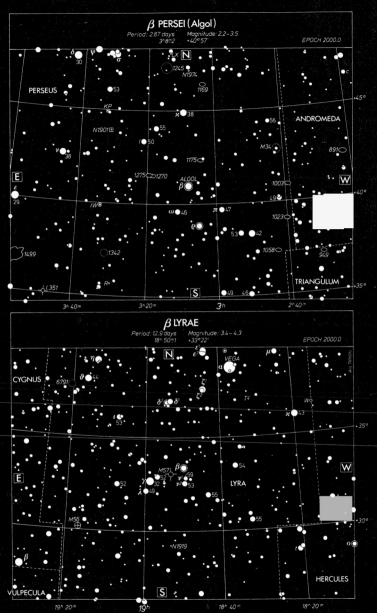

Fig. 40. Algol (beta Persei) and beta Lyrae. (Wil Tirion) Note: Decimal points are omitted from magnitudes on this chart, to avoid confusion with points representing stars.

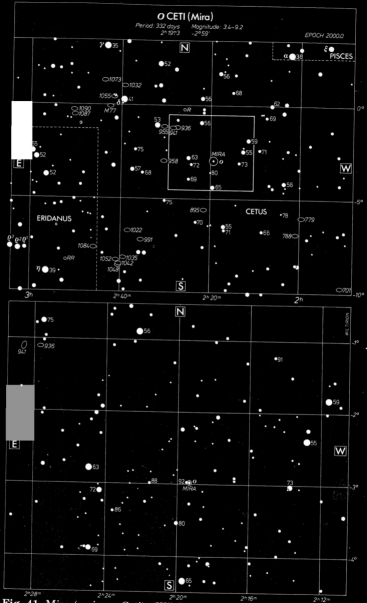

Fig. 41. Mira (omicron Ceti). (Wil Tirion) Note: Decimal points are omitted from magnitudes on this chart, to avoid confusion with points representing stars.

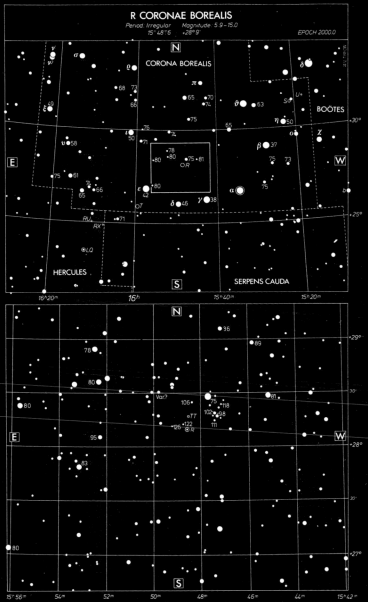

Fig. 42. R Coronae Borealis. (Wil Tirion) Note: Decimal points are omitted from magnitudes on this chart, to avoid confusion with points representing stars.

161

7

Atlas of the Sky

On the following pages are 52 Atlas Charts that together cover the entire sky, accompanied by 5 charts showing close-ups of areas of special interest. A visual key to the Atlas Charts appears on the endpapers; Table 14 (p. 169) also outlines the region of sky covered by each chart.

The charts have been drawn with high precision by Wil Tirion to show positions of celestial objects as they will appear in epoch 2000.0, that is, at the beginning of the year 2000. The charts contain 25,000 stars down to visual magnitude 7.5, and about 2,500 deep-sky objects (star clusters, nebulae, galaxies, and other objects among and beyond the stars). A key to the symbols for these objects and a list of lower-case Greek letters (which are used to label the brightest stars) appear below every chart.

The Greek letters were used to name the brightest stars in constellations by Johann Bayer in his sky atlas of 1603. On the whole, the stars were lettered in order of brightness, with α (alpha) usually being the brightest star in a constellation. This rule was not always followed; for example, the stars in the Big Dipper (which is part of the constellation Ursa Major) are lettered in order around the bowl and handle. A few stars are shown on the charts with their Flamsteed numbers. These numbers, from the 1725 catalogue of John Flamsteed, were assigned to stars in order of their position in the sky (increasing right ascension) at that time, rather than brightness. Note: Only stars are labelled with Flamsteed numbers; the bulk of celestial objects that have been assigned numbers, but not letters, are deep-sky objects listed in Dreyer's *New General Catalogue*. Some of the fainter stars are labelled with lower-case Roman (regular) or italic letters.

Astronomers mark positions in the sky using coordinates comparable to those we use to plot positions on earth. In astronomy, *right ascension* is celestial longitude, analogous to terrestrial longitude. *Declination* is celestial latitude, analogous to terrestrial latitude. The Atlas Charts in this chapter are marked in right ascension and declination. Right ascension is marked in hours, minutes, and seconds (abbreviated ^{h}, ^{m}, and ^{s}) of time, with each 24 hours representing a full rotation of 360°. Declination is marked in degrees, minutes, and seconds (°, ′, and ″) north (+) or south (−) of the celestial equator. If you are familiar with right ascension and declination, you can refer to Table 14 (p. 169) to see at a glance which constellations are covered on each chart. (The constellations are labelled on the charts.)

If you are looking at an Atlas Chart, and want to know if the objects in it are visible at a given time, find the equivalent sidereal time in Appendix Table A-10. An object is crossing your meridian (is highest in the sky) when its right ascension matches your sidereal time. You might also choose to simply locate the constellation on the Monthly Sky Maps in Chapter 3.

The positions of celestial objects take account of *precession* — the drifting of the direction of the earth's axis of rotation and thus of the celestial coordinate system. Current positions of stars and deep-sky objects and constellation boundaries differ from the positions for the year 2000.0 (as shown on the charts) by less than one-half degree. For further details on right ascension, declination, precession, and matters of time and calendars, turn to Chapter 14. Formulae needed to calculate the exact amount of precession, should you want to do so, appear on p. 415.

Constellation figures are outlined and the official constellation boundaries are marked with dotted lines. The line connecting stars is dotted in a few cases where a star that has traditionally been part of an asterism or constellation is no longer considered an official part of it; for example, one of the four stars marking the Great Square of Pegasus is really in the constellation Andromeda (see Atlas Charts 9 and 20).

Stellar magnitudes (see p. 32) are represented by dots of different sizes. Each size represents one whole magnitude, including a range of $\frac{1}{2}$ magnitude on each side of the integer (whole number) given, as shown in the key beneath each chart. The relative brightness of stars is shown as they appear to the eye rather than as they appear on a photographic plate, which is usually sensitive to a different range of colors (wavelengths).

Moving stars — stars whose motion across the sky over centuries can be detected quite easily — are especially close to our solar system. Two special charts, one for Proxima Centauri (the nearest star) opposite Chart 50 and one for Barnard's Star (the fourth nearest star) opposite Chart 30, show stars whose angular motion across the sky, called *proper motion,* is especially large.

Double stars (Table 10, pp. 148–149) are marked with a horizontal line through the dot that represents the total magnitude of the components. Only visual binaries — those systems in which two or more stars can be seen through a telescope (as opposed to stars that have been identified as binaries by studying their spectrum or in some other way) — are marked as doubles. The members of double star systems are plotted as separate stars, though, if they are separated by more than one minute of arc. A few double stars are listed on the charts using the letter A (for Aitken) fol-

lowed by their numbers in the Aitken Double Star (ADS) catalogue of 1932. Sometimes several members of a binary system
share the same letter but each component is marked with a different superscript, as in ι^1 (iota-1) and ι^2 (iota-2).

   Variable stars (Tables 11 and 12, pp. 155, 157) whose maximum brightness is within the range of magnitude covered on these
charts are indicated by a dot surrounded by a concentric circle to
show the range of brightness; no inside dot appears if the minimum brightness is fainter than magnitude 7.5. Only variables
whose brightness varies by more than 0.1 magnitude are marked.
(The system for naming variable stars is described in Chapter 6.)

   Deep-sky objects. In 1784, Charles Messier compiled a list
(Table 13) of fuzzy objects in the sky, so that he would not be
confused by them when searching for comets. A few additional
objects were added later. The list turns out to contain many of the
most interesting objects in the sky that are accessible to amateur
observers. Most deep-sky objects are marked with their numbers in
the Messier list, in J. L. E. Dreyer's *New General Catalogue*
(NGC) of 1888, or in the supplemental *Index Catalogues* (ICs) of
1895 and 1908. Messier numbers are preceded by an M, NGC numbers are written (on the charts) without prefixes, and IC numbers
are preceded by the letter *I* and a period. A few objects are also
labelled with their popular names.

   The **Milky Way** is outlined by a dotted line on the charts.

   Open star clusters (also known as *galactic clusters*) are
marked with dotted circles, whose size signifies the size of each
cluster. A special chart, following Chart 10, shows the stars of the
Pleiades at a larger scale. Some star clusters are often referred to
by their numbers in special catalogues, such as the Trumpler Catalogue (*e.g.*, Tr 1) or the Melotte Catalogue (*e.g.*, Mel 71).

   Globular star clusters are marked with symbols showing three
size ranges.

   Planetary nebulae, shells of gas given off when stars like the
sun end their lives, are marked with circles bearing four external
spikes.

   Pulsars, tiny neutron stars that give off regular pulses of radio
waves, representing the death of stars somewhat more massive
than the sun, and **black holes,** representing the death of extremely massive stars, also have their own symbols. The most
likely black-hole candidate is Cygnus X-1 on Atlas Chart 19. The
name Cygnus X-1 means that it was the first x-ray source to be
discovered in the constellation Cygnus, a discovery that was not
possible until observations could be made from x-ray telescopes
orbiting above the earth's atmosphere. Even though pulsars and
black holes cannot usually be seen in visible light (only two pulsars
have been detected optically, even with large telescopes), these
objects are so fascinating that it is interesting to know that one is
nearby when you are observing the sky.

Novae — sudden brightenings of stars — are listed with the year in which they went off, such as Nova 1934 (for the nova visible in that location in the year 1934). We now know that novae are events that take place in binary star systems containing a white dwarf. Since they were called novae at the time they occurred, two **supernovae** are given the same symbol. These supernovae — the one Tycho saw in 1572 (Atlas Chart 1) and the one Kepler saw in 1604 (Atlas Chart 41) — represented the explosive deaths of massive stars. Their remnants are not easy to see; the Crab Nebula and S147 in Taurus (Atlas Chart 11) and the Veil Nebula in Cygnus (Atlas Chart 19) are much more interesting supernova remnants to see with small telescopes.

The shapes of other **nebulae,** regions of gas and dust in space, are drawn on the charts when they are more than 10 minutes of arc across; a square signifies nebulae smaller than 10 minutes of arc across. A special chart following Atlas Chart 24 shows at an enlarged scale the complex of stars and nebulae in Orion's sword (including the Orion Nebula) that are such beautiful objects for observers to see or photograph.

Galaxies are drawn as ovals, in four size ranges. They are labelled down to magnitude 13.0 and fainter. The central part of the nearest cluster of galaxies, the Virgo Cluster, is featured on a larger-scale chart, 27A (p. 261).

A few selected **quasars** are marked with triangles. Quasars are farther from us than most galaxies marked on the charts, and all are optically fainter than the normal limit of our charts. In parts of the spectrum other than the visible, though, some of the quasars can be relatively bright. Quasars are probably the result of high-energy events going on in the cores of certain galaxies, and may show that giant black holes with perhaps a million times the mass of the sun are located in those places (see p. 125). Some of the quasars are listed by their commonly used *3C numbers,* that is, their numbers in the Third Cambridge Catalogue of radio sources. Others are named for their position in other catalogues, such as OQ 172 (one of the farthest known objects in the universe), where the O stands for the Ohio State University Catalogue, the Q stands for a certain band of right ascension, and 172 is the number assigned to this quasar in the catalogue. Similarly, PKS 0405 −123 stands for "Parkes at right ascension 04^h05^m declination $−12.3°$," where Parkes is the site of the Australian National Radio Observatory.

Radio astronomers — astronomers using radio telescopes — originally named objects emitting radio waves with a capital letter and the name of the constellation, such as Taurus A for the first radio source discovered in Taurus. We now know that Taurus A is the Crab Nebula, named for its shape. It is also known as M1, because it was the first object in Messier's list (see Table 13). The center of our galaxy coincides with the radio source Sagittarius A.

The Atlas Charts are in the following order: the north polar charts come first, then the midnorthern charts, equatorial charts, midsouthern charts, and south polar charts. In order to minimize distortion on these charts, Tirion has used a conic projection for the polar regions and intermediate declinations, and a cylindrical projection for the equatorial region.

Numbers set in triangles at the borders of each map direct you to adjacent charts. (The adjacent charts in each group are in order of right ascension.) North is up, west is to the right, south is down, and east is to the left on the charts.

Keep a life list of the Messier objects you have seen, recording the date on which you observed each one and sketching what it looked like. In March each year, it is possible (though difficult) to see all the Messier objects in a single night. The table below and the Atlas Charts will help you find these objects, whether you want to track them all down in a "Messier marathon" or pursue them at a more leisurely pace.

Table 13. Messier Catalogue

M	NGC	r.a. 2000.0 h m	dec. °	Visual Magnitude	Description	Atlas Chart
1	1952	05 34.5	+22 01	11.3	Crab Nebula	11
2	7089	21 33.5	−00 49	6.3	Globular cluster	32
3	5272	13 42.2	+28 23	6.2	Globular cluster	15
4	6121	16 23.6	−26 31	6.1	Globular cluster	41
5	5904	15 18.5	+02 05	6	Globular cluster	29
6	6405	17 40.0	−32 12	6	Open cluster	41
7	6475	17 54.0	−34 49	5	Open cluster	41
8	6523	18 03.7	−24 23		Lagoon Nebula	42
9	6333	17 19.2	−18 31	7.6	Globular cluster	30
10	6254	16 57.2	−04 06	6.4	Globular cluster	29, 30
11	6705	18 51.1	−06 16	7	Open cluster	30
12	6218	16 47.2	−01 57	6.7	Globular cluster	29
13	6205	16 41.7	+36 28	5.8	Globular cluster	17
14	6402	17 37.6	−03 15	7.8	Globular cluster	30
15	7078	21 30.0	+12 10	6.3	Globular cluster	32
16	6611	18 18.9	−13 47	7	Open cluster, nebula	30
17	6618	18 20.8	−16 10	7	Omega Nebula	30
18	6613	18 19.9	−17 08	7	Open cluster	30
19	6273	17 02.6	−26 16	6.9	Globular cluster	41
20	6514	18 02.4	−23 02		Trifid Nebula	41, 42
21	6531	18 04.7	−22 30	7	Open cluster	41, 42
22	6656	18 36.4	−23 54	5.2	Globular cluster	42
23	6494	17 56.9	−19 01	6	Open cluster	30, 41

Note: Magnitudes based on a table in the *Observer's Handbook 1980* of the Royal Astronomical Society of Canada.

Table 13 (contd.). Messier Catalogue

M	NGC	r.a. 2000.0 h m	dec. °	Visual Magnitude	Description	Atlas Chart
24	6603	18 18.4	−18 25	6	Open cluster	30
25	IC4725	18 31.7	−19 14	6	Open cluster	30
26	6694	18 45.2	−09 24	9	Open cluster	30
27	6853	19 59.6	+22 43	8.2	Dumbbell Nebula	18, 19
28	6626	18 24.6	−24 52	7.1	Globular cluster	42
29	6913	20 24.0	+38 31	8	Open cluster	18, 19
30	7099	21 40.4	−23 11	7.6	Globular cluster	43
31	224	00 42.7	+41 16	3.7	Andromeda Galaxy	9
32	221	00 42.7	+40 52	8.5	Elliptical galaxy	9
33	598	01 33.8	+30 39	5.9	Spiral galaxy (Sc)	9
34	1039	02 42.0	+42 47	6	Open cluster	10
35	2168	06 08.8	+24 20	6	Open cluster	11, 12
36	1960	05 36.3	+34 08	6	Open cluster	11
37	2099	05 53.0	+32 33	6	Open cluster	11, 12
38	1912	05 28.7	+35 50	6	Open cluster	11
39	7092	21 32.3	+48 26	6	Open cluster	19
40	WNC4	12 22.2	+58 05	9, 9.3	Double star	
41	2287	06 47.0	−20 44	6	Open cluster	36
42	1976	05 35.3	−05 23		Orion Nebula	24
43	1982	05 35.5	−05 16		Orion Nebula; smaller part	24
44	2632	08 40.0	+20 00	4	Praesepe; open cluster	13, 25
45	—	03 47.5	+24 07	2	The Pleiades; open cluster	10, 11
46	2437	07 41.8	−14 49	7	Open cluster	25
47	2422	07 36.6	−14 29	5	Open cluster	25
48	2548	08 13.8	−05 48	6	Open cluster	25
49	4472	12 29.8	+08 00	8.9	Elliptical galaxy	27A
50	2323	07 03.0	−08 21	7	Open cluster	24, 25
51	5194	13 29.9	+47 12	8.4	Whirlpool Galaxy.	15
52	7654	23 24.2	+61 36	7	Open cluster	1
53	5024	13 12.9	+18 10	7.7	Globular cluster	15
54	6715	18 55.1	−30 28	7.7	Globular cluster	42
55	6809	19 40.0	−30 57	6.1	Globular cluster	42
56	6779	19 16.6	+30 11	8.3	Globular cluster	18
57	6720	18 53.6	+33 02	9.0	Ring Nebula	18
58	4579	12 37.7	+11 49	9.9	Spiral galaxy (SBb)	27A
59	4621	12 42.0	+11 39	10.3	Elliptical galaxy	27A
60	4649	12 43.7	+11 33	9.3	Elliptical galaxy	27A
61	4303	12 21.9	+04 28	9.7	Spiral galaxy (Sc)	27A
62	6266	17 01.2	−30 07	7.2	Globular cluster	41
63	5055	13 15.8	+42 02	8.8	Spiral galaxy (Sb)	15
64	4826	12 56.7	+21 41	8.7	Spiral galaxy (Sb)	15
65	3623	11 18.9	+13 06	9.6	Spiral galaxy (Sa)	27
66	3627	11 20.3	+13 00	9.2	Spiral galaxy (Sb)	27
67	2682	08 51.3	+11 48	7	Open cluster	25

Table 13 (contd.). Messier Catalogue

M	NGC	r.a. 2000.0 h m	dec. ° '	Visual Magni- tude	Description	Atlas Chart
68	4590	12 39.5	− 26 45	8	Globular cluster	39
69	6637	18 31.4	− 32 21	7.7	Globular cluster	42
70	6681	18 43.2	− 32 17	8.2	Globular cluster	42
71	6838	19 53.7	+ 18 47	6.9	Globular cluster	19, 31
72	6981	20 53.5	− 12 32	9.2	Globular cluster	31, 32
73	6994	20 59.0	− 12 38		Open cluster	31, 32
74	628	01 36.7	+ 15 47	9.5	Spiral galaxy (Sc)	9, 22
75	6864	20 06.1	− 21 55	8.3	Globular cluster	42, 43
76	650	01 42.2	+ 51 34	11.4	Planetary nebula	1, 2, 9, 10
77	1068	02 42.7	− 00 01	9.1	Spiral galaxy (Sb)	22
78	2068	05 46.7	+ 00 04		Emission nebula	24
79	1904	05 24.2	− 24 31	7.3	Globular cluster	35
80	6093	16 17.0	− 22 59	7.2	Globular cluster	41
81	3031	09 55.8	+ 69 04	6.9	Spiral galaxy (Sb)	4, 5
82	3034	09 56.2	+ 69 42	8.7	Irregular galaxy (Irr)	4, 5
83	5236	13 37.7	− 29 52	7.5	Spiral galaxy (Sc)	39
84	4374	12 25.1	+ 12 53	9.8	Elliptical galaxy	27A
85	4382	12 25.4	+ 18 11	9.5	Spiral galaxy (S0)	15, 27
86	4406	12 26.2	+ 12 57	9.8	Elliptical galaxy	27A
87	4486	12 30.8	+ 12 23	9.3	Elliptical galaxy	27A
88	4501	12 32.0	+ 14 25	9.7	Spiral galaxy (Sb)	27A
89	4552	12 35.7	+ 12 33	10.3	Elliptical galaxy	27A
90	4569	12 36.8	+ 13 10	9.7	Spiral galaxy (Sb)	27A
91	4548	12 35.4	+ 14 30	9.5	Spiral galaxy or M58?	27A
92	6341	17 17.1	+ 43 08	6.3	Globular cluster	17
93	2447	07 44.6	− 23 53	6	Open cluster	36
94	4736	12 50.9	+ 41 07	8.1	Spiral galaxy (Sb)	15
95	3351	10 44.0	+ 11 42	9.9	Barred spiral galaxy	26
96	3368	10 46.8	+ 11 49	9.4	Spiral galaxy (Sa)	26
97	3587	11 14.9	+ 55 01	11.1	Owl Nebula	5, 14
98	4192	12 13.8	+ 14 54	10.4	Spiral galaxy (Sb)	27A
99	4254	12 18.8	+ 14 25	9.9	Spiral galaxy (Sc)	27A
100	4321	12 22.9	+ 15 49	9.6	Spiral galaxy (Sc)	15, 27A
101	5457	14 03.5	+ 54 21	8.1	Spiral galaxy (Sc)	6, 15, 16
102	—	—	—		M101	6, 15, 16
103	581	01 33.1	+ 60 42	7	Open cluster	1, 2
104	4594	12 40.0	− 11 42	8	Sombrero Galaxy	27
105	3379	10 47.9	+ 12 43	9.5	Elliptical galaxy	26
106	4258	12 19.0	+ 47 18	9	Spiral galaxy (Sb)	14, 15
107	6171	16 32.5	− 13 03	9	Globular cluster	29
108	3556	11 11.6	+ 55 40	10.5	Spiral galaxy (Sb)	5, 14
109	3992	11 57.7	+ 53 22	10.6	Barred spiral galaxy	5, 14, 15

Note: Magnitudes based on a table in the *Observer's Handbook 1980* of the Royal Astronomical Society of Canada.

Table 14. Regions Covered by the Atlas Charts

Each chart covers roughly the area between the coordinates given below. The right ascension column in the middle can be used with either the left-hand or the right-hand columns of declination:

Constellation for + declination zone	Chart declination +50° to +90°	right ascension	Chart declination −50° to −90°
Cas, Cep	1	$22\frac{1}{2}^{h}$–$1\frac{1}{2}^{h}$	45
Cam, Cas, Cep, Per	2	$1\frac{1}{2}^{h}$–$4\frac{1}{2}^{h}$	46
Cam, Lyn	3	$4\frac{1}{2}^{h}$–$7\frac{1}{2}^{h}$	47
Cam, Dra, UMa	4	$7\frac{1}{2}^{h}$–$10\frac{1}{2}^{h}$	48
Dra, UMa, UMi	5	$10\frac{1}{2}^{h}$–$13\frac{1}{2}^{h}$	49
Dra, UMa, UMi	6	$13\frac{1}{2}^{h}$–$16\frac{1}{2}^{h}$	50
Dra, UMi	7	$16\frac{1}{2}^{h}$–$19\frac{1}{2}^{h}$	51
Cep, Cyg, Dra, Lac	8	$19\frac{1}{2}^{h}$–$22\frac{1}{2}^{h}$	52

	declination +20° to +50°		declination −20° to −50°
And, Psc, Tri	9	0^{h}–2^{h}	33
And, Ari, Per, Tau, Tri	10	2^{h}–4^{h}	34
Aur, Per, Tau	11	4^{h}–6^{h}	35
Aur, Gem, Lyn, Tau	12	6^{h}–8^{h}	36
Can, LMi, Lyn, UMa	13	8^{h}–10^{h}	37
Leo, LMi, UMa	14	10^{h}–12^{h}	38
Com, CVn	15	12^{h}–14^{h}	39
Boo, CrB	16	14^{h}–16^{h}	40
Her	17	16^{h}–18^{h}	41
Cyg, Lyr, Vul	18	18^{h}–20^{h}	42
Cyg, Vul	19	20^{h}–22^{h}	43
And, Lac, Peg	20	22^{h}–24^{h}	44

			declination −20° to +20°
Aqr, Cet, Peg, Psc		23^{h}–1^{h}	21
Ari, Cet, Psc		1^{h}–3^{h}	22
Eri, Tau		3^{h}–5^{h}	23
CMa, Lep, Mon, Ori		5^{h}–7^{h}	24
CMi, Cnc, Hya, Mon, Pup		7^{h}–9^{h}	25
Leo, Hya, Sex		9^{h}–11^{h}	26
Com, Crt, Crv, Leo, Vir		11^{h}–13^{h}	27
Boo, Lib, Vir		13^{h}–15^{h}	28
Her, Lib, Oph, Sco, Ser		15^{h}–17^{h}	29
Her, Oph, Sct, Ser, Sgr		17^{h}–19^{h}	30
Aql, Cap, Del, Sge, Sgr		19^{h}–21^{h}	31
Aqr, Cap, Equ, Peg		21^{h}–23^{h}	32

Note: Constellation abbreviations are shown in Table A-1. The Atlas Charts where each constellation is found are also listed in Chapter 5.

ATLAS CHART 1. Bubble Nebula, Tycho's Supernova
The Milky Way in Cassiopeia, with its rich fields of clouds, gas and dust, and star clusters, is an interesting area to scan with low-power telescopes or binoculars. Several open clusters are easy to find because they lie close to bright stars. NGC 457 is an especially bright open cluster in the same field of view as ϕ (phi) Cas, which lies southwest of δ (delta) Cas in the W of Cassiopeia. Even a small telescope gives a good view of the stars in this cluster; more stars are visible with telescopes of higher power. Close by is the open cluster NGC 436. Since binoculars or telescopes with small apertures show only a few stars here, NGC 436 is more suitable for viewing with larger telescopes. M103 (NGC 581) is a fan-shaped, 7th-magnitude cluster (C.Pl. 29) located northeast of δ (delta) Cas.

About 5° northwest of β (beta) Cas, near the border of Cepheus, is M52 (NGC 7654), a rich, 7th-magnitude open cluster. Close to M52 is the Bubble Nebula, NGC 7635, which has a high total brightness even though it is spread over so much sky that its average surface brightness is low. Southwest of β (beta) Cas, about halfway between ρ (rho) and σ (sigma) Cas, is the open cluster NGC 7789. Its diameter is about the same as the moon's. You can detect this cluster with binoculars, and telescopes resolve many of its 1,000 stars.

Northeast of β (beta) Cas is κ (kappa) Cas, with the two open clusters NGC 133 and NGC 146 nearby. NGC 146 has about 50 stars within a 6 arc min diameter; this cluster should be viewed with high power because of the richness of the Milky Way in this part of the sky.

Tycho's Supernova of 1572 appeared slightly to the northwest of NGC 146, growing so bright that it was visible to the naked eye for about six months. Since the appearance of a new star showed that the sky changed, contrary to the Ptolemaic theory, this object was important for the acceptance of Copernicus' heliocentric theory. The remnant of the supernova is now only faintly visible with large telescopes, appearing as thin filaments that form an incomplete ring 8 arc minutes across. Radio telescopes, however, detect strong signals.

Fig. 43. The Bubble Nebula (NGC 7635) in Cassiopeia. (Lick Observatory photo)

ADAPTED FROM "SKY ATLAS 2000.0" BY WIL TIRION

ATLAS CHART 2. Polaris, double cluster in Perseus
Polaris, α (alpha) Ursa Minoris, the North Star, lies about 1° from
the true north pole and circles the pole once every 24 hours. Polaris
is a double star, with a 9th-magnitude companion at a distance of
18 arc min. The pair can be resolved by a good small telescope.

In the Milky Way in Perseus, we find the famous "double cluster," also known as *h* and χ (chi) Persei, and marked on the chart
as NGC 869 and NGC 884. The double cluster is a favorite of amateurs because it is so easy to observe. Binoculars reveal it as a hazy
patch. A small telescope used with a low-power, wide-angle eyepiece (field of 1°) shows both clusters. Each cluster has a diameter
of about 70 light-years. NGC 869 is the younger of the two, only
about 10 million years old, which makes it one of the youngest
clusters known; it lies over 7000 light-years away from us.

About 1½° east is the smaller open cluster NGC 957. At lower
right in Perseus is M76 (NGC 650), a planetary nebula of magnitude 11.4. Near the chart's right border, at 58°, η (eta) Cas is a
well-known binary. Its components show a beautiful color contrast, sometimes reported as gold and purple or as yellow and red.

As we move from Cassiopeia to Camelopardalis, we go further
from the Milky Way and into less interesting regions of the sky.

Fig. 44. The double open cluster in Perseus, h and chi Persei (NGC 869
and 884). (Lick Observatory photo)

MAP FROM "SKY ATLAS 2000.0" BY WIL TIRION

MAGNITUDES		OPEN CLUSTERS				QUASAR	△	GREEK ALPHABET			
●	>-0.4		>10', TO SCALE	<10'		PULSAR	☆	α	Alpha	ν	Nu
●	-0.4-+0.5	GLOBULAR CLUSTERS				BLACK HOLE	Y	β	Beta	ξ	Xi
●	0.6-1.5		>10'	5'-10'	<5'	MILKY WAY		γ	Gamma	o	Omicron
●	1.6-2.5							δ	Delta	π	Pi
●	2.6-3.5	PLANETARY NEBULAE				GALACTIC EQUATOR		ε	Epsilon	ϱ	Rho
●	3.6-4.5		>1'	0.5'-1'	<0.5'		70°	ζ	Zeta	σ	Sigma
●	4.6-5.5							η	Eta	τ	Tau
●	5.6-6.5	BRIGHT DIFFUSE NEB.			□	ECLIPTIC		θ ϑ	Theta	υ	Upsilon
●	6.6-7.5		>10', TO SCALE	<10'			100°	ι	Iota	φ φ	Phi
DOUBLE or MULTIPLE	●-					CONSTELLATION BOUNDARIES		κ ϰ	Kappa	χ	Chi
VARIABLE	◎ ○	GALAXIES	○ ○ ○ ○					λ	Lambda	ψ	Psi
			>30' 20'-30' 10'-20' <10'					μ	Mu	ω	Omega

ATLAS CHART 3. Camelopardalis Camelopardalis, the Giraffe, has very few bright stars. In Lynx, a trio of multiple stars lies just right of 7^h and just below +60°. The multiple system farthest to the right is an interesting triple. The double star A6012, near 7^{h}23^m +55°, is surrounded by several faint companions.

The galaxy NGC 2403 (7^{h}34^m +65°40′) is one of the nearest spiral galaxies outside of our Local Group of galaxies. Binoculars show it as a large hazy spot, while large telescopes hint at spiral structure. NGC 2403 is 8 million light-years away and 37,000 light-years across. The galaxy is easily seen even with small telescopes at low power, and the view with larger telescopes is striking.

Further down in Camelopardalis, at 4^{h}08^m +62°, is the open cluster NGC 1502, readily visible through small telescopes. Larger telescopes show it to be tightly packed with many stars of differing magnitudes. On nights with good seeing, the cluster is visible even through binoculars.

Six degrees above NGC 1502 is the large beautiful spiral galaxy IC 342. Though it can be seen with small telescopes, larger ones are needed to show any spiral structure. The planetary nebula NGC 1501, located in Camelopardalis slightly below NGC 1502, is about 1 arc min in diameter and slightly oval. Because it is dim, at least a medium aperture is needed for a good view.

Fig. 45. Camelopardalis, the Giraffe (drawn backwards), from the star atlas of Hevelius (1690).

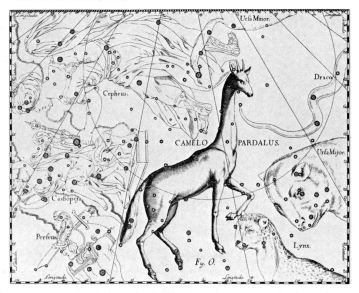

MAGNITUDES		OPEN CLUSTERS				QUASAR	△	GREEK ALPHABET			
●	>-0·4		>10'	TO SCALE	<10'	PULSAR	⯭	α Alpha		ν Nu	
●	-0·4 - 0·5					BLACK HOLE	⯭	β Beta		ξ Xi	
●	0·6 - 1·5	GLOBULAR CLUSTERS	⊕	⊕	⊛	MILKY WAY		γ Gamma		ο Omicron	
●	1·6 - 2·5		>10'	5'-10'	<5'			δ Delta		π Pi	
●	2·6 - 3·5	PLANETARY NEBULAE	◇	◇	◇	GALACTIC EQUATOR	70°	ε Epsilon		ϱ Rho	
●	3·6 - 4·5		>1'	0·5'-1'	<0·5'			ζ Zeta		σ Sigma	
●	4·6 - 5·5	BRIGHT DIFFUSE NEB.			□	ECLIPTIC	100°	η Eta		τ Tau	
●	5·6 - 6·5		>10'. TO SCALE		<10'			θ ϑ Theta		υ Upsilon	
	6·6 - 7·5							ι Iota		φ φ Phi	
DOUBLE or MULTIPLE	●─●					CONSTELLATION BOUNDARIES		κ ϰ Kappa		χ Chi	
		GALAXIES	◯	◯	◯			λ Lambda		ψ Psi	
VARIABLE	◎ ○		>30'	20'-30'	10'-20' <10'			μ Mu		ω Omega	

ATLAS CHART 4. M81 and M82   The galaxies M81 (NGC 3031) and M82 (NGC 3034) appear at the center of this chart, just short of 10^h and below $70°$. They are members of the Ursa Major cluster of galaxies. M81 is one of the prettiest spirals in the sky. M82 (C.Pl. 22) appears as though gas is exploding from it, but there has been continual controversy over whether we are seeing gas exploding or light reflected from relatively stationary gas. These galaxies are two of the most easily observed galaxies in the sky, and are visible through a small telescope or a good pair of binoculars. Nearby are NGC 3077 and NGC 2976, two other members of the cluster that are easily visible with low-power instruments.

QSO 0957 + 561 is the "double quasar," the first known example of a "gravitational lens." The light and radio waves from this quasar are bent by the gravity of an intervening galaxy as they pass it, making us see at least two images separated by only 6 arc sec. The optical objects are 17th magnitude and have a redshift of 1.41 (141%), making them some of the most distant ones known.

Fig. 46. M81 (NGC 3031), a type Sb spiral galaxy in Ursa Major, with M82 (NGC 3034) above it, NGC 3077 to its lower left, and NGC 2976 at bottom right. (Palomar Observatory photo)

MAGNITUDES	OPEN CLUSTERS	QUASAR	GREEK ALPHABET

MAGNITUDES

●	>-0.4
●	-0.4-+0.5
●	0.6-1.5
●	1.6-2.5
●	2.6-3.5
•	3.6-4.5
•	4.6-5.5
•	5.6-6.5
•	6.6-7.5

DOUBLE or
MULTIPLE ●—●

VARIABLE ◉ ○

OPEN CLUSTERS ○ ○ ○ ○ >10', TO SCALE <10'

GLOBULAR CLUSTERS ⊕ ⊕ ∘ >10' 5'-10' <5'

PLANETARY NEBULAE ◇ ◇ ◇ >1' 0.5'-1' <0.5'

BRIGHT DIFFUSE NEB. ◻ >10', TO SCALE <10'

GALAXIES ○ ○ ○ ○ > 30' 20'-30' 10'-20' <10'

QUASAR △

PULSAR ✶

BLACK HOLE ⅄

MILKY WAY

GALACTIC EQUATOR 70°

ECLIPTIC 100°

CONSTELLATION BOUNDARIES

GREEK ALPHABET

α Alpha	ν Nu
β Beta	ξ Xi
γ Gamma	ο Omicron
δ Delta	π Pi
ε Epsilon	ϱ Rho
ζ Zeta	σ Sigma
η Eta	τ Tau
θ ϑ Theta	υ Upsilon
ι Iota	φ φ Phi
κ ϰ Kappa	χ Chi
λ Lambda	ψ Psi
μ Mu	ω Omega

ATLAS CHART 5. Big Dipper, Owl Nebula At lower right in Ursa Major — about $2\frac{1}{2}°$ southeast of β (beta) UMa — is M97 (NGC 3587), the Owl Nebula (C.Pl. 27), one of the largest of the planetary nebulae. Small telescopes show a featureless circular disk, but the image in large telescopes suggests the face of an owl. Within 1° of M97 is the galaxy M108 (NGC 3556), shown in C.Pl. 32.

Mizar, ζ (zeta) UMa, is the second star from the end of the handle of the Big Dipper and is located near $13^{h}30^{m}$ +55°. Mizar is one of the most interesting doubles in the entire sky. Its companion, Alcor (labelled g on the chart), lies about 12 arc min away (a distance equal to about $\frac{1}{3}$ of the moon's diameter) and is just barely visible with the naked eye. Both components of Mizar, called Mizar A and Mizar B, are in turn double stars, though their components cannot be resolved by telescopes. The motions of the components show up in spectra. The American Indians referred to Mizar and Alcor as the Horse and Rider.

M109 (NGC 3992), a bright barred spiral galaxy in Ursa Major, lies near γ (gamma) Ursa Majoris, at 12^{h} and above +50°. Another easily observed galaxy, NGC 3953, lies nearby. These and other galaxies are part of the Ursa Major cluster of galaxies. Another cluster of galaxies can be observed in Draco.

An interesting variable to observe is RY in Draco, at 13^{h} +65°, which varies with a 6-month period.

Fig. 47. The Owl Nebula, M97 (NGC 3587) in Ursa Major. (Mt. Wilson and Las Campanas Observatories photo)

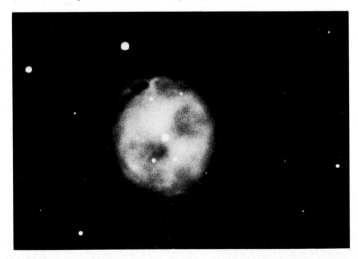

CEP

CAMELOPARDALIS

DRACO

URSA MINOR

DRACO

URSA MAJOR

CANES VENATICI

FROM "SKY ATLAS 2000.0" BY WIL TIRION

Owl neb.

MAGNITUDES		OPEN CLUSTERS		QUASAR	△	GREEK ALPHABET			
	>-0.4		>10' TO SCALE <10'	PULSAR		α Alpha		ν Nu	
	-0.4-0.5			BLACK HOLE		β Beta		ξ Xi	
	0.6-1.5	GLOBULAR CLUSTERS	>10' 5'-10' <5'	MILKY WAY		γ Gamma		o Omicron	
	1.6-2.5					δ Delta		π Pi	
	2.6-3.5	PLANETARY NEBULAE	>1' 0.5'-1' <0.5'	GALACTIC EQUATOR		ϵ Epsilon		ϱ Rho	
	3.6-4.5				70°	ζ Zeta		σ Sigma	
	4.6-5.5					η Eta		τ Tau	
	5.5-6.5	BRIGHT DIFFUSE NEB.	>10' TO SCALE <10'	ECLIPTIC	100°	θ ϑ Theta		υ Upsilon	
	6.6-7.5					ι Iota		ϕ φ Phi	
DOUBLE or MULTIPLE				CONSTELLATION BOUNDARIES		κ $\varkappa$ Kappa		χ Chi	
		GALAXIES	>30' 20'-30' 10'-20' <10'			λ Lambda		ψ Psi	
VARIABLE						μ Mu		ω Omega	

ATLAS CHART 6. M101, Little Dipper   M101 (NGC 5457) is an exceptionally beautiful spiral galaxy in Ursa Major, near $14^h + 50°$. It is seen face-on, and at almost 8th magnitude, it is one of the brightest galaxies in the sky (C.Pl. 28). M101's spiral arms can barely be seen on a very clear night through a medium-sized telescope, but because the galaxy is so spread out, it is hard to find unless conditions are perfect. M102 was one of Messier's few mistakes; it was really a second reference to M101. Many other members of the Ursa Major cluster of galaxies, such as NGC 5907 (C.Pl. 44), also appear on this chart, near $15^h 20^m + 56°$.

The Little Dipper in Ursa Minor lies in the central region of this chart, prominently marked by β (beta) UMi, γ (gamma) UMi, η (eta) UMi, and ζ (zeta) UMi. β and γ UMi are also known as Kochab and Pherkad, respectively, the "guards" of the Little Dipper. Few other objects of special interest are in this region. One double star worth observing is κ (kappa) Boötis, near $14^h 13^m + 52°$.

Fig. 48. M101 (NGC 5457), a type Sc spiral galaxy in Ursa Major. (Palomar Observatory photo)

ADAPTED FROM "SKY ATLAS 2000.0" BY WIL TIRION

MAGNITUDES	OPEN CLUSTERS				QUASAR	△	GREEK ALPHABET			
● >-0.4		○ >10', TO SCALE	○	⚬ <10'	PULSAR	⚡	α Alpha		ν Nu	
● -0.4 - +0.5					BLACK HOLE	Y	β Beta		ξ Xi	
● 0.6-1.5	GLOBULAR CLUSTERS	⊕ >10'	⊕ 5'-10'	⊕ <5'	MILKY WAY		γ Gamma		ο Omicron	
● 1.6-2.5							δ Delta		π Pi	
● 2.6-3.5	PLANETARY NEBULAE	◇ >1'	◇ 0.5'-1'	◇ <0.5'	GALACTIC EQUATOR	70°	ε Epsilon		ϱ Rho	
● 3.6-4.5						—+—	ζ Zeta		σ Sigma	
● 4.6-5.5							η Eta		τ Tau	
● 5.5-6.5	BRIGHT DIFFUSE NEB.		▱ >10', TO SCALE	▫ <10'	ECLIPTIC	100°	θ ϑ Theta		υ Upsilon	
● 6.6-7.5						— + —	ι Iota		φ φ Phi	
DOUBLE or MULTIPLE ●—●					CONSTELLATION BOUNDARIES		κ ϰ Kappa		χ Chi	
VARIABLE ◉ ○	GALAXIES ⬭ >30'	⬭ 20'-30'	⬬ 10'-20'	⬬ <10'			λ Lambda		ψ Psi	
							μ Mu		ω Omega	

ATLAS CHART 7. Draco   The four stars in the head of Draco — ξ, ν, β, and γ Dra (xi, nu, beta, and gamma Dra) — form a conspicuous asterism, the Lozenge, at the bottom of this chart. They are not far from the bright star Vega (see Atlas Chart 18) and are close to the foot of Hercules. The Lozenge and other areas of Draco contain several interesting doubles. Binoculars resolve the star ν (nu) Dra in the Lozenge, and a good small telescope resolves the nearby star μ (mu) Dra, at 17^h $+55°$.

Several other doubles appear toward the center of the chart, including φ (phi) Dra, near 18^h20^m $+71°$, and the pair at 18^h $+80°$, whose members have Flamsteed numbers 40 and 41 Dra. Due east (left) of φ (phi) Dra is ε (epsilon) Dra, a good double to observe at moderate magnifications since its companions are only about 3 arc sec apart.

The planetary nebula NGC 6543 (C.Pl. 48), at magnitude 8.8, is one of the brightest in the sky. It is located almost in the center of this chart, about halfway between δ (delta) and ζ (zeta) Dra. NGC 6543 was the first planetary nebula to be observed with a spectroscope; the fact that emission lines were present settled the controversy over whether planetaries were numerous stars or — as they turned out to be — clouds of diffuse gas. A small telescope at moderate magnification shows the disk, but more powerful telescopes are needed to show the internal structure, a bright irregular helix. Since NGC 6543 is a circumpolar object for most observers, you can view it throughout the year.

The quasar 3C 351 lies at 17^h $+60°$. It is magnitude 15.3 and has a redshift of .371 (37.1%), making it about 7 billion light-years away.

Fig. 49. Draco, the Dragon, snakes across the sky in this drawing from the star atlas of Hevelius (1690). The constellation is drawn backwards from the way it appears in the sky because Hevelius drew the celestial sphere as it would appear from the outside.

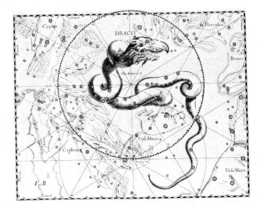

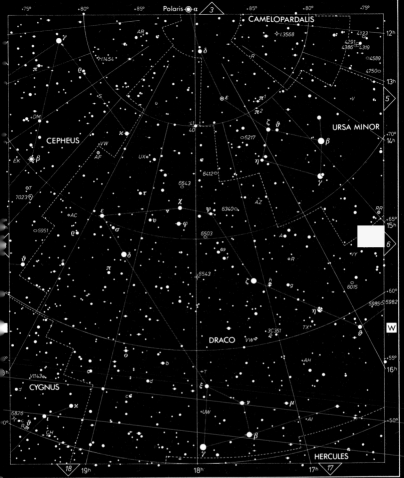

ADAPTED FROM "SKY ATLAS 2000.0" BY WIL TIRION

MAGNITUDES		OPEN CLUSTERS	QUASAR △	GREEK ALPHABET

MAGNITUDES
- -1 > -0.4
- 0 -0.4 - +0.5
- 1 0.6 - 1.5
- 2 1.6 - 2.5
- 3 2.6 - 3.5
- 4 3.6 - 4.5
- 5 4.6 - 5.5
- 6 5.6 - 6.5
- 7 6.6 - 7.5

DOUBLE or MULTIPLE

VARIABLE ◉ ○

OPEN CLUSTERS >10', TO SCALE <10'

GLOBULAR CLUSTERS >10' 5'-10' <5'

PLANETARY NEBULAE >1' 0.5'-1' <0.5'

BRIGHT DIFFUSE NEB. >10', TO SCALE <10'

GALAXIES >30' 20'-30' 10'-20' <10'

QUASAR △
PULSAR ☿
BLACK HOLE Y
MILKY WAY
GALACTIC EQUATOR —— 70° ——
ECLIPTIC —— 100° ——
CONSTELLATION BOUNDARIES

GREEK ALPHABET

α	Alpha	ν	Nu
β	Beta	ξ	Xi
γ	Gamma	ο	Omicron
δ	Delta	π	Pi
ε	Epsilon	ϱ	Rho
ζ	Zeta	σ	Sigma
η	Eta	τ	Tau
θ ϑ	Theta	υ	Upsilon
ι	Iota	φ φ	Phi
κ ϰ	Kappa	χ	Chi
λ	Lambda	ψ	Psi
μ	Mu	ω	Omega

ATLAS CHART 8. delta Cephei The Milky Way passes through Cepheus and Cygnus in this chart, and presents a very rich star field for observing. The region contains a number of open clusters, planetary nebulae, and gaseous nebulae. Look in Cepheus, near $21^h50^m +58°$, for the variable star μ (mu) Cep, one of the reddest stars visible to the naked eye. It is sometimes called the Garnet Star because of its deep red color. Just south of μ (mu) Cep is a large faint structure, IC 1396, a region of extended nebulosity. 1^h15^m to the right of μ (mu) Cep and slightly north, at about $20^h30^m +60°$, are the open cluster NGC 6939 and the spiral galaxy NGC 6946 (C.Pl. 51). Above μ (mu) Cep, near $22^h +64°$, is one of the best doubles in this constellation, ξ (xi) Cep.

Near the galactic equator, at $22^h30^m +58°$, is the famous star δ (delta) Cephei, a variable star whose discovery led to the first measurements of the distances to galaxies and thus to our subsequent understanding of the immense scale of the universe (p. 153). This star is the prototype of the *Cepheid variables* (p. 152). δ Cephei has a 5.4-day period, during which it varies by about one magnitude in brightness. By observing δ Cephei every night, you can easily detect the variation. Small telescopes also reveal that δ Cephei is a double star.

Toward the top of the chart, near $21^h30^m + 78°$, is a red variable, S Cep, which is visible through small telescopes. Closer to Polaris is the open cluster NGC 188, one of the oldest open clusters known, at 14 billion years of age. About 70 stars can be seen in this cluster with a medium-sized telescope at low power.

This completes the set of 8 Atlas Charts covering the region from $+50°$ to the north celestial pole. The next 12 Atlas Charts cover the region from $+20°$ to $+50°$.

Fig. 50. Cepheus (drawn backwards), from the star atlas of Hevelius.

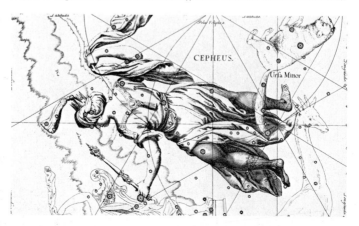

MAGNITUDES		OPEN CLUSTERS				QUASAR	△		GREEK ALPHABET		
-1	>-0.4		>10'	TO SCALE	<10'	PULSAR		α	Alpha	ν	Nu
0	-0.4-+0.5	GLOBULAR CLUSTERS				BLACK HOLE		β	Beta	ξ	Xi
1	0.6-1.5		>10'	5-10'	<5'	MILKY WAY		γ	Gamma	o	Omicron
2	1.6-2.5							δ	Delta	π	Pi
3	2.6-3.5	PLANETARY NEBULAE				GALACTIC EQUATOR		ε	Epsilon	ϱ	Rho
4	3.6-4.5		>1'	0.5'-1'	<0.5'		70°	ζ	Zeta	σ	Sigma
5	4.6-5.5							η	Eta	τ	Tau
6	5.6-6.5	BRIGHT DIFFUSE NEB.				ECLIPTIC		$\theta\ \vartheta$	Theta	υ	Upsilon
7	6.6-7.5		>10'	TO SCALE	<10'		100°	ι	Iota	$\phi\ \varphi$	Phi
DOUBLE or MULTIPLE								$\kappa\ \varkappa$	Kappa	χ	Chi
		GALAXIES				CONSTELLATION BOUNDARIES		λ	Lambda	ψ	Psi
VARIABLE			>30'	20'-30'	10'-20' <10'			μ	Mu	ω	Omega

ATLAS CHART 9. M31 (Andromeda Galaxy) M31 (NGC 224), the Great Galaxy in Andromeda (C.Pl. 9), lies in the center of the chart, near 0ʰ45ᵐ +41°. At a distance of 1.5 million light-years from earth, M31 is the largest neighboring galaxy. It is visible to the naked eye as a hazy glow. It appears in telescopes as a yellowish oval glow, brighter toward the center; the dust lane below its center may be seen or photographed. If you slowly sweep the field with medium power you will see more detail in outer regions of this spiral galaxy. M31 has two faint elliptical companions, M32 (NGC 221) and NGC 205, sometimes known as M110. M32 usually appears as a bright roundish haze, while NGC 205 looks a bit larger though not as bright.

At left in Triangulum is the face-on spiral galaxy M33 (NGC 598), the brightest spiral in the northern sky (Fig. 51, C.Pl. 10) except for M31. M33 covers an area about the same size as the moon but can be difficult to find. (Try testing a telescope on M33 before attempting to find dimmer objects.) With a low-power, wide-field telescope, the galaxy appears as an oval glow covering a large area. On nights of particularly good seeing, you can observe it with binoculars and may even be able to just barely detect it with the naked eye.

3C 48, one of the first quasars discovered, lies at about 1ʰ40ᵐ +34°, above M33. It has a magnitude of 16.2 and a redshift of 0.367 (36.7%), making it about 7 billion light-years away.

M76 (NGC 650), a prominent planetary nebula (C.Pl. 20), lies near the boundary of Perseus and Andromeda. M76 is difficult to find, though it is easiest to observe in the early fall. Many consider it to be the faintest of all the Messier objects. Because it has two distinct components, M76 looks like a smaller version of the Dumbbell Nebula and is sometimes called the Barbell Nebula.

Some noteworthy double stars in this area of the sky are: ι (iota) Tri (yellow and blue) at left center; A1457 (white and green), above and to the right of β (beta) Ari; and γ (gamma) And, at 2ʰ +43° (orange and yellow). γ (gamma) Ari is one of the most beautiful doubles in the sky, with contrasting orange and green components.

Fig. 51. M33 (NGC 598), a type Sc spiral galaxy in Triangulum. (The Kitt Peak National Observatory)

MAGNITUDES

-1   ●   >−0·4
0   ●   −0·4–·05
1   ●   0·6–1·5
2   ●   1·6–2·5
3   ●   2·6–3·5
4   ●   3·6–4·5
5   •   4·6–5·5
6   ·   5·6–6·5
7   ·   6·6–7·5

DOUBLE or MULTIPLE ●—●

VARIABLE ◉ ○

OPEN CLUSTERS   ○ ○ ◌ ✧
    >10' TO SCALE <10'

GLOBULAR CLUSTERS   ⊕   ⊕   ⊙
    >10' 5–10' <5'

PLANETARY NEBULAE   ◇   ◇   ◇
    >1' 0·5–1' <0·5'

BRIGHT DIFFUSE NEB. ⬭   □
    >10' TO SCALE <10'

GALAXIES   ⬭ ○ ○ ○
    >30' 20'–30' 10'–20' <10'

QUASAR   △
PULSAR   ⚡
BLACK HOLE   ⴲ
MILKY WAY   ∿
GALACTIC EQUATOR   —70°—┼—
ECLIPTIC   ——100°——
CONSTELLATION BOUNDARIES   ·—·—·

GREEK ALPHABET

α	Alpha	ν	Nu
β	Beta	ξ	Xi
γ	Gamma	ο	Omicron
δ	Delta	π	Pi
ε	Epsilon	ϱ	Rho
ζ	Zeta	σ	Sigma
η	Eta	τ	Tau
θ ϑ	Theta	υ	Upsilon
ι	Iota	φ φ	Phi
κ κ	Kappa	χ	Chi
λ	Lambda	ψ	Psi
μ	Mu	ω	Omega

ADAPTED FROM "SKY ATLAS 2000.0" BY WIL TIRION

ATLAS CHART 10. The Pleiades, Algol Algol, β (beta) Per, is sometimes called the Demon Star (see p. 150). It is the most famous of the eclipsing binaries, and the easiest of these variable stars to observe. The eclipses occur approximately every 69 hours, dropping the total brightness of the system from about magnitude 2.3 to a minimum of magnitude 3.5 within a five-hour period; the system returns to its normal brightness after about 20 minutes. The eclipses are fun to observe with the naked eye (see Fig. 40 and Table 9 in Chapter 6).

M34 (NGC 1039) is an open cluster in Perseus, about halfway between the two bright stars Algol (β Per) and γ (gamma) And. It appears to the naked eye as a hazy object. About 80 stars are loosely grouped in M34, and many of these are blue-white doubles. M34 is beautiful in binoculars and small telescopes with wide fields.

M45, the Pleiades, one of the most spectacular and most obvious open clusters in the sky, is in Taurus. Six or seven of its stars, arranged roughly in the shape of a little dipper, are visible to the naked eye. Binoculars clearly show numerous faint stars (see Fig. 52 below), and even a small telescope shows more than 100 stars. Wisps of nebulosity, reflection nebulae, cover the group and are especially bright near the more luminous stars (see C.Pl. 15). The nebulous background shows clearly on long-exposure photographs, but you will probably not be able to see it visually.

North of the Pleiades, in Perseus, are NGC 1499 — the California Nebula (C.Pl. 35) — a large, faint nebula even more difficult to find than M76; and IC 348, a bright diffuse nebula. NGC 1220, a small open cluster that appears in binoculars as a hazy patch, is at the top center of the chart, near γ (gamma) Per. NGC 1528, a large open cluster visible to the naked eye on clear nights, is at the top left of the chart.

Fig. 52. A close-up of the Pleiades. (Wil Tirion)

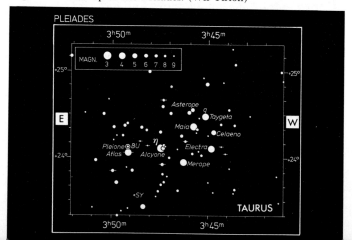

CAS

CAM

ANDROMEDA

TRIANGULUM

W

PERSEUS

PSC

ARIES

TAURUS

9

W

MAGNITUDES						
●	>-0.4				
●	-0.4 - +0.5				
●	0.6 - 1.5				
●	1.6 - 2.5				
●	2.6 - 3.5				
●	3.6 - 4.5				
●	4.6 - 5.5				
·	5.6 - 6.5				
·	6.6 - 7.5				

DOUBLE or
MULTIPLE

VARIABLE ◉ ○

OPEN CLUSTERS	○ ○ ○ ○
	>10' TO SCALE <10'

GLOBULAR CLUSTERS	⊕ ⊕ ⊕
	>10' 5'-10' <5'

PLANETARY NEBULAE	◇ ◇ ◇
	>1' 0.5'-1' <0.5'

BRIGHT DIFFUSE NEB.	▭
	>10' TO SCALE <10'

GALAXIES	◯ ◯ ○ ○
	>30' 20'-30' 10'-20' <10'

QUASAR △

PULSAR ☼

BLACK HOLE ⅄

MILKY WAY

GALACTIC EQUATOR
70°

ECLIPTIC
100°

CONSTELLATION
BOUNDARIES

GREEK ALPHABET			
α	Alpha	ν	Nu
β	Beta	ξ	Xi
γ	Gamma	ο	Omicron
δ	Delta	π	Pi
ε	Epsilon	ϱ	Rho
ζ	Zeta	σ	Sigma
η	Eta	τ	Tau
θ ϑ	Theta	υ	Upsilon
ι	Iota	φ φ	Phi
κ ϰ	Kappa	χ	Chi
λ	Lambda	ψ	Psi
μ	Mu	ω	Omega

ATLAS CHART 11. Crab Nebula, Hyades, Aldebaran

With binoculars or a rich-field telescope, sweep along the Milky Way through Perseus, Auriga, and Gemini. Many rich star fields, including numerous nebulae and open clusters, will reward your search. M36, M37, and M38 are all open clusters in Auriga. In the same region of Auriga, several gaseous nebulae of interest can be seen. IC 405 is sometimes called the Flaming Star; IC 410 is a cluster with surrounding nebulosity.

The most famous object on the chart is M1 (NGC 1952), the Crab Nebula in Taurus. The Crab is the remnant of a supernova, a star whose explosion the Chinese recorded in 1054 A.D. The Crab is still rapidly expanding, with gas travelling outward at about 1000 km per second. Small telescopes disappointingly show only an oval nebulosity; increasing the magnification reveals more shape. Only with a large aperture might you detect some of the filamentary structure that appears in photographs (C.Pl. 1).

A spectacular open cluster, the Hyades, outlines the bull's face in Taurus. It appears at the bottom edge of the chart, near $4^h30^m +15°$. The Hyades is an older and looser cluster than the Pleiades and contains fewer stars. Use low power or good binoculars to see this extended V-shaped group; it is so large that it is not circled on the chart. The stars in this cluster are approximately 140 light-years away and formed about a billion years ago. Their white color contrasts with the red of the star Aldebaran, α (alpha) Tau. Aldebaran is much closer to us than the Hyades.

Northeast of the Hyades, near $4^h45^m +18°$, is the open cluster NGC 1647; it includes about 50 stars spread across an area about the size of the moon's diameter.

In Auriga, at about $5^h10^m +45°$, is Capella, α (alpha) Aur, the sixth brightest star in the sky. Near Capella, ε (epsilon) Aur is an eclipsing variable with a period of 27 years, one of the longest periods known. Below ε Aur is ζ (zeta) Aur, another eclipsing variable with a fairly long period of $2\frac{2}{3}$ years.

Fig. 53. Taurus, the Bull (drawn backwards), from the star atlas of Hevelius (1690).

MAGNITUDES		OPEN CLUSTERS				QUASAR		GREEK ALPHABET			
-1	>-0.4		>10'	TO SCALE	<10'	PULSAR		α	Alpha	ν	Nu
0	-0.4 - +0.5							β	Beta	ξ	Xi
1	0.6 - 1.5	GLOBULAR CLUSTERS				BLACK HOLE		γ	Gamma	ο	Omicron
2	1.6 - 2.5		>10'	5'-10'	<5'	MILKY WAY		δ	Delta	π	Pi
3	2.6 - 3.5							ε	Epsilon	ϱ	Rho
4	3.6 - 4.5	PLANETARY NEBULAE				GALACTIC EQUATOR		ζ	Zeta	σ	Sigma
5	4.6 - 5.5		>1'	0.5'-1'	<0.5'			η	Eta	τ	Tau
6	5.6 - 6.5	BRIGHT DIFFUSE NEB.				ECLIPTIC		θ ϑ	Theta	υ	Upsilon
7	6.6 - 7.5		>10'	TO SCALE	<10'			ι	Iota	φ φ	Phi
DOUBLE or MULTIPLE						CONSTELLATION BOUNDARIES		κ ϰ	Kappa	χ	Chi
		GALAXIES						λ	Lambda	ψ	Psi
VARIABLE			>30'	20'-30'	10'-20' <10'			μ	Mu	ω	Omega

ATLAS CHART 12. Castor and Pollux Gemini is marked
by the heavenly twins, Castor and Pollux. The fainter of the two,
Castor, α (alpha) Gem, is one of the most beautiful double stars
easily resolved in small telescopes. The two stars, of magnitudes 2.0
and 2.8, slowly revolve around each other with a period of 477
years; a third, fainter star called Castor C is also present. Castor is
actually a sextuple system, since each of its three components has
been found by spectroscopy to be double. Castor C is not only a
spectroscopic double but also an eclipsing binary, and is known as
YY Gem. Pollux, β (beta) Gem, is brighter than Castor by about ½
magnitude and is more yellowish.

Southwest of Pollux is δ (delta) Gem, a bright double star.
Nearby is NGC 2392, the Eskimo or Clown Face Nebula (Fig. 54).
The distinctive greenish "face" of this planetary nebula is visible
only in large telescopes. ζ (zeta) Gem, to the right of the Eskimo
Nebula, near $7^h +20°$, is one of the brightest Cepheid variables. It
changes in brightness by more than ½ magnitude during its 10-day
period.

Toward the other side of the chart, near $6^h15^m +22°$, is η (eta)
Gem, a variable red giant. Slightly above η Gem is M35 (NGC
2168), a beautiful open cluster, and its companion NGC 2158. M35
is about 40 arc min in diameter and contains about 120 stars. It is
sometimes visible to the naked eye as a hazy, faint patch; any
telescope or even binoculars will reveal a fine open cluster. It can
be seen best with an eyepiece with at least a 1° field. High power
tends to reduce the beauty of M35, but is fine for observing its
companion about ½° to the southwest, NGC 2158. NGC 2158 is a
very rich open cluster, with about 40 stars crowded into an area
only 4 arc sec across. Although M35 and NGC 2158 appear close to
each other, they are completely unrelated physically: M35 is only
2200 light-years away, while NGC 2158 is 16,000 light-years away,
near the edge of our galaxy.

NGC 2419 in Lynx, about 7° north of Castor, is so far from the
center of our galaxy that it has been called an "intergalactic
tramp." On a clear dark night, you can see it in a small telescope.

Fig. 54. The Eskimo Nebula (NGC 2392), a planetary nebula. The
"face" is visible only in large telescopes. (Lick Observatory photo)

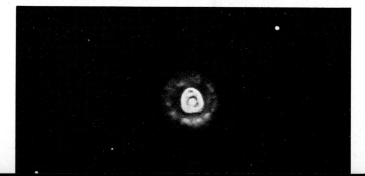

MAGNITUDES

-1	● > -0.4
0	● -0.4 - +0.5
1	● 0.6 - 1.5
2	● 1.6 - 2.5
3	● 2.6 - 3.5
4	● 3.6 - 4.5
5	● 4.6 - 5.5
6	• 5.6 - 6.5
7	· 6.6 - 7.5

DOUBLE or MULTIPLE

VARIABLE ◉ ○

OPEN CLUSTERS ○ ○ ⊙ ⊙
>10'; TO SCALE <10'

GLOBULAR CLUSTERS ⊕ ⊕ ●
>10' 5'-10' <5'

PLANETARY NEBULAE ◇ ◇ ◇
>1' 0.5'-1' <0.5'

BRIGHT DIFFUSE NEB. □
>10'; TO SCALE <10'

GALAXIES ⬭ ⬭ ⬬
>30' 20'-30' 10'-20' <10'

QUASAR △

PULSAR ☿

BLACK HOLE Y

MILKY WAY ∿

GALACTIC EQUATOR ─70°─
─┼─

ECLIPTIC ─100°─
─┼─

CONSTELLATION BOUNDARIES

GREEK ALPHABET

α	Alpha	ν	Nu
β	Beta	ξ	Xi
γ	Gamma	o	Omicron
δ	Delta	π	Pi
ε	Epsilon	ϱ	Rho
ζ	Zeta	σ	Sigma
η	Eta	τ	Tau
$\theta\ \vartheta$	Theta	υ	Upsilon
ι	Iota	$\phi\ \varphi$	Phi
$\kappa\ \varkappa$	Kappa	χ	Chi
λ	Lambda	ψ	Psi
μ	Mu	ω	Omega

ATLAS CHART 13. Praesepe The four stars γ, η, θ, and δ (gamma, eta, theta, and delta) Cancri (Cnc) at lower right outline the irregular shape that forms the body of Cancer, the Crab. Within the crab's body lies the open cluster M44 (NGC 2632), also called Praesepe or the Beehive Cluster (Fig. 55). M44 covers more than 80 arc min — an area almost three times the diameter of the moon — and is one of the most spectacular clusters in the sky. Although M44 is visible to the naked eye only as a hazy patch, a small telescope or binoculars will resolve it into its component stars, many of which are multiples. It is estimated to be 520 light-years from us. Praesepe means "the manger"; the stars γ (gamma) and δ (delta) Cnc nearby are known as "the asses," from a story in Greek mythology.

To the west of M44 is ζ (zeta) Cnc (at about $8^h12^m +18°$), an unusual triple star with three yellow components. This system can be viewed in medium-sized or large telescopes. $ι^1$ (iota[1]) Cnc (at about $8^h47^m +29°$) is an interesting orange-green double, outstanding in even a small telescope because of the beautiful color contrast between its components. Located slightly to the northeast, $ι^2$ (iota[2]) Cnc is a triple.

R Leo Minoris is a long-period variable, ranging in brightness from magnitude 6.0 to magnitude 13.3 with a period of 372 days. Its deep red color, especially when near maximum, makes this star an interesting telescopic object.

A noteworthy galaxy in Lynx is NGC 2683 at $8^h55^m +33°$, a spiral seen almost edge-on. Near the boundary between Lynx (Lyn) and Leo Minor, at $9^h20^m +35°$, are several other galaxies, easy for beginners to find by sweeping near α (alpha) Lyn. The brightest of these galaxies is NGC 2859, a barred spiral in Leo Minor. NGC 2903 (C.Pl. 38) is an isolated spiral in Leo.

Fig. 55. M44 (NGC 2632), also known as Praesepe, the Beehive Cluster. (Hopkins Observatory, Williams College)

26

MAGNITUDES

-1	> -0.4
0	-0.4 - -0.5
1	0.6 - 1.5
2	1.6 - 2.5
3	2.6 - 3.5
4	3.6 - 4.5
5	4.6 - 5.5
6	5.6 - 6.5
7	6.6 - 7.5

DOUBLE or MULTIPLE

VARIABLE

OPEN CLUSTERS
>10' TO SCALE <10'

GLOBULAR CLUSTERS
>10' 5-10' <5'

PLANETARY NEBULAE
>1' 0.5'-1' <0.5'

BRIGHT DIFFUSE NEB.
>10' TO SCALE <10'

GALAXIES
>30' 20'-30' 10'-20' <10'

QUASAR

PULSAR

BLACK HOLE

MILKY WAY

GALACTIC EQUATOR
70°

ECLIPTIC
100°

CONSTELLATION BOUNDARIES

GREEK ALPHABET

α	Alpha	ν	Nu
β	Beta	ξ	Xi
γ	Gamma	ο	Omicron
δ	Delta	π	Pi
ε	Epsilon	ϱ	Rho
ζ	Zeta	σ	Sigma
η	Eta	τ	Tau
θ ϑ	Theta	υ	Upsilon
ι	Iota	φ φ	Phi
κ ϰ	Kappa	χ	Chi
λ	Lambda	ψ	Psi
μ	Mu	ω	Omega

ADAPTED FROM "SKY ATLAS 2000.0" BY WIL TIRION

ATLAS CHART 14. Leo, M64 Leo Minor and Leo contain a number of interesting doubles. One of the most striking is γ (gamma) Leo, one of the stars in the sickle that makes up Leo, actually a double with magnitudes 2.3 and 3.5. This golden-yellow pair is easily resolved and provides a good view even in small telescopes. ζ (zeta) Leo, located in the sickle above γ Leo, is a bright pair of stars far enough apart to be detected with binoculars.

Nearby, at about $11^h +25°$, is another double star, 54 Leo, that can be resolved with very low power into a pair with magnitudes 4.5 and 6.3. The star Denebola — β (beta) Leo — at the tip of the lion's tail and just below the bottom of the chart at left, is a beautiful blue-orange pair. However, it is merely an optical double — the stars are actually far apart in space. The pulsar CP1133 nearby was one of the original four pulsars discovered by Jocelyn Bell and Antony Hewish in 1968.

To the lower left of the chart's center is the double star ξ (xi) Ursa Majoris; it is only 26 light-years away and is a good object to observe with small telescopes. The 60-year elliptical orbit of the two visual components, magnitudes 4.4 and 4.9, is tilted 57° to our line of sight. The pair reached their maximum separation of 2.9 arc min as seen from earth in 1980. Each of these two components is in turn a double. The stars in one of the pairs are separated by only 0.5 A.U. — half the distance between the earth and the sun — and orbit each other in 1.8 years. The stars in the other pair are separated by only the distance from the earth to the moon and orbit each other every 4 days.

Since this chart includes an area close to the north galactic pole (which is on the next chart), we are looking as far away as possible from the Milky Way and can thus see many galaxies. Scanning the nearby region of Leo Minor reveals a hard-to-see group of galaxies (Fig. 56). NGC 3158, at $10^h20^m +38°$, is the brightest of this group and should be visible in a medium-sized telescope.

We will discuss the cluster of galaxies at upper left with the next chart. M97 and M108 are discussed with Chart 5.

Fig. 56. Four galaxies in Leo, from Chart 14 (lower right): NGC 3193 (type E2, upper left), NGC 3187 (type SBc, upper right), NGC 3190 (type Sb, upper center, with dust lane), and NGC 3185 (type SBa, bottom right). (Palomar Observatory photo)

MAGNITUDES		OPEN CLUSTERS				QUASAR	△	GREEK ALPHABET			
	> -0·4		>10'	TO SCALE	<10'	PULSAR		α Alpha		ν Nu	
	-0·4 - +0·5					BLACK HOLE		β Beta		ξ Xi	
	0·6-1·5	GLOBULAR CLUSTERS	⊕	⊕	⊕	MILKY WAY		γ Gamma		ο Omicron	
	1·6-2·5		>10'	5'-10'	<5'			δ Delta		π Pi	
	2·6-3·5					GALACTIC EQUATOR		ε Epsilon		ϱ Rho	
	3·6-4·5	PLANETARY NEBULAE	◇	◇	◇		70°	ζ Zeta		σ Sigma	
	4·6-5·5		>1'	0·5'-1'	<0·5'			η Eta		τ Tau	
	5·6-6·5					ECLIPTIC		θ ϑ Theta		υ Upsilon	
	6·6-7·5	BRIGHT DIFFUSE NEB.			□		100°	ι Iota		φ φ Phi	
			>10'	TO SCALE	<10'			κ ϰ Kappa		χ Chi	
DOUBLE or MULTIPLE						CONSTELLATION BOUNDARIES		λ Lambda		ψ Psi	
VARIABLE	◎ ○	GALAXIES						μ Mu		ω Omega	

ATLAS CHART 15. Whirlpool Galaxy Characteristic of this part of the sky is the large number of galaxies, most of them millions of light-years distant. Relatively close to us is one of the brightest galaxies, M51 (NGC 5194) — also known as the Whirlpool Galaxy (C.Pl. 16) — at upper left-center in Canes Venatici. Located a few degrees south of the handle of the Big Dipper, M51 is a face-on spiral about 15 million light-years distant, with a companion galaxy NGC 5195. Other galaxies nearby to the south and west include M63 (NGC 5055), M94 (NGC 4736), and M106 (NGC 4258), all spirals of type Sc that can be observed with small telescopes. M106 is shown in C.Pl. 31. Far beyond are the clusters of galaxies in Ursa Major and Canes Venatici.

At $12^h52^m +27°$ in Coma Berenices, we come to the region around the north galactic pole. Since we are looking away from the Milky Way at this point, there is minimum interference from gas and dust in our galaxy, and many galaxies are visible here. M64 (NGC 4826), at lower center in Coma Berenices, is a spiral known as the Black-eye Galaxy because the dark dust lane across its center is so prominent. The galaxy is relatively bright and can be detected in binoculars. Also located here are the spiral galaxy M100 (NGC 4321) — shown in C.Pl. 26 — and the elliptical galaxy M85 (NGC 4382). Ten degrees to the north are NGC 4559 and NGC 4565 (C.Pl. 43), an edge-on spiral tilted only 4° from our line of sight. These galaxies are all in the Virgo Cluster of galaxies (see Charts 27 and 28). Much farther away is the Coma Cluster of galaxies, in which only NGC 4889 and NGC 4874, near $13^h +28°$, are bright enough to be seen by most amateur observers.

In this area of the sky we also find several globular clusters. At magnitude 4.5, the globular cluster M3 (NGC 5272) — located at $13^h40^m +28°$ in Canes Venatici — is one of the brightest in the sky. Two other globular clusters, M53 (NGC 5024) and NGC 5053, appear lower in the sky in Coma Berenices. M53, near α (alpha) Com, is twice as far away as M3 and therefore twice as large, since both star clusters have roughly the same apparent size in the sky.

Mel 111 (at $12^h20^m +27°$ in Coma Berenices) is a large, diffuse open cluster of 5th- to 10th-magnitude stars.

This part of the sky is best placed for viewing in the spring.

Fig. 57. The Whirlpool Galaxy, M51 (NGC 5194), a type Sb spiral, with the irregular companion galaxy NGC 5195 at the end of one of its spiral arms. (The Kitt Peak National Observatory)

MAGNITUDES		OPEN CLUSTERS			QUASAR	△	GREEK ALPHABET			
	>-0.4		>10', TO SCALE <10'		PULSAR	☆	α Alpha		ν	Nu
	-0.4--0.5	GLOBULAR CLUSTERS			BLACK HOLE	Υ	β Beta		ξ	Xi
	0.6-1.5		>10' 5'-10' <5'		MILKY WAY		γ Gamma		o	Omicron
	1.6-2.5						δ Delta		π	Pi
	2.6-3.5	PLANETARY NEBULAE			GALACTIC EQUATOR		ϵ Epsilon		ϱ	Rho
	3.6-4.5		>1' 0.5'-1' <0.5'			70° +	ζ Zeta		σ	Sigma
	4.6-5.5						η Eta		τ	Tau
	5.6-6.5	BRIGHT DIFFUSE NEB.		□	ECLIPTIC		θ ϑ Theta		υ	Upsilon
	6.6-7.5		>10', TO SCALE <10'			100° +	ι Iota		ϕ φ	Phi
DOUBLE or MULTIPLE	•-•-						κ $\varkappa$ Kappa		χ	Chi
		GALAXIES			CONSTELLATION BOUNDARIES		λ Lambda		ψ	Psi
VARIABLE	◉ ○		>30' 20'-30' 10'-20' <10'				μ Mu		ω	Omega

ATLAS CHART 16. Arcturus At lower right is the third brightest star in the sky, Arcturus, α (alpha) Boötis. It is 36 light-years away and about 25 times the diameter of the sun. ε (epsilon) Boo (at $14^h45^m +27°$) is a double star with widely spaced, contrasting orange and green components that orbit each other with a period of 153 years. Further to the northeast, μ (mu) Boo (at $15^h25^m +37°$) is a widely spaced, white-orange pair with fainter companions.

To the south, in Corona Borealis (the Northern Crown), η (eta) CrB is a close double, with components orbiting each other in about 42 years. ζ (zeta) CrB is a wide pair of blue stars, separated from each other by over 6 arc sec. σ (sigma) CrB is also a wide pair. γ (gamma) CrB is a very close pair, orbiting in about 100 years. o (omicron) CrB has several faint companions and a period of 215 years.

R Coronae Borealis (R CrB) is one of the most unusual stars in the sky. It stays at 6th magnitude for a period ranging from months to years and then precipitously drops to 11th magnitude or even fainter for a period that lasts only weeks (see light curve below). At the same time, its spectrum of absorption lines is obscured and a spectrum of emission lines appears. R CrB is a carbon-rich star, and apparently it sometimes throws off clouds of carbon soot that obscure its photosphere, making spectral lines from its higher and relatively hotter chromosphere appear as emission lines. The clouds of soot have been detected in the infrared. A finding chart for R CrB appears on p. 161.

Fig. 58. Light curve for R Coronae Borealis. (AAVSO)

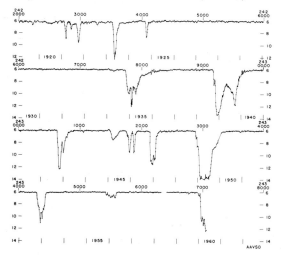

CHART FROM "SKY ATLAS 2000.0" BY WIL TIRION

ATLAS CHART 17. M13   At the center of the chart in Hercules we find the Keystone, an asterism formed by the stars η, ζ, ε, and π (eta, zeta, epsilon, and pi) Her. About $\frac{1}{3}$ of the way from η (eta) to ζ (zeta) Her is the spectacular globular cluster, M13 (NGC 6205), shown in C.Pl. 4. M13 lies about 25,000 light-years away. It is faintly visible to the naked eye as a fuzzy mothball of the 4th magnitude. You can resolve the outer edges of M13 into its member stars with a medium-sized telescope.

About $\frac{1}{2}°$ northeast of M13 is the faint spiral galaxy NGC 6207, sometimes visible in the same field at low power.

At $17^h15^m +42°$, about 7° north of π (pi) Her, is the globular cluster M92 (NGC 6341), about 1 magnitude fainter than M13 but still a good object for small telescopes. A third globular cluster, NGC 6229 (at $16^h45^m +47°$), lies about 10° above M13; it is much smaller and fainter (10th magnitude) than the other two clusters.

By sweeping with low power near β (beta) Her at the bottom of the chart, you can find NGC 6210, a bluish-green planetary nebula. NGC 6210 has a fairly bright inner ring with a faint outer ring that measures 20 by 43 arc sec. The nebula can be detected with a small telescope, although a larger telescope is needed to distinguish the rings. The planetary nebula NGC 6058 (at $16^h +41°$) is more difficult to locate.

δ (delta) Her is an optical double for which many different colors have been reported. ρ (rho) Her is a pair of blue-white stars that differ in brightness by 1 magnitude. Only 30 light-years away from us is ζ (zeta) Her, a pair with magnitudes 2.9 and 5.5. All lie on the constellation outline drawn.

The quasar 3C 345 is 16th magnitude and has a redshift of 0.6 (60%), placing it about 8 billion light-years away.

Fig. 59. Hercules (drawn backwards), from the star atlas of Hevelius.

AGNITUDES		OPEN CLUSTERS		QUASAR	$\triangle$	GREEK ALPHABET			
	>-0.4		>10' TO SCALE <10'	PULSAR		α Alpha		ν Nu	
	-0.4 - +0.5			BLACK HOLE		β Beta		ξ Xi	
	0.6 - 1.5	GLOBULAR CLUSTERS		MILKY WAY		γ Gamma		o Omicron	
	1.6 - 2.5		>10' 5'-10' <5'			δ Delta		π Pi	
	2.6 - 3.5	PLANETARY NEBULAE		GALACTIC EQUATOR		ε Epsilon		ϱ Rho	
	3.6 - 4.5		>1' 0.5'-1' <0.5'			ζ Zeta		σ Sigma	
	4.6 - 5.5					η Eta		τ Tau	
	5.6 - 6.5	BRIGHT DIFFUSE NEB.		ECLIPTIC		θ ϑ Theta		υ Upsilon	
	6.6 - 7.5		>10' TO SCALE <10'			ι Iota		ϕ φ Phi	
OUBLE or ULTIPLE				CONSTELLATION BOUNDARIES		κ $\varkappa$ Kappa		χ Chi	
		GALAXIES				λ Lambda		ψ Psi	
ARIABLE			>30' 20'-30' 10'-20' <10'			μ Mu		ω Omega	

ATLAS CHART 18. Ring Nebula, Dumbbell Nebula The brightest star on this chart is Vega, α (alpha) Lyrae. Vega is a pure white star of 1st magnitude and is dazzlingly beautiful in a telescope. With ε (epsilon) and ζ (zeta) Lyr, Vega forms a small equilateral triangle. A person with acute eyesight can barely detect the two components of ε (epsilon) Lyr, each of which is about 5th magnitude. Even a small telescope separates the pair and splits each component into a second pair, making four white stars in all.

β (beta) Lyr, a spectroscopic double and an eclipsing binary, varies between magnitudes 3.4 and 4.3 every 12.9 days. The variations are easy to follow by comparing the magnitude of this double with that of γ (gamma) Lyr, magnitude 3.2 (see Fig. 40, p. 159). Other nearby stars in Lyra to compare are η (eta) at magnitude 4.4, θ (theta) at 4.5, ζ^1 (zeta^1) at 4.4, κ (kappa) at 4.3, λ (lambda) at 4.9, and ι (iota) at 5.3.

Albireo — β (beta) Cygni — at the nose of the Swan, is one of the prettiest doubles. Binoculars split it into orange and blue components. The components of δ (delta) Cygni, whose orbital period is 300 years, differ by 3 magnitudes.

RR Lyrae (at 19^h25^m $+43°$) is the prototype of a class of variable stars with regular periods of less than one day (p. 152).

At the center of the chart, located about halfway between β and γ Lyr, is M57 (NGC 6720), the Ring Nebula (C.Pl. 17). This nebula, the most famous of the planetary nebulae, has a shell of gas expanding from a central star. The shell represents the outer quarter of the star's mass and bares the core of the star, which appears bluish because it is so hot. This is the end that our sun will reach one day, in about 5 billion years. A small telescope shows the nebula clearly, although a larger one is necessary to reveal structural details. The colors show up only in long-exposure photographs.

To the south and east of M57, near 19^h20^m $+30°$, is the globular cluster M56 (NGC 6779). This small, highly concentrated cluster is easily visible with small telescopes. Further south and east in Vulpecula, near 20^h $+23°$, is M27 (NGC 6853), the second largest planetary nebula, about 3 light-years in diameter. Because of its shape, M27 is commonly called the Dumbbell Nebula (C.Pl. 8). Though its total brightness is high, M27 is spread over such a large area that the average brightness of its surface remains fairly low. A dark sky is needed to view it well. Low power shows its shape and greenish color; higher power begins to show some structure. To the west are the pulsars PSR 1937 $+215$ and CP 1919, which can be detected only in the radio spectrum. PSR 1937 $+215$ gives off over 600 pulses of radio waves per second.

Below M27, in Sagitta, is the globular cluster M71 (NGC 6838), which is rich in stars despite its relatively small diameter.

On the boundary between Lyra and Hercules, at 18^h20^m $+46°$, is the remnant of Nova Herculis 1934.

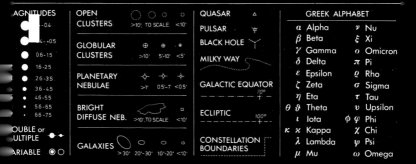

MAGNITUDES		OPEN CLUSTERS				QUASAR		GREEK ALPHABET			
	−0.4		>10', TO SCALE	<10'		PULSAR		α Alpha		ν Nu	
	−0.5	GLOBULAR CLUSTERS	⊕	⊕	·	BLACK HOLE		β Beta		ξ Xi	
	0.6-1.5		>10'	5'-10'	<5'	MILKY WAY		γ Gamma		o Omicron	
	1.6-2.5							δ Delta		π Pi	
	2.6-3.5	PLANETARY NEBULAE		0.5'-1'	<0.5'	GALACTIC EQUATOR		ε Epsilon		ϱ Rho	
	3.6-4.5		>1'					ζ Zeta		σ Sigma	
	4.6-5.5							η Eta		τ Tau	
	5.6-6.5	BRIGHT DIFFUSE NEB.				ECLIPTIC		θ ϑ Theta		υ Upsilon	
	6.6-7.5		>10', TO SCALE	<10'				ι Iota		ϕ φ Phi	
DOUBLE or MULTIPLE						CONSTELLATION BOUNDARIES		κ $\varkappa$ Kappa		χ Chi	
VARIABLE		GALAXIES		20'-30'	10'-20' <10'			λ Lambda		ψ Psi	
			>30'					μ Mu		ω Omega	

ATLAS CHART 19. Deneb, North America Nebula, Veil Nebula, Cygnus Loop In the center of the chart near the bright star Deneb, α (alpha) Cyg, is the North America Nebula (NGC 7000), named because of its shape. Although this nebula and its neighbor, the Pelican Nebula (IC 5067–5070), are extremely hard to observe telescopically, long-exposure photography makes them visible. The Pelican's beak is facing the North America Nebula (C.Pl. 53). Another region of nebulosity lies around γ (gamma) Cyg, southwest of Deneb, near $20^h20^m +40°$.

On the 20^h meridian is Cygnus X-1, invisible in ordinary light but the brightest x-ray source in Cygnus. The x-rays probably come from gas swirling around a black hole; the presence of the black hole is inferred from its gravitational effect on the 15th-magnitude star HDE226868 that is observed at that location.

Southwest of this region, near $21^h +30°$, is another area of diffuse nebulae. NGC 6960 appears close to the double star 52 Cygni, whose orange and blue components are separated by $6\frac{1}{2}$ arc sec. NGC 6960, 6992, 6995, and 6979 are jointly known as the Veil Nebula, which makes up most of the Cygnus Loop, a faint circular nebula about $2\frac{1}{2}°$ across. The Cygnus Loop is the remnant of a supernova that exploded over 100,000 years ago. The Veil Nebula is visible through a wide range of optical instruments, from binoculars to large telescopes, but its structure and beautiful colors show only in long-exposure photographs (C.Pl. 52).

Within the triangle formed by α, γ, and ε (alpha, gamma, and epsilon) Cyg is the Northern Coalsack, an opaque patch of nebulosity that obscures the Milky Way behind it.

Near the upper right of the chart, at $19^h40^m +50°$, is the planetary nebula NGC 6826. This nebula is often called the Blinking Nebula because its central star, which is relatively bright, appears to blink on and off if you look rapidly toward and away from it.

M39 (NGC 7092) is an open cluster in northern Cygnus (near $21^h30^m +48°$), with about 30 stars scattered across its 7 light-year diameter. This cluster is a fine object to view with binoculars or low-power, wide-field telescopes.

Near $21^h40^m +43°$ is the variable star SS Cygni, which rises sharply to its maximum about once every 50 days, on a somewhat irregular schedule (Fig. 34, p. 152). It is very interesting to follow as it rapidly increases in brightness from 12th to 8th magnitude. You can see its brightness change markedly in two or three hours during certain parts of its cycle. SS Cyg is too faint to be shown on this chart, but its position is marked on a finding chart (Fig. 39) in Chapter 6.

To the southwest, near $21^h20^m +39°$, is 61 Cyg, the first star whose distance from the sun was measured (from its parallax). 61 Cyg is an orange pair of stars only 10.9 light-years away from us.

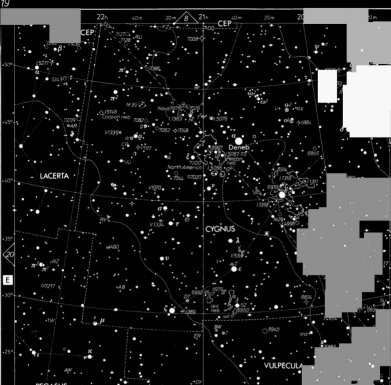

ADAPTED FROM "SKY ATLAS 2000.0" BY WIL TIRION

MAGNITUDES	
-1	>-0.4
0	-0.4–-0.5
1	0.6–1.5
2	1.6–2.5
3	2.6–3.5
4	3.6–4.5
5	4.6–5.5
6	5.6–6.5
7	6.6–7.5

DOUBLE or MULTIPLE

VARIABLE

OPEN CLUSTERS   >10' TO SCALE <10'

GLOBULAR CLUSTERS   >10' 5'–10' <5'

PLANETARY NEBULAE   >1' 0.5'–1' <0.5'

BRIGHT DIFFUSE NEB.   >10' TO SCALE <10'

GALAXIES   > 30' 20'–30' 10'–20' <10'

QUASAR

PULSAR

BLACK HOLE

MILKY WAY

GALACTIC EQUATOR

ECLIPTIC

CONSTELLATION BOUNDARIES

GREEK ALPHABET

α	Alpha	ν	Nu
β	Beta	ξ	Xi
γ	Gamma	o	Omicron
δ	Delta	π	Pi
ε	Epsilon	ϱ	Rho
ζ	Zeta	σ	Sigma
η	Eta	τ	Tau
θ ϑ	Theta	υ	Upsilon
ι	Iota	φ φ	Phi
κ ϰ	Kappa	χ	Chi
λ	Lambda	ψ	Psi
μ	Mu	ω	Omega

ATLAS CHART 20. Pegasus As we move away from the
Milky Way into Lacerta, Andromeda, and Pegasus, the star fields
thin out. The bright star β (beta) Peg — below the center of the
chart, near 23ʰ +27° — marks the northwest corner of the Great
Square of Pegasus. β Peg, also called Scheat, is a red-giant star
that varies with an irregular period from magnitude 2.4 to 3.0. One
corner of the square is not in Pegasus; α (alpha) Andromedae
(And) marks the northeast corner of the Square. About halfway
between these two stars is the double star 78 Peg, a close pair with
magnitudes 5 and 8. The third star in the square, α (alpha) Peg, is
drawn just outside the chart boundary at the bottom center; the
fourth star does not show on this chart.

About 5° to the northwest of β Peg is η (eta) Peg. Follow the line
from β to η Peg about 7° further to a pair of double stars, π¹ (pi¹)
and π². π² is an optical double — two yellow stars at different distances from earth that are apparently separated in the sky by only
15 minutes of arc. The pair is especially suitable for viewing with
binoculars.

Just to the right of the center of the chart, about 5° north and
slightly west of η Peg, is the spiral galaxy NGC 7331 (Fig. 60). This
galaxy can be found by sweeping the area with low power.

The planetary nebula NGC 7662 lies near 23ʰ26ᵐ +42.5° in
Andromeda, about 3° southwest of ι (iota) And. One of the best
planetaries for observing, NGC 7662 appears round, slightly ring-
like, and bluish-green. About 3 arc min south of NGC 7662 is NGC
7640, a barred spiral galaxy that is fairly easy to locate.

Sweeping the northern area of Lacerta with a low-power telescope reveals several deep-sky objects. Binoculars give a good view
of the open cluster NGC 7243, west of α (alpha) Lac, near
22ʰ15ᵐ +50°. Large telescopes show about 50 of the cluster's stars.
About 4° southwest is another open cluster, NGC 7209, smaller
and fainter than NGC 7243. Many of its stars are 9th to 10th magnitude; they provide a beautiful view in a medium-sized telescope.
East of β (beta) Lac is a third open cluster, NGC 7296.

Fig. 60. NGC 7331, a type Sb spiral galaxy in Pegasus. The dust
lane surrounding its disk shows
clearly. Three other spirals are
visible beyond it. (The Kitt Peak
National Observatory)

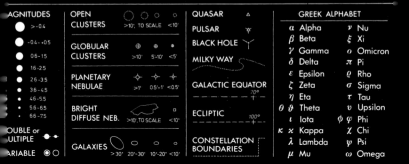

MAGNITUDES

- ● >-0.4
- ● -0.4 - +0.5
- ● 0.6 - 1.5
- ● 1.6 - 2.5
- ● 2.6 - 3.5
- ● 3.6 - 4.5
- ● 4.6 - 5.5
- ● 5.6 - 6.5
- ● 6.6 - 7.5

DOUBLE or MULTIPLE ●
VARIABLE ◉ ○

OPEN CLUSTERS	◌	○	◌	◌
	>10'	TO SCALE		<10'

GLOBULAR CLUSTERS	⊕	⊕	●
	>10'	5'-10'	<5'

PLANETARY NEBULAE	◇	◇	◇
	>1'	0.5'-1'	<0.5'

BRIGHT DIFFUSE NEB. ▭ >10' TO SCALE ▫ <10'

GALAXIES	◯	○	○	○
	>30'	20'-30'	10'-20'	<10'

QUASAR △
PULSAR ✶
BLACK HOLE ⅄
MILKY WAY 〰
GALACTIC EQUATOR ⊦ 70°
ECLIPTIC ⊦ 100°
CONSTELLATION BOUNDARIES ⸽

GREEK ALPHABET

α	Alpha	ν	Nu
β	Beta	ξ	Xi
γ	Gamma	ο	Omicron
δ	Delta	π	Pi
ε	Epsilon	ϱ	Rho
ζ	Zeta	σ	Sigma
η	Eta	τ	Tau
θ ϑ	Theta	υ	Upsilon
ι	Iota	φ φ	Phi
κ ϰ	Kappa	χ	Chi
λ	Lambda	ψ	Psi
μ	Mu	ω	Omega

ATLAS CHART 21. Pisces   The next group of star charts shows the region of sky surrounding the celestial equator. Near the top of this chart are γ (gamma) and α (alpha) Peg, which are the bottom two stars of the Great Square of Pegasus (Atlas Chart 20).

In Pisces, in the middle of the right-hand side of the chart, the stars ι, θ, γ, κ, and λ (iota, theta, gamma, kappa, and lambda) Psc form a distinctive asterism shaped like a small, irregular pentagon. This asterism, called the Circlet, marks the western fish in Pisces, and lies directly south of the Great Square of Pegasus. The Circlet covers an area about 7° by 5° and also includes fainter stars.

Due west of δ (delta) Psc is UU Psc, a wide pair with components of 6th and 8th magnitudes.

Below the Circlet, at −10°, is the triple system ψ^1, ψ^2, ψ^3 (psi¹, psi², psi³) in Aquarius. About $4\frac{1}{2}$° south of ψ^3 Aqr is the double star 94 Aqr, a pair of white and yellow-white stars with magnitudes 5.3 and 7.5. Even a small telescope at low power resolves this wide pair.

To the southeast, R Aqr lies south of ω^1 (omega¹) and ω^2 (omega²) Aqr. R Aqr is a long-period (Mira-type) variable that changes in brightness from about 6th to 12th magnitude and back to 6th magnitude again during its 385-day period. The double star 107 Aqr can be found 4° south and slightly east of R Aqr. The white and yellow-white components of 6th and 7th magnitude are separated by about 6 arc sec.

In Cetus we find ϕ^1, ϕ^2, and ϕ^3 (phi¹, phi², and phi³) about 7° north of β (beta) Cet. ϕ^1 and ϕ^2 form an equilateral triangle with the planetary nebula NGC 246. The planetary looks like a dim oval through small telescopes but looks like a ring with a 10th-magnitude central star through large telescopes.

Fig. 61. Pisces, the Fish, from the star atlas of Hevelius (1690). The constellation is drawn backwards from the way it appears in the sky because Hevelius drew the celestial sphere as it would appear from the outside.

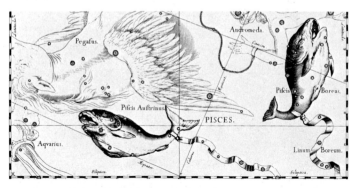

MAGNITUDES		OPEN		QUASAR	△	GREEK ALPHABET			
-1	>-0.4	CLUSTERS	>10' TO SCALE <10'	PULSAR	⚡	α	Alpha	ν	Nu
0	-0.4 -+0.5			BLACK HOLE	⌄	β	Beta	ξ	Xi
1	0.6-1.5	GLOBULAR				γ	Gamma	ο	Omicron
2	1.6-2.5	CLUSTERS	>10' 5-10' <5'	MILKY WAY		δ	Delta	π	Pi
3	2.6-3.5					ε	Epsilon	ϱ	Rho
4	3.6-4.5	PLANETARY		GALACTIC EQUATOR		ζ	Zeta	σ	Sigma
5	4.6-5.5	NEBULAE	>1' 0.5'-1' <0.5'		70°	η	Eta	τ	Tau
6	5.6-6.5					θ ϑ	Theta	υ	Upsilon
7	6.6-7.5	BRIGHT		ECLIPTIC	100°	ꝯ ι	Iota	φ φ	Phi
		DIFFUSE NEB.	>10' TO SCALE <10'			κ ϰ	Kappa	χ	Chi
DOUBLE or						λ	Lambda	ψ	Psi
MULTIPLE	•-•	GALAXIES		CONSTELLATION		μ	Mu	ω	Omega
VARIABLE	◎ ○		>30' 20'-30' 10'-20' <10'	BOUNDARIES					

ATLAS CHART 22. Cetus, M77 Just above the center of the chart, the "cords" of Pisces — the "cords" holding the two fish together — converge to form an acute angle at α (alpha) Psc, also called El Rischa, the Knot. With components of magnitudes 4.2 and 5.2 separated by about 2 arc sec, α Psc can be observed with a medium-sized telescope.

The cords point to the lower left toward Mira, o (omicron) Ceti, the variable star about halfway down the neck of Cetus, the Whale. The first variable discovered, in 1596, Mira ("the Wonderful") is the prototype of long-period variables (p. 152). At its maximum magnitude of about 2.0, it is the brightest star in this constellation, shining with a deep red color. During its 332-day cycle, Mira slowly fades to fainter than 9th magnitude, far below naked-eye visibility, then gradually returns to maximum. (See Figs. 35 and 41 in Chapter 6.)

About 6° — three fingers' width — northeast of Mira is δ (delta) Cet, and about ½° southeast of this star is M77 (NGC 1068), one of the brighter spiral galaxies. M77's bright nucleus can be seen with a small telescope, but observations of its face-on spiral structure require a large telescope.

α (alpha) Cet, at left center, is an optical double. The two components, separated by 16 arc min, can be resolved with binoculars.

ζ (zeta) Cet lies near $1^h50^m -10°$, toward the bottom of the neck of Cetus. About 5° northeast of ζ Cet is NGC 615, a spiral galaxy. This field contains two of the stars closest to us. The flare star UV Ceti near the bottom of the chart is in the fifth nearest star system. It is too faint to see easily when at its normal 13th magnitude, but sometimes it brightens within about five minutes to almost 6th magnitude, a change of nearly 250 times in brightness. To its upper left is τ (tau) Ceti, the 20th closest star to earth. τ Cet, magnitude 3.5, is the nearest solar-type star (that is, a star with the same type of spectrum, temperature, and other properties as our sun). It is thus one of the prime objects in the search for life elsewhere in the universe.

Fig. 62. M77 (NGC 1068), a type Sb spiral galaxy in Cetus. At magnitude 10.5, it is a difficult object to detect in small telescopes. (Lick Observatory photo)

MAGNITUDES		OPEN CLUSTERS				QUASAR		GREEK ALPHABET			
●	>-0.4		>10'	TO SCALE	<10'	△		α	Alpha	ν	Nu
●	-0.4 - +0.5							β	Beta	ξ	Xi
●	0.6-1.5	GLOBULAR CLUSTERS	⊕	⊕	⊕	PULSAR		γ	Gamma	o	Omicron
●	1.6-2.5		>10'	5'-10'	<5'	BLACK HOLE		δ	Delta	π	Pi
●	2.6-3.5					MILKY WAY		ε	Epsilon	ϱ	Rho
●	3.6-4.5	PLANETARY, NEBULAE	◇	◇	◇			ζ	Zeta	σ	Sigma
●	4.6-5.5		>1'	0.5'-1'	<0.5'	GALACTIC EQUATOR		η	Eta	τ	Tau
●	5.6-6.5						70°	θ ϑ	Theta	υ	Upsilon
●	6.6-7.5	BRIGHT DIFFUSE NEB.		>10' TO SCALE	□ <10'	ECLIPTIC	100°	ι	Iota	φ φ	Phi
DOUBLE or MULTIPLE	●● ○○							κ ϰ	Kappa	χ	Chi
VARIABLE	◉ ○	GALAXIES	○	○	○	CONSTELLATION BOUNDARIES		λ	Lambda	ψ	Psi
			> 30'	20'-30'	10'-20' <10'			μ	Mu	ω	Omega

ATLAS CHART 23. Aldebaran, Hyades As we move toward the western edge of the Milky Way in Taurus and Eridanus, the star fields become richer. On winter evenings, sweep southward with binoculars from the Hyades (at the top of this chart; see also Chart 11) through the shield of Orion.

Just north of the Hyades is T Tauri (at $4^h22^m +19°32'$, but unmarked because it is too faint to be included on our chart), one of the most interesting stars in the sky. T Tauri varies in brightness because it is extremely young and has not yet settled down to a steady existence. It fluctuates between 9th and 13th magnitudes. T Tauri stars, the youngest stars, are being studied by optical, infrared, and radio techniques. T Tauri is near a diffuse nebula, NGC 1554–55, known as Hind's Variable Nebula. Over the last century, the nebula faded from view, then brightened considerably.

Slightly northwest of μ (mu) Eri, at $4^h46^m -3°$, is NGC 1637, a galaxy visible with medium-sized telescopes. The planetary nebula NGC 1535 lies 12° southwest, nearer γ (gamma) Eri, at lower center; it consists of a bright inner ring and a faint outer ring with an 11.5-magnitude central star. NGC 1535 is difficult to find because of its isolated position. Far beyond it is a quasar.

At $4^h15^m -8°$ is the triple system o^2 (omicron²) Eri, also known as 40 Eridani. It is less than 16 light-years from us. It contains a 9th-magnitude white dwarf, 40 Eri B, the easiest white dwarf to observe with a small telescope. Astronomers have been able to measure the mass of the white dwarf by studying this system and have verified the theoretical prediction that its mass is about that of the sun. The system has also provided a test of Einstein's theory of relativity, since the white dwarf's spectral lines are redshifted by gravity, as Einstein predicted. 11th-magnitude 40 Eri C, which forms a visual binary with the white dwarf, is a small red star, one of the least massive stars known. The pair is separated from the bright (4th-magnitude) 40 Eri A by 82 arc sec.

About 6° northwest of o^1 (omicron¹), just to the right of the chart's center, lies w Eri. This star is also known as 32 Eri (double star number ADS 2850). Its greenish and yellowish components are 5th and 6th magnitude.

The cluster of galaxies at the bottom edge of the chart includes NGC 1300, a fine example of a barred spiral galaxy.

(*Atlas Charts continue on p. 250.*)

Fig. 63. T Tauri, with Hind's Variable Nebula (NGC 1555) 45 arc seconds to its west (right). Another nebulous patch surrounds the star itself and is known as Burnham's Nebula. (Lick Observatory photo)

TAURUS · ARIES · CETUS · ORION · ERIDANUS · LEPUS

Aldebaran · Hyades

PKS 0405-123

FROM "SKY ATLAS 2000.0" BY WIL TIRION

MAGNITUDES		OPEN CLUSTERS				QUASAR	△		GREEK ALPHABET			
●	>-0.4		>10', TO SCALE	<10'		PULSAR		α	Alpha	ν	Nu	
●	-0.4 - +0.5	GLOBULAR CLUSTERS				BLACK HOLE		β	Beta	ξ	Xi	
●	0.6 - 1.5		>10'	5'-10'	<5'	MILKY WAY		γ	Gamma	ο	Omicron	
●	1.6 - 2.5							δ	Delta	π	Pi	
●	2.6 - 3.5	PLANETARY NEBULAE				GALACTIC EQUATOR		ε	Epsilon	ϱ	Rho	
●	3.6 - 4.5		>1'	0.5'-1'	<0.5'		70°	ζ	Zeta	σ	Sigma	
•	4.6 - 5.5	BRIGHT DIFFUSE NEB.				ECLIPTIC		η	Eta	τ	Tau	
•	5.6 - 6.5		>10', TO SCALE	<10'			100°	θ ϑ	Theta	υ	Upsilon	
•	6.6 - 7.5					CONSTELLATION BOUNDARIES		ι	Iota	φ φ	Phi	
DOUBLE or MULTIPLE		GALAXIES						κ ϰ	Kappa	χ	Chi	
			>30'	20'-30'	10'-20'	<10'		λ	Lambda	ψ	Psi	
VARIABLE								μ	Mu	ω	Omega	

Notes on Color Plates
and Astronomical Photography

This color plate section includes both amateur and professional photographs. Nebulae and other deep-sky objects from the Messier Catalogue and Dreyer's New General Catalogue are shown first, in numerical order. The rest of the photographs appear in roughly the same order in which their subjects are described in this *Field Guide*: the moon, planets, comets, meteors, and the sun. The plates can be enjoyed without knowing the details of how they were made, but for those who are interested in trying some photography on their own, a few notes follow. (See also pp. 386 and 408–411, for more suggestions on photographing comets and solar eclipses.)

Astronomical photography requires certain special considerations. In order to capture images of objects in the nighttime sky, the camera's shutter must be left open longer than for most daytime exposures. The earth's rotation has an interesting effect on these time exposures: unless the camera is mounted on a sturdy tracking mount, the images of stars and other objects show up as streaks or curved trails (C.Pl. 70) instead of points of light or distinct shapes. You can take excellent photographs of bright comets, constellations, planetary conjunctions, and solar eclipses with a camera on a stationary tripod, but images of fainter objects (which require longer exposures) will come out blurred.

Another problem occurs when you try to photograph very faint objects: at very low light levels, film seems to "forget" that it has been struck by light. Unless you use special black-and-white films designed to compensate for this problem (which is called reciprocity failure), you will find that you will need even longer exposure times for very faint objects than you would expect.

Color film also responds less predictably during long exposures: the sensitivity of each layer in the film becomes different for a given exposure. The effect can be lessened by chilling the film (a technique often used in the color photos in this section), or by using special filters to compensate for color shifts, or both. Another approach is to make three separate exposures through different color filters onto black-and-white film, and then combine them in the darkroom using similar color filters. For technical notes on the equipment and techniques used to produce the color plates in this section, turn to p. 249.

216

C.Pl. 1. The Crab Nebula, M1 in Taurus (Chart 11), the remnant of the 1054 A.D. supernova. (© 1980 Ben Mayer)

217

C.Pl. 2. The Lagoon Nebula, M8 (Chart 42). (© 1978 Ben Mayer)

C.Pl. 3. The open cluster M11 in Scutum (Chart 30). (© 1980 Ben Mayer)

C.Pl. 4. The globular cluster M13 in Hercules, and the spiral galaxy NGC 6207 at upper left (Chart 17). (© 1980 Ben Mayer)

C.Pl. 5. The open cluster and nebula M16, often known as the Eagle Nebula, in Serpens (Chart 30). (© 1980 Ben Mayer)

C.Pl. 6. The Omega Nebula, M17, in Sagittarius (Chart 30). (Meade Instruments Corp.)

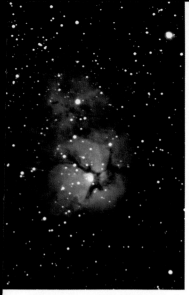

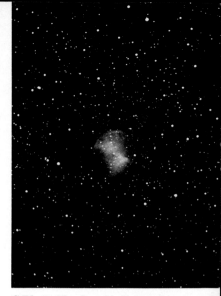

C.Pl. 7. The reddish Trifid Nebula, M20, and a separate blue reflection nebula (Charts 41, 42). (© 1979 Ben Mayer)

C.Pl. 8. The Dumbbell Nebula, M27, a planetary nebula in Vulpecula (Charts 18, 19). (© 1979 Ben Mayer)

219

C.Pl. 9. The Andromeda Galaxy, M31 (Chart 9), with the elliptical galaxies NGC 205 (above it) and M32. (© 1959 California Institute of Technology)

C.Pl. 10. The inner part of the spiral galaxy M33 in Triangulum, seen face-on (Chart 9). (© 1979 Ben Mayer)

C.Pl. 11. *(opposite)* A wide-field view of Orion, showing his belt, sword, and two nebulae (Chart 24). The Horsehead Nebula (IC 434) is below zeta Orionis, the first star (at left) in the belt; the Orion Nebula (M42 and M43) is farther south, in the sword. (© 1979 Ben Mayer)

C.Pl. 12. *(above)* The Horsehead Nebula and zeta Orionis (Chart 24). North is at left. (© 1980 Royal Observatory, Edinburgh)

C.Pl. 13. *(below)* The Orion Nebula, including M42 (the main nebula) and M43 (Chart 24). (Meade Instruments Corp.)

C.Pl. 15. The Pleiades, M45, an open cluster in Taurus containing many hot, young stars (Chart 10). This long exposure shows starlight reflecting off dust around many of the stars, making reflection nebulae. The bright star at left is Alcyone. (© 1961 California Institute of Technology)

C.Pl. 14. The constellation Orion, with Betelgeuse (top left), Rigel (lower right), and the Orion Nebula. (Jay M. Pasachoff)

C.Pl. 16. The Whirlpool Galaxy, M51, in Canes Venatici (Chart 15). (© 1980 Ben Mayer)

C.Pl. 17. The Ring Nebula, M57, a planetary nebula in Lyra (Chart 18). (© 1980 Ben Mayer)

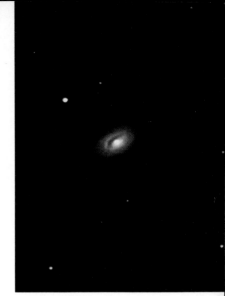

C.Pl. 18. The spiral galaxy M61 in Virgo (Chart 27A). (© 1980 James D. Wray, McDonald Observatory)

C.Pl. 19. The spiral galaxy M64 in Coma Berenices (Chart 15). (© 1980 Ben Mayer)

C.Pl. 20. The planetary nebula M76 in Perseus (Charts 2, 9, 10). (© 1980 Ben Mayer)

C.Pl. 21. The diffuse nebulae M78 and NGC 2071 in Orion (Chart 24). (© 1980 Ben Mayer)

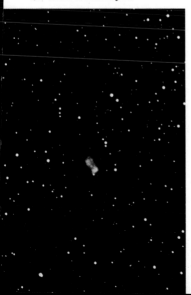

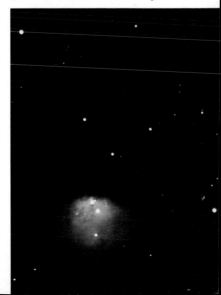

C.Pl. 22. The irregular galaxy M82 in Ursa Major (Charts 4, 5). (Lick Observatory photo)

C.Pl. 23. The spiral galaxy M83 in Hydra (Chart 39). (© 1979 Ben Mayer)

224

C.Pl. 24. The elliptical galaxy M87 in Virgo, with its jet of gas (Chart 27A). (© James D. Wray, McDonald Observatory)

C.Pl. 25. The spiral galaxy M90 in Virgo (Chart 27A). (© 1979 Ben Mayer)

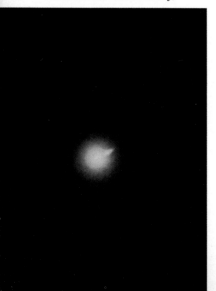

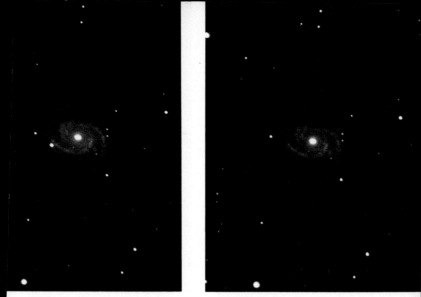

C.Pl. 26. The spiral galaxy M100 in Coma Berenices, with (left) and without (right) a supernova (Charts 15, 27A). (© 1979 Ben Mayer)

225

C.Pl. 27. The Owl Nebula, M97, a faint planetary nebula in Ursa Major (Chart 5). (© 1980 Ben Mayer)

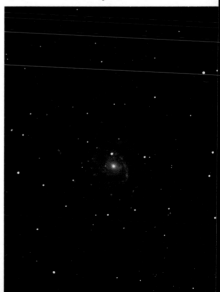

C.Pl. 28. The face-on spiral galaxy M101 in Ursa Major (Chart 6). (© 1980 Ben Mayer)

C.Pl. 29. The open cluster M103 in Cassiopeia (Chart 1). (© 1980 Ben Mayer)

C.Pl. 30. The Sombrero Galaxy, M104, in Virgo, a spiral galaxy seen edge-on (Chart 27). (© 1979 Ben Mayer)

226

C.Pl. 31. The spiral galaxy M106 in Canes Venatici (Chart 15). (© James D. Wray, McDonald Observatory)

C.Pl. 32. The edge-on spiral galaxy M108 in Ursa Major (Charts 5, 14). (© 1980 Ben Mayer)

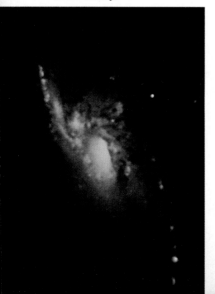

C.Pl. 33. The spiral galaxy NGC 253 (Chart 33) with its light-absorbing dust lanes. (© 1980 Anglo-Australian Telescope Board)

C.Pl. 34. The Large Magellanic Cloud, with its reddish Tarantula Nebula, NGC 2070 (Chart 47). (© 1977 Hans Vehrenberg)

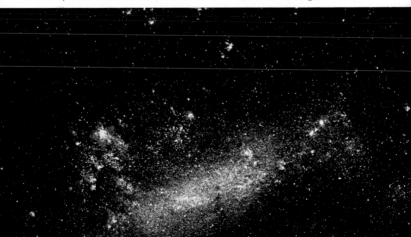

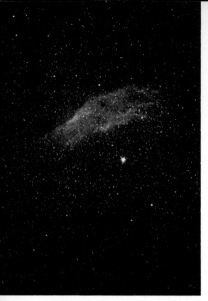

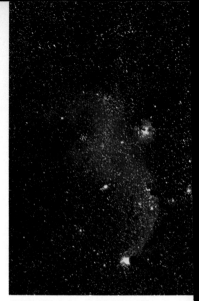

C.Pl. 35. The California Nebula, NGC 1499 (Chart 10), named because of its shape. (© 1977 Hans Vehrenberg)

C.Pl. 37. The Rosette Nebula, NGC 2237 – 39, in Monoceros, around the cluster NGC 2244 (Chart 24). (© 1965 California Institute of Technology)

C.Pl. 36. *(above)* IC 2177 (Chart 25), with nearby open clusters. (© 1977 Hans Vehrenberg)

C.Pl. 39. *(opposite)* The Cone Nebula, NGC 2264 (Chart 24). (© 1981 Anglo-Australian Telescope Board)

C.Pl. 38. *(below)* The spiral galaxy NGC 2903 in Leo (Chart 13). (© James D. Wray, McDonald Observatory)

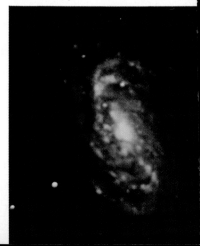

C.Pl. 40. The Eta Carinae Nebula, NGC 3372 (Chart 49). (© 1977 Hans Vehrenberg)

230

C.Pl. 41. NGC 2997, a spiral galaxy in Antlia (Chart 37). (© 1980 Anglo-Australian Telescope Board)

C.Pl. 42. NGC 5128, a galaxy wrapped with a tremendous dust lane (Chart 39), corresponds to the strong radio source Centaurus A. North is at upper left. (© 1979 R.J. Dufour)

C.Pl. 43. The edge-on spiral galaxy NGC 4565 in Coma Berenices (Chart 15). (U.S. Naval Observatory)

C.Pl. 44. The spiral galaxy NGC 5907 in Draco (Chart 6), seen edge-on. (U.S. Naval Observatory)

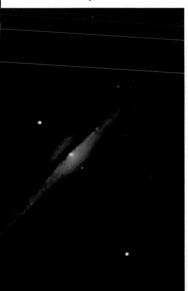

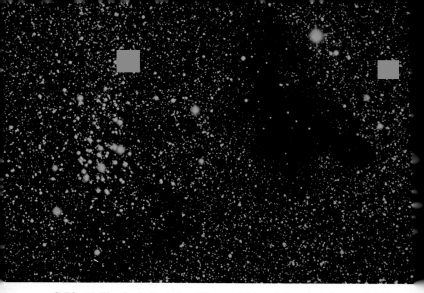

C.Pl. 45. The dark dust cloud Barnard 86 and the open cluster NGC 6520 in Sagittarius (Chart 42). (© 1980 Anglo-Australian Telescope Board)

C.Pl. 46. Dust and gas in Sagittarius, including the bluish reflection nebulae NGC 6589 and NGC 6590. The reddish glow is from the nebulae IC 1283 and IC 1284 (Charts 30, 42). (© 1981 Anglo-Australian Telescope Board)

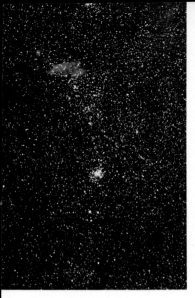

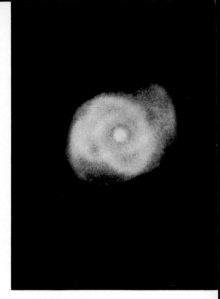

C.Pl. 47. The open cluster NGC 6231 (center) and the nearby nebula IC 4628 in Scorpius (Chart 41). (© 1977 Hans Vehrenberg)

C.Pl. 48. The planetary nebula NGC 6543 in Draco (Chart 7). (Lick Observatory photo)

C.Pl. 49. The spiral galaxy NGC 6744 in Pavo (Chart 51). (© 1979 R.J. Dufour)

C.Pl. 50. The faint planetary nebula NGC 6781 in Aquila (Chart 31). (© 1965 California Institute of Technology)

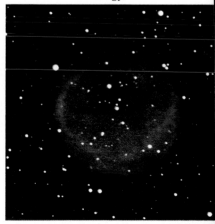

C.Pl. 51. NGC 6946, a spiral galaxy in Cepheus (Chart 8). (U.S. Naval Observatory)

234

C.Pl. 52. The Veil Nebula in Cygnus (Chart 19). NGC 6992–95 are at left and NGC 6960 is at right in this supernova remnant. (© 1977 Hans Vehrenberg)

C.Pl. 53. *(opposite)* The North America Nebula, NGC 7000, glows red because of the red emission from hydrogen; dark dust absorbs light in the region that corresponds to the Gulf of Mexico. The Pelican Nebula, IC 5067–70, is to the right (Chart 19). (© 1977 Hans Vehrenberg)

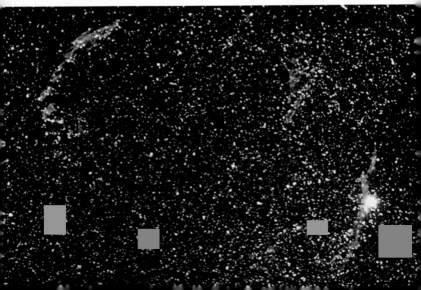

C.Pl. 54. The Helix Nebula, NGC 7293, is the nearest planetary to us, at 400 light-years (Chart 44). The red is from nitrogen and hydrogen; the greenish color is from oxygen that has lost two electrons. (© 1979 Anglo-Australian Telescope Board)

236

C.Pl. 55. Gas and dust around rho Ophiuchi (in the blue reflection nebula). The bright star Antares is surrounded by yellow and red nebulosity. M4, the nearest globular cluster to us, is nearby. North is at left (Chart 41). (© 1979 Royal Observatory, Edinburgh)

C.Pl. 56. An Apollo 17 astronaut on the moon. (NASA)

C.Pl. 57. Multiple exposures of a total lunar eclipse, taken over a 3-hour period. The totality exposures (1–4 seconds) were longer, to show the faint reddish glow, than the exposures for the partial phases (1/125–1/8 second). (Suginami Science Education Center)

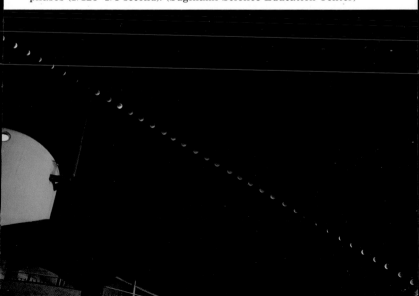

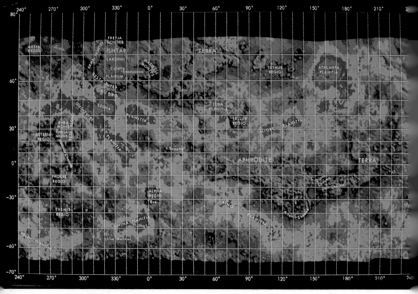

C.Pl. 58. A radar map of Venus, made from the Pioneer Venus spacecraft; the radar could penetrate the clouds that enshroud Venus. (NASA/Ames, U.S. Geological Survey, and M.I.T.)

C.Pl. 59. A view from the Viking 1 lander of its surroundings on Mars, including reddish rocks and a pink sky caused by reddish dust suspended in the atmosphere. The sampler arm (at right) dug soil samples that were tested for signs of life. (JPL/NASA)

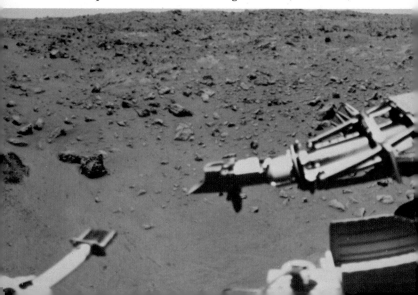

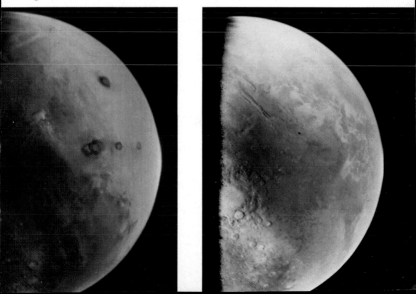

C.Pl. 60. A series of photographs of Mars taken from earth, showing surface features and the planet's rotation. North is at top. (International Planetary Patrol/Lowell Observatory)

C.Pl. 61. Viking 1 views of Mars. *(left)* Olympus Mons is the large volcano at top right, with three other volcanoes, the Tharsis Mountains, below it; *(right)* above center is part of the giant canyon, longer than the continental U.S. is wide. (JPL/NASA)

C.Pl. 62. One of the best ground-based photos of Jupiter. The Great Red Spot shows clearly. North is at top. (Lunar and Planetary Laboratory, University of Arizona)

C.Pl. 63. Jupiter, as seen from the Voyager 1 spacecraft. Io appears as a tiny reddish-brown ball suspended in front of the planet, and Europa is at far right. (JPL/NASA)

C.Pl. 64. Voyager photos of Jupiter's Galilean moons, to scale: Io (top left), made reddish by sulfur from its volcanoes; smooth, icy Europa; Ganymede, the largest solar-system moon (lower left); and Callisto (bottom right), with bright craters. (JPL/NASA)

241

C.Pl. 65. Computer processing of Voyager data unrolls Jupiter's dark belts and bright zones. (JPL/NASA)

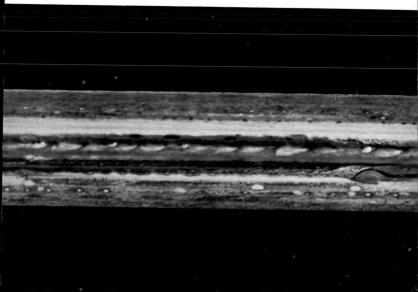

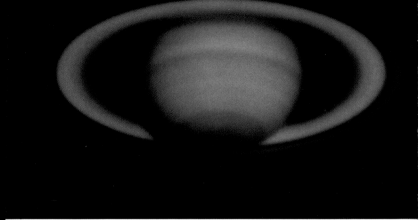

242

C.Pl. 66. *(above)* One of the best ground-based photos of Saturn, taken with a 61-in. telescope. North is at top. (Catalina Observatory, University of Arizona)

C.Pl. 67. *(below)* As Voyager 1 left Saturn and looked back, the planet appeared as a crescent inside its rings. (JPL/NASA)

C.Pl. 68. *(opposite)* A montage of Saturn and some of its moons, photographed from the Voyager 1 and 2 spacecraft. Clockwise from upper left, the moons are Titan (reddish), Iapetus (with one dark side), Tethys, Mimas (a small moon with a giant crater), Enceladus (lower center), Dione, and Rhea. (JPL/NASA)

C.Pl. 69. Comet West (1976) photographed over Los Angeles with a camera on a tripod (2-sec. exposure, *f*/2.5, ASA 200 film). (© 1976 Ben Mayer)

244

C.Pl. 70. Star trails over the Kitt Peak National Observatory (3-hour exposure; tripod-mounted camera). (Richard E. Hill)

C.Pl. 71. A multiple exposure of the sun's position over a year, making an analemma (see p. 417). (Dennis di Cicco)

C.Pl. 72. The green flash, a greenish solar edge appearing at sunset (see p. 402). (George V. Coyne, S.J., Vatican Observatory)

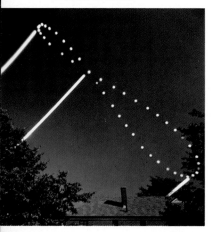

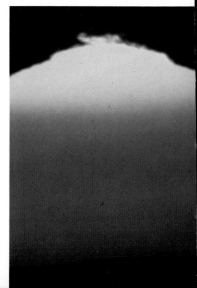

C.Pl. 73. The diamond-ring effect, including the ghostly white inner corona and the pinkish prominences, at the 1983 total solar eclipse. (Jay M. Pasachoff)

C.Pl. 74. A conjunction of the moon, Venus, and Jupiter, showing the dark side of the moon illuminated by earthshine — sunlight that reflects off the earth to the moon. (Jay M. Pasachoff)

C.Pl. 75. Meteors in the Leonid meteor shower show up as straight streaks across the curved star trails. (Don Pearson)

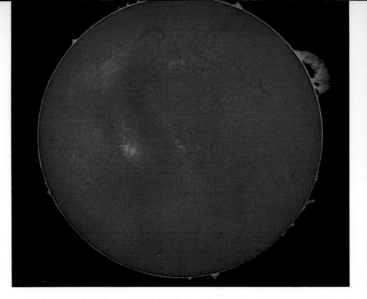

C.Pl. 76. The sun through an H-alpha filter. Printed over the center of an image showing prominences around the sun's edge is a less-exposed image, which shows dark filaments snaking across the solar disk (and some minor blotches from the filter). (© 1982 DayStar Filter Corp.)

C.Pl. 77. A prominence photographed through an H-alpha filter. (R. J. Poole/DayStar Filter Corp.)

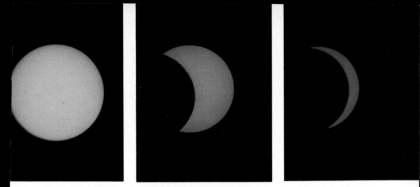

C.Pl. 78. Partial phases of the 1980 total solar eclipse, observed from India through a dense filter. (Jay M. Pasachoff)

C.Pl. 79. The total solar eclipse of 1980, photographed from India with a 500-mm lens on a standard Nikon. No filter was used. (Jay M. Pasachoff)

C.Pl. 80. A solar spectrum observed during a total solar eclipse. The arcs (which are white because they are overexposed) show the spectrum of the chromosphere around the sun's edge; the greenish circle shows emission from iron in the solar corona at a temperature of 2,000,000°C. (William C. Atkinson)

C.Pl. 81. The view during the total solar eclipse of 1980, showing the overexposed corona surrounding the dark, barely visible disk of the moon in the midday sky. The distant horizon is bright because the eclipse is not total there. (Jay M. Pasachoff)

C.Pl. 82. The diamond-ring effect—the last bit of solar photosphere shining through a valley on the edge of the moon—at the total solar eclipse of 1973. Scattering in the lens adds to the luster of the "diamond." (Jay M. Pasachoff)

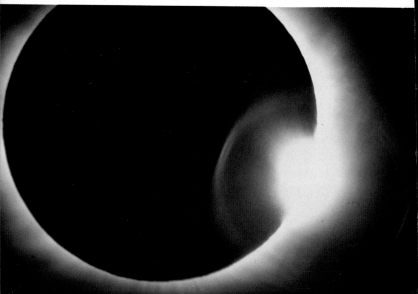

(*continued from p. 216*)

Here are a few technical notes on the equipment and techniques used to produce the color plates in this section:

Amateur Photographs:

Ben Mayer: Celestron 14-in. Schmidt-Cassegrain telescope and 8-in. Schmidt optics, Byers mount; chilled color film, ASA/ISO 400 film, exposures 30–120 minutes.

Meade: Meade 8-in. Schmidt-Cassegrain telescope; ordinary color film.

Hans Vehrenberg: 14-in. Celestron Schmidt telescope, Zeiss mount; separate black-and-white images combined in the darkroom.

Professional Photographs:

Anglo-Australian Telescope Board: 3.9-m (153-in.) telescope; black-and-white photos taken through filters chosen to respond to spectra of nebulae; images combined in the darkroom by David Malin.

Royal Observatory, Edinburgh: 1.2-m (48-in.) Schmidt telescope; same technique as the one used with the Anglo-Australian telescope.

James D. Wray: 0.9-m (36-in.) McDonald Observatory telescope; photographs taken through filters using an electronic intensifier, dye-transfer printing.

Lick Observatory: 120-in. telescope; black-and-white images taken through filters chosen for accurate reproduction of nebular spectra, dye-transfer printing.

R.J. Dufour: 4-m (156-in.) Cerro Tololo telescope; black-and-white photographs taken through filters, printed by projection through filters.

Palomar Observatory: 200-in. and 48-in. Schmidt telescopes; color film exposed or printed through color filters.

U.S. Naval Observatory: 40-in telescope; chilled color film.

Jay M. Pasachoff: Photos taken with Nikon equipment (including wide-angle or telephoto lenses up to 500 mm), usually on a stationary tripod; C.Pl. 82 was taken through a Celestron telescope.

ATLAS CHART 24. Orion Nebula   The entire region of Orion and Monoceros is interesting to sweep with a low-power telescope or binoculars. The region of Orion is one of the most fascinating areas in the sky, because it contains both the youngest known stars and many beautiful nebulae.

Orion's belt, three stars in a line, is one of the easiest asterisms to locate (C.Pl. 14). In Orion's shoulder is Betelgeuse, α (alpha) Ori, a supergiant star whose reddish color is apparent to the naked eye. The 10th brightest star in the sky, Betelgeuse varies in brightness by about 1 magnitude during its 5.7-year period. Rigel, the bright bluish star β (beta) Ori, marks Orion's heel. Rigel, a blue-white double, is the 7th brightest star in the sky. Because its primary component is so bright, the secondary is hard to find.

From Orion's belt extends his sword. M42 (NGC 1976), the Great Nebula in Orion, is a magnificent luminous cloud surrounding θ^1 (theta[1]) Ori in the sword (C.Pl. 13). It is often simply called the Orion Nebula. Even the smallest optical aid reveals the wispy structure of this nebula. Small telescopes readily resolve θ^1 into four components, which are called the Trapezium. The four stars (magnitudes 5.1, 6.7, 6.7, and 7.9) are very hot and provide the energy to illuminate the nebula. Larger telescopes also show two additional companions of 11th magnitude and reveal more of the nebula's structure. The colors, though, show only with time exposures. Even a short time exposure with a tripod-mounted 35 mm

Fig. 64. (*left*) The Orion Nebula region. (Lick Observatory photo)
Fig. 65. (*right*) A chart of the Orion Nebula region. (Wil Tirion)

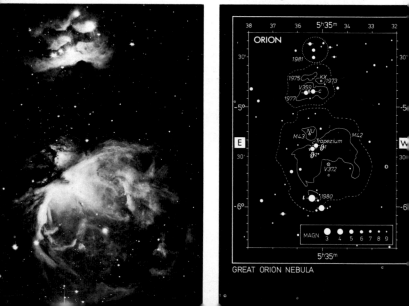

GREAT ORION NEBULA

MAGNITUDES	OPEN CLUSTERS				QUASAR		GREEK ALPHABET			
>-0.4		>10' TO SCALE	<10'		PULSAR		α Alpha		ν Nu	
-0.4 - +0.5	GLOBULAR CLUSTERS				BLACK HOLE		β Beta		ξ Xi	
0.6-1.5		>10'	5'-10'	<5'	MILKY WAY		γ Gamma		o Omicron	
1.6-2.5							δ Delta		π Pi	
2.6-3.5	PLANETARY NEBULAE				GALACTIC EQUATOR		ε Epsilon		ϱ Rho	
3.6-4.5		>1'	0.5'-1'	<0.5'			ζ Zeta		σ Sigma	
4.6-5.5							η Eta		τ Tau	
5.6-6.5	BRIGHT DIFFUSE NEB.				ECLIPTIC		$\theta\,\vartheta$ Theta		υ Upsilon	
6.6-7.5		>10', TO SCALE		<10'			ι Iota		$\phi\,\varphi$ Phi	
DOUBLE or TRIPLE					CONSTELLATION BOUNDARIES		$\kappa\,\varkappa$ Kappa		χ Chi	
VARIABLE	GALAXIES						λ Lambda		ψ Psi	
		>30'	20'-30'	10'-20' <10'			μ Mu		ω Omega	

camera, using an ordinary 50 mm or 35 mm lens, shows the nebula as red, and allows the principal stars in Orion to be picked out. Photographs taken through equatorially driven telescopes look very different depending on the exposure time; relatively short exposures show only the bright inner core of the nebula, while longer ones overexpose the inner parts but show the nebula's outer contours.

The Orion Nebula lies about 1500 light-years from the sun. The Nebula is a blister on the side of the Orion Molecular Cloud that is closer to us. In the Orion Molecular Cloud, dust shields molecules from being broken apart by the ultraviolet light from the hot stars. Astronomers study the molecules with radio telescopes; objects that are probably stars in formation are studied in the infrared.

NGC 1973–75–77 refers to the nebulosity and open cluster around the northern end of Orion's sword. NGC 1981 is an open cluster slightly farther north, with about 12 widely scattered members. NGC 1980 is a faint nebula around ι (iota) Ori, south of the Orion Nebula.

Extending downward from the star at the far left of Orion's belt — ζ (zeta) Ori — is IC 434, which contains the famous Horsehead Nebula (C.Pl. 12). The horsehead shape is formed by dark absorbing gas and dust that blocks our view of a bright reddish emission nebula beyond. Northeast of the Horsehead, the nebula NGC 2023 is partially covered by the same dust cloud. Many beautiful photographs of the Orion Nebula and the Horsehead Nebula have been taken by amateur astronomers.

M78 (NGC 2068) is a wispy cloud north of Orion's belt (C.Pl. 21), found by sweeping the area with low power. A very long exposure shows Barnard's Loop, which rings the northeast side of the Orion complex.

With low power, sweep the triangle formed by β (beta) Ori, η (eta) Ori (about 5° up toward Orion's belt), and β (beta) Eridani (Eri). The triangle contains many faint irregular patches of nebulosity. IC 2118, the Witch Head Nebula, lies about $1\frac{1}{2}$° west of Rigel. In one of Orion's shoulders, γ (gamma) Ori has some faint nebulosity around it.

Continue sweeping northeast from the Orion and Horsehead Nebulae into Monoceros. About 2° east of ε (epsilon) Mon is NGC 2244, a loose open cluster visible to the naked eye and resolvable into its member stars with good binoculars. NGC 2244 has about 24 member stars in an area 40 arc min across. Around NGC 2244 is NGC 2237, the Rosette Nebula. It is relatively large — 1° across — and hard to locate. With low power it often shows just as a halo or glow around the open cluster; photographed in color, it looks spectacular (C.Pl. 37).

Nearby, further northeast, NGC 2264 is a nebulous cluster around S Mon, a variable double star. The cluster contains about 20 stars in an area close to the size of the full moon. The Cone Nebula (Fig. 66, C.Pl. 39), like the Rosette, is difficult to find, but

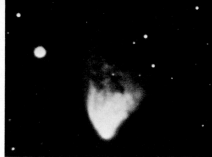

Fig. 66. (*left*) The Cone Nebula and S Monocerotis. (© 1979 Anglo-Australian Telescope Board)

Fig. 67. (*above*) Hubble's Variable Nebula (NGC 2261). (Palomar Observatory photo)

can be located with a medium-sized telescope.

To the south is NGC 2261, Hubble's Variable Nebula (Fig. 67). This nebula is so interesting that it was the first object photographed with the huge 5-m telescope when it opened at Palomar Observatory. The nebula reflects light from the irregular variable R Mon, which varies from 10th to 13th magnitude. The nebula has the shape of a comet's tail. The shape sometimes changes in a matter of weeks. It is hard to observe with small telescopes.

Moving farther south in the sky, near 6^{h}50^m 0°, is the open cluster NGC 2301, which consists of several overlapping groups of stars. This cluster has about 50 stars, including many bright ones.

Toward the bottom of the chart in Canis Major is Sirius, α (alpha) CMa, the Dog Star. It gave its name to the "dog days" of summer. At magnitude −1.46, it is the brightest star in the sky.

In Orion's sword, ι (iota) and θ^2 (theta^2) are interesting doubles, as is η (eta) to the northwest. ζ (zeta) Ori, the easternmost star in Orion's belt, is a triple system with blue-white components of magnitudes 2, 4, and 10. σ (sigma) Ori, 1° south of ζ, is a multiple with at least 5 components. The two brightest components, of magnitudes 3.8 and 6.6, are separated by only ¼ arc sec and so are difficult to resolve, but a small telescope can separate the other components. Two other components are 8th- and 10th-magnitude stars; they are about 12 arc sec away, on opposite sides of the brighter pair. Another companion (6th magnitude) is 40 arc sec away.

The components of the double star ε (epsilon) Mon, 7° southeast of Betelgeuse, show as blue and gold. The two components are separated by about 13 arc min and are accompanied by a third companion. Low power should be used to find ε Mon because the surrounding star field is so rich.

About 12° south is β (beta) Mon, an easily resolved triple that is visible to the naked eye. Its three blue components are of magnitudes 4.6, 5.1, and 5.4.

ATLAS CHART 25. Procyon A number of open clusters appear in this region, which is worth sweeping with low power. The most spectacular is the Beehive Cluster (M44), called Praesepe, which is described opposite Chart 13. Southeast of ζ (zeta) Mon, near 8h13m −6° in Hydra, the rich open cluster M48 (NGC 2548) is bright enough to be visible to the naked eye but difficult to locate. On the right of the chart, near 7h03m −8° in Monoceros, M50 (NGC 2323) is a rich open cluster visible with a small telescope. M67 (NGC 2682) is an open cluster west of α (alpha) Cancri.

A beautiful open cluster in Puppis is M46 (NGC 2437), somewhat fainter than M67 but containing more stars. On the northern edge of M46 is the planetary nebula NGC 2438, an irregular patchy ring. Nearby is M47 (NGC 2422), an open cluster that is faintly visible to the naked eye because of its many 6th-magnitude stars. Both M46 and M47 are only about 20 million years old. NGC 2539 is another rich open cluster in Puppis. Low power should be used to view the open cluster NGC 2506 in Monoceros, slightly northwest of NGC 2539.

Near 7h40m −18° in Puppis is the planetary nebula NGC 2440, best viewed with high power.

At the lower right of the chart (at −12°), extending from Canis Major into Monoceros, is IC 2177, a nebulous patch more than 3° across (C.Pl. 36). This nebula is called the Eagle Nebula in older references, though that name is now usually applied to M16 (NGC 6611) instead (Chart 30). Toward the upper tip of IC 2177 are two open clusters, NGC 2335 and NGC 2343.

Procyon, α (alpha) Canis Minoris, is the 8th brightest star in the sky (magnitude 0.38) and one of the closest stars to us (about 11 light-years away). Procyon has a faint (11th-magnitude) white-dwarf companion, the second closest white dwarf to us.

Fig. 68. Argo Navis, the Ship, from the star atlas of Bayer (1603). It has since been divided into the constellations Carina (the Keel), Puppis (the Stern), Pyxis (the Compass), and Vela (the Sails). (Smithsonian Institution Libraries)

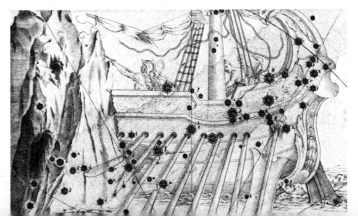

FROM "SKY ATLAS 2000.0" BY WIL TIRION

GNITUDES		OPEN CLUSTERS	○ ○ ○ ○ >10', TO SCALE <10'	QUASAR	△	GREEK ALPHABET

●	>-0.4
●	-0.4-+0.5
●	0.6-1.5
●	1.6-2.5
●	2.6-3.5
●	3.6-4.5
●	4.6-5.5
●	5.6-6.5
●	6.6-7.5

OPEN CLUSTERS   ○ ○ ○ ○   >10', TO SCALE <10'

GLOBULAR CLUSTERS   ⊕ ⊕ ⊕   >10' 5'-10' <5'

PLANETARY NEBULAE   ◇ ◇ ◇   >1' 0.5'-1' <0.5'

BRIGHT DIFFUSE NEB.   >10', TO SCALE <10'

GALAXIES   ○ ○ ○ ○   >30' 20'-30' 10'-20' <10'

QUASAR   △

PULSAR   ☿

BLACK HOLE   Υ

MILKY WAY

GALACTIC EQUATOR   ——— 70° ———

ECLIPTIC   ——— 100° ———

CONSTELLATION BOUNDARIES

	GREEK ALPHABET		
α	Alpha	ν	Nu
β	Beta	ξ	Xi
γ	Gamma	ο	Omicron
δ	Delta	π	Pi
ε	Epsilon	ϱ	Rho
ζ	Zeta	σ	Sigma
η	Eta	τ	Tau
θ ϑ	Theta	υ	Upsilon
ι	Iota	φ φ	Phi
κ ϰ	Kappa	χ	Chi
λ	Lambda	ψ	Psi
μ	Mu	ω	Omega

UBLE or LTIPLE   ●—●

RIABLE   ◉ ○

ATLAS CHART 26. Regulus As we move away from the obscuring dust clouds of the Milky Way, the more distant galaxies become visible. In Leo we see an outer edge of the huge cluster of galaxies — the Virgo Cluster — that extends through Virgo, Coma Berenices, and Corvus. Three of the brighter galaxies in Leo are M95 (NGC 3351), M96 (NGC 3368), and M105 (NGC 3379), located about 30 million light-years from the Milky Way Galaxy. M95 (magnitude 9.9) is a barred spiral and M96 (magnitude 9.4) is a regular spiral; both are relatively large. M105 is a small elliptical galaxy of magnitude 9.5. If you use a wide-angle, medium-power telescope to sweep about 9° east of Regulus, you can see these galaxies in the same field of view.

Regulus, α (alpha) Leo, lies directly on the ecliptic. Regulus is at the base of the "sickle" that many people see as part of the constellation's shape; it marks the lion's heart. The 21st brightest star in the sky at magnitude 1.35, Regulus is visible for most of the year, and can be seen in the eastern sky in the spring. It has a faint companion that can sometimes be detected with binoculars.

γ (gamma) Leo, the brightest star in the sickle, is a binary located about 8° northeast of Regulus. One of the best doubles in the sky, γ Leo consists of two yellow components of magnitudes 2.1 and 3.4 that orbit with a period of 407 years. The components are fairly close, so they are difficult to resolve with a small telescope. The gap between the pair is currently widening, though; these stars will be about 5 arc sec apart by the year 2100.

This chart includes Wolf 359, the third closest star to the sun, after the α (alpha) Centauri system and Barnard's Star. At magnitude 13.5, Wolf 359 is too faint to be marked on this chart, but it is located near VY Leo, at $10^h56^m +7°$. It is one of the least luminous stars known, but has been discovered because its rapid proper motion makes its closeness (only 7.8 light-years from earth) apparent.

The radiant of the Leonid meteor shower (p. 394), which reaches its maximum frequency on November 17 each year, lies about 2° northwest of γ Leo. This shower peaks every 33 years, as the earth passes through the part of a defunct comet's orbit where the remaining particles from the comet are bunched. The last spectacular display of Leonids occurred in 1966; the next might occur in 1998 or 1999.

In the Lion's foreleg, R Leo is one of the brightest of the long-period variables; it ranges from magnitude 11.6 to 4.4 over a period of 312 days. Located about 5° west of Regulus, R Leo forms a triangle with two westward stars of magnitudes 9.0 and 9.6.

Although the constellation Hydra is not particularly rich in stars, sweeping the region with low-power binoculars does show some objects of interest. NGC 3242 is a planetary nebula located about 1.8° south of μ (mu) Hydra at the lower left of the chart. In a small telescope the planetary appears as a pale blue disk; a large telescope reveals a bright inner ring and a faint outer shell.

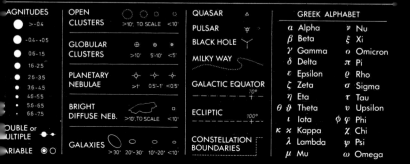

ATLAS CHART 27. Sombrero Galaxy, Virgo Cluster of Galaxies Here we find the greatest concentration of galaxies in the Virgo Cluster of galaxies that extends north into Canes Venatici and south into Corvus. The Virgo Cluster is about 65 million light-years away, and is the closest to the Milky Way Galaxy of the large clusters of galaxies. Many of these galaxies are shown on an expanded chart (27A) of the area outlined in the upper-left region of this chart.

Just above the outlined area is M85 (NGC 4382), an elliptical galaxy in Coma Berenices, at $12^h25^m +18°$. One of the brightest galaxies in the Virgo Cluster, M85 shows in a small telescope as a bright oval patch.

In the Leo Cluster of galaxies at upper right, M65 (NGC 3623) and M66 (NGC 3627) are two bright spiral galaxies about 20 arc min apart, located about halfway between θ (theta) and ι (iota) Leo. Both galaxies appear similar to the Andromeda Galaxy. At magnitude 9.2, M66 is about 0.4 magnitude brighter than M65. M65 and M66 can be seen in the same field of view with low-power telescopes. On a clear night, you may be able to glimpse both galaxies with a good pair of binoculars. Two nearby spirals, NGC 3628 and NGC 3593, may be visible in the same telescopic field.

On the border of Virgo and Corvus, near $12^h40^m -12°$, is M104 (NGC 4594), another of the brighter members of the cluster. M104 is called the Sombrero Galaxy (C.Pl. 30) because of its distinctive large center and the dark lane of dust that runs along the equatorial plane of this spiral galaxy, seen only 6° from edge-on. These features are hard to see in small telescopes, though the galaxy can be readily detected as a hazy oval patch. In larger telescopes, the dust lane can be detected under good observing conditions.

In Corvus, about 4° southwest of γ (gamma) Corvi, near $12^h -18°$, are a pair of peculiar galaxies, NGC 4038 and NGC 4039, known as the Antennae. Each of these elliptical galaxies appears to have a long curving tail; computer models show that the tails may have been formed by gravity as a result of a collision of the two galaxies. The Antennae are a source of radio waves and lie about 90 million light-years away.

One of the finest double stars in the sky and a favorite of amateurs is γ (gamma) Vir (at $12^h42^m -1°$). Its two almost equal components (magnitudes 3.6 and 3.7) are separated by about 4 arc sec.

Located about 4° south of σ (sigma) Leo, near the ecliptic (at about $11^h27^m +3°$), is τ (tau) Leo, a pair of yellow and blue stars of 5th and 7th magnitudes, separated by 90 arc sec. δ (delta) Crv, whose components are magnitudes 3 and 8.5, is also easily resolved.

COMA BERENICES
CHART 27-A
LEO
VIRGO
SEX
W
HYA
CORVUS
CRATER

D FROM "SKY ATLAS 2000.0" BY WIL TIRION

AGNITUDES		OPEN				QUASAR	△	GREEK ALPHABET		
●	>-0.4	CLUSTERS	◯ ◯	○	◌			α Alpha		ν Nu
			>10' TO SCALE	<10'		PULSAR	※	β Beta		ξ Xi
●	-0.4 - +0.5							γ Gamma		o Omicron
●	0.6 - 1.5	GLOBULAR	⊕	⊕	⊛	BLACK HOLE	⅄	δ Delta		π Pi
●	1.6 - 2.5	CLUSTERS	>10'	5'-10'	<5'	MILKY WAY		ε Epsilon		ϱ Rho
●	2.6 - 3.5							ζ Zeta		σ Sigma
●	3.6 - 4.5	PLANETARY	✧	◇	◦	GALACTIC EQUATOR		η Eta		τ Tau
●	4.6 - 5.5	NEBULAE	>1'	0.5'-1'	<0.5'			θ ϑ Theta		υ Upsilon
●	5.6 - 6.5							ι Iota	φ φ Phi
●	6.6 - 7.5	BRIGHT				ECLIPTIC		κ ϰ Kappa		χ Chi
		DIFFUSE NEB.	>10' TO SCALE	<10'				λ Lambda		ψ Psi
DUBLE or ULTIPLE	●—					CONSTELLATION BOUNDARIES		μ Mu		ω Omega
ARIABLE	◉ ○	GALAXIES	◯ ○	○	○					
			>30' 20'-30'	10'-20'	<10'					

ATLAS CHART 27A. Virgo Cluster of Galaxies This chart shows the heart of the Virgo Cluster. About 75% of the brightest galaxies in this cluster are spirals.

Since the Virgo Cluster is the closest rich cluster of galaxies to us, it is the easiest to observe. In small telescopes most of the galaxies appear as faint patches of light. But on a clear dark night in the summer, over 100 galaxies can be viewed with a medium-sized telescope. It is best to first locate a galaxy with low power and then switch to higher magnifications.

Some outstanding galaxies in Coma Berenices include: M98 (NGC 4192) at $12^h13^m +15°$, a bright, elongated spiral seen nearly edge-on; M99 (NGC 4254) — about 1.5° southeast of M98, near $12^h19^m +14°$ — a bright, round spiral seen face-on; and M88 (NGC 4501) — about 4° east of M99, near $12^h32^m +14°$ — a spiral that can be located by searching for the two faint stars near its edge. M100 (NGC 4321), near $12^h23^m +16°$, is the largest spiral in the Virgo Cluster (C.Pl. 26).

In Virgo, some noteworthy galaxies are: M49 (NGC 4472), M58 (NGC 4579), M59 (NGC 4621), M60 (NGC 4649), M61 (NGC 4303), M84 (NGC 4374), M86 (NGC 4406), M89 (NGC 4552), and M90 (NGC 4569). M49, one of the brightest ellipticals, is fairly easy to find near $12^h38^m +12°$. M61, about 6° southwest of M49, near $12^h22^m +4°30'$, is a beautiful face-on spiral (C.Pl. 18). M84 and M86 are nearly identical ellipticals separated by less than 20 arc min. M90 is an edge-on spiral with a bright nucleus (C.Pl. 25).

One of the brighter and most interesting galaxies in the Virgo Cluster is M87 (NGC 4486), a huge peculiar elliptical galaxy at $12^h31^m +13'$. The best photographs taken with large telescopes show the more than 1000 globular clusters that surround this galaxy (Fig. 20, p. 122). Short-exposure photographs taken with large telescopes have also revealed an unusual jet of gas (C.Pl. 24). M87 has been identified as the radio source Virgo A, and many astronomers think that a giant black hole containing millions of times more mass than the sun lies at its center.

Near $12^h30^m +2°$, about $3\frac{1}{2}°$ northeast of η (eta) Vir, is 3C 273, the brightest and one of the first known quasars. The easiest quasar to observe, 3C 273 ranges in magnitude from about 12.2 to 13.0. Depending on its magnitude and on observing conditions, 3C 273 can sometimes be glimpsed in medium to large telescopes with the aid of an accurate finding chart. It appears almost starlike, not nebulous, and is thus "quasi-stellar." From the enormous redshift in its spectrum, astronomers have calculated that this bluish object is receding at the tremendous speed of 50,000 km per second, 16% of the speed of light. Hubble's Law tells us that 3C 273 is out among the distant galaxies.

DOUBLE or MULTIPLE

VARIABLE

QUASAR

GALAXIES
>30'   20'-30'   10'-20'   <10'

MAGNITUDES
3   4   5   6   7   >8.1

ATLAS CHART 28. Arcturus, Spica   Arcturus, α (alpha)
Boötis, at magnitude 0.0 the third brightest single star in the sky,
and 1st-magnitude Spica, α (alpha) Virginis, are the most prominent objects in this region. Arcturus is a cool red giant. Spica, the
16th brightest star, is extremely hot and therefore bluish. It is an
eclipsing spectroscopic binary that drops about one magnitude
from its usual magnitude of 1.0 during a four-day period.

In the Virgo Cluster of galaxies, NGC 5364 is a spiral galaxy of
11th magnitude (Fig. 69). NGC 5746 (at $14^h45^m +2°$) is an edge-on
spiral of 10th magnitude, best seen with small telescopes; its companion, NGC 5740, a little over $\frac{1}{4}°$ south, is a spiral of magnitude
11.5. To the left of the chart, the galaxies NGC 5806 (spiral), NGC
5813 (elliptical), and NGC 5831 (elliptical) are visible in medium-
sized telescopes. The dim galaxies NGC 5838, NGC 5854, and NGC
5864 are slightly further east. The nearby spiral NGC 5846 is
brighter. Another spiral galaxy located slightly to the southeast,
NGC 5850, is dimmer but about twice as large as NGC 5846.

Of the double stars shown on this chart, ζ (zeta) Boo, at upper
left, is a beautiful pair of white stars with magnitudes 4.5 and 4.6,
which orbit each other with a period of 130 years. They can be
resolved with a medium-sized telescope. π (pi) Boo, above ζ Boo, is
another double, more easily resolved.

S Vir, about halfway between Spica and ζ (zeta) Vir, is a long-
period variable ranging from 6th to 13th magnitude, with an average period of 377 days. As it nears maximum brightness, its red
color makes it an interesting telescopic object.

This chart contains two of the most famous quasars. 3C 279, one
of the brightest quasars, has reached an absolute magnitude as
high as −31, which is 10,000 times brighter than the Andromeda
Galaxy. 3C 279 is on the ecliptic at extreme right. OQ 172 — at
$14^h45^m +10°$, at upper left — has all its spectral lines redshifted by
3.53 times their original wavelength (353%); it is one of the farthest
known objects in the universe. Though both quasars are extremely
bright intrinsically, they are so far away that they can be seen only
through large telescopes.

Fig. 69. NGC 5364 in Virgo, a spiral galaxy, type Sc. (The Kitt
Peak National Observatory)

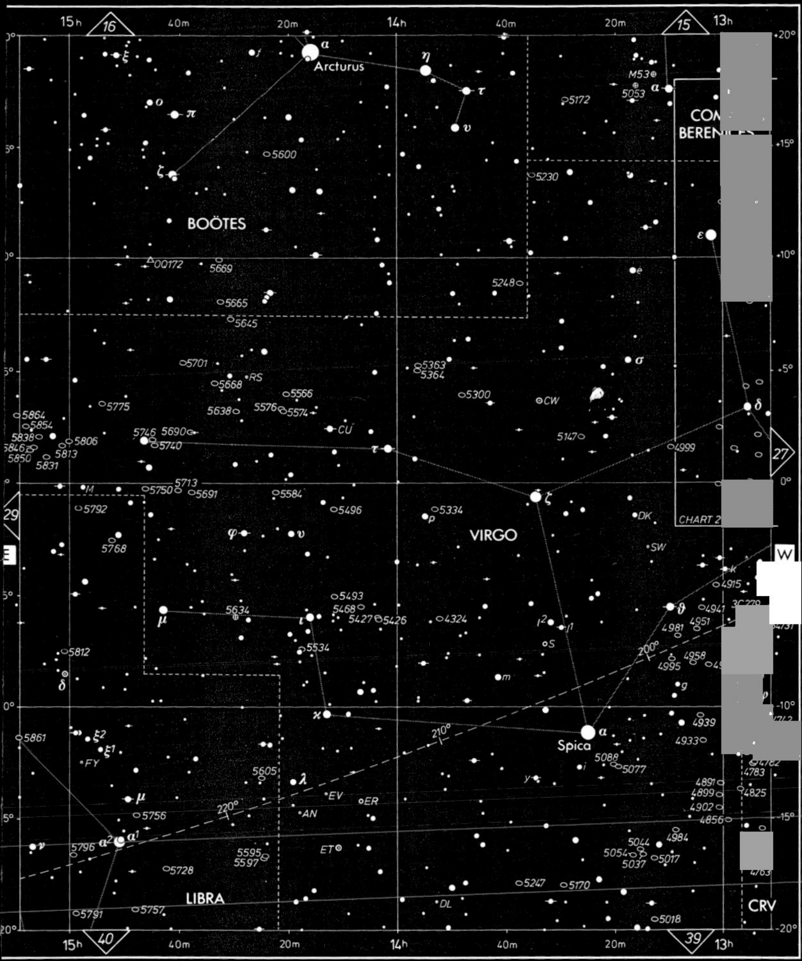

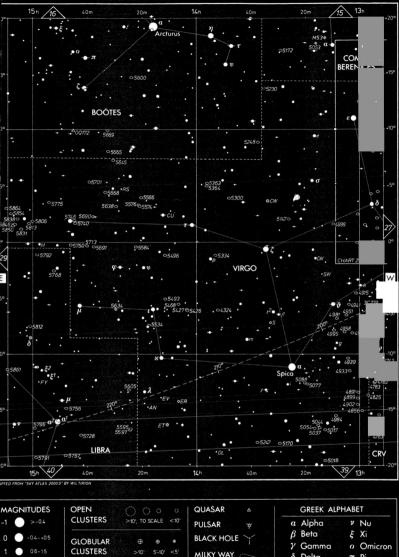

...APTED FROM "SKY ATLAS 2000.0" BY WIL TIRION

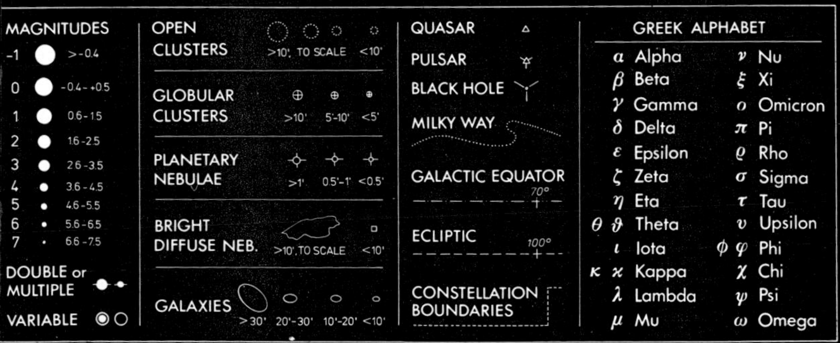

ATLAS CHART 29. Globular Clusters M5, M10, M12   In Serpens, the globular cluster M5 (NGC 5904) is one of the finest in the sky, surpassed only by 47 Tuc, ω (omega) Cen, and possibly M13. During the summer months it is easily located in the sparse area near $15^h18^m +2°$, slightly northwest of the naked-eye star 5 Ser. Under very good conditions, M5 is also visible to the naked eye. With binoculars, it appears as a hazy star, while small telescopes show it as a bright, circular glow. Larger telescopes begin to resolve its stars. Nearby, 5 Ser is a double with 5th- and 10th-magnitude components that are separated by 11 arc sec.

M10 and M12, two very similar globular clusters in Ophiuchus, are also outstanding. Both brighter than 7th magnitude, they are visible to the naked eye. Located about 5° northeast of ζ (zeta) Oph, near $16^h57^m -4°$, M10 (NGC 6254) is a rich cluster with a compressed center. Small telescopes show M10 as an oval with a diameter of about 12 arc min, and medium-sized telescopes begin to resolve its stars. M12 (NGC 6218) lies about 3° northwest of M10 and is a bit larger. Since M12 has a relatively loose structure, its stars are more readily resolved.

M107 (NGC 6171), a third globular cluster in Ophiuchus, lies about 3° south of ζ (zeta) Oph. At magnitude 9 and about 8 arc min across, M107 is much fainter and smaller than M10 or M12.

See Chart 28 for the galaxies in Virgo shown here west of M5.

During the summer months scan the region of Serpens Caput (the Serpent's Head); the tail, Serpens Cauda, is on Chart 30. Use binoculars or a low-power telescope near β (beta) Ser. δ (delta) Ser is a fine double with components of magnitudes 4.1 and 5.2, easily resolved by a small telescope. About 1° east of β (beta) Ser near $15^h46^m +15°$, R Ser is a long-period variable, ranging from magnitude 5.6 to 14.0 over an average period of 356 days. Its color near maximum, when it can sometimes be seen with the naked eye, is crimson.

S Her, a long-period variable located near $16^h50^m +15°$ in Hercules, ranges from magnitude 5.9 to 13.6 in 307 days.

Fig. 70. M5 (NGC 5904), a globular cluster in Serpens. (The Kitt Peak National Observatory)

ADAPTED FROM "SKY ATLAS 2000.0" BY WIL TIRION

MAGNITUDES			OPEN CLUSTERS				QUASAR	△		GREEK ALPHABET			
−1	●	>−0.4		>10'	TO SCALE	<10'	PULSAR	⚡		α	Alpha	ν	Nu
0	●	−0.4 −+0.5	GLOBULAR CLUSTERS				BLACK HOLE	Y		β	Beta	ξ	Xi
1	●	0.6−1.5		⊕	⊕	✦	MILKY WAY	～～		γ	Gamma	o	Omicron
2	●	1.6−2.5		>10'	5−10'	<5'				δ	Delta	π	Pi
3	●	2.6−3.5	PLANETARY NEBULAE	✧	✧	✧	GALACTIC EQUATOR			ε	Epsilon	ϱ	Rho
4	●	3.6−4.5		>1'	0.5−1'	<0.5'		70°		ζ	Zeta	σ	Sigma
5	●	4.6−5.5					ECLIPTIC			η	Eta	τ	Tau
6	·	5.6−6.5	BRIGHT DIFFUSE NEB.			□		100°		$\theta \vartheta$	Theta	υ	Upsilon
7	·	6.6−7.5		>10'	TO SCALE	<10'				ι	Iota	$\phi \varphi$	Phi
										$\kappa \varkappa$	Kappa	χ	Chi
DOUBLE or MULTIPLE	◉ ○		GALAXIES	⬭	⬬	○	CONSTELLATION BOUNDARIES	⌐ ¬		λ	Lambda	ψ	Psi
VARIABLE	◉ ○			>30'	20'−30'	10'−20' <10'				μ	Mu	ω	Omega

ATLAS CHART 30. Omega Nebula The star fields begin to increase in brilliance toward the Milky Way in Aquila, Scutum, and Sagittarius. The area provides good views in binoculars or low-power telescopes.

The open clusters NGC 6633 in Ophiuchus (at 18^h28^m $+7°$) and IC 4756 in Serpens (at 18^h40^m $+6°$) are easily resolved at relatively low power. About 5° southwest of ζ (zeta) Aquilae at upper left, you can see the open cluster NGC 6709 with medium power. Another rich field lies about 1° northeast of β (beta) Oph, in the open cluster IC 4665 (at 17^h47^m $+6°$).

The rich open cluster M11 (NGC 6705), the Wild Duck, is located near 18^h50^m $-6°$ on the northern edge of the Scutum star cloud, a bright region of the Milky Way. M11 is magnificent in both small and large telescopes (C.Pl. 3); it can sometimes be seen with the naked eye. A medium-sized telescope resolves its edges, while larger telescopes resolve hundreds of its stars. About $3\frac{1}{2}°$ south of M11 is the open cluster M26 (NGC 6694), which can also be observed easily in small telescopes.

In Sagittarius, M23 (NGC 6494) is a large open cluster that appears prominent against a dark nebula. Its bright stars are outstanding even when viewed through binoculars. The open clusters M18 (NGC 6613), M24 (NGC 6603), and M25 (IC 4725) also lie among the dark nebulae in Sagittarius. M24 is actually a part of the Milky Way and includes the open cluster NGC 6603. Nearby are the reflection nebulae NGC 6589 and NGC 6590 (C.Pl. 46).

An interesting planetary, NGC 6572, is in Ophiuchus. Use high power to observe the globular clusters M9 (NGC 6333), NGC 6356, and NGC 6342.

In Serpens, near 18^h20^m $-14°$, is M16 (NGC 6611), an easily located nebulous cluster (C.Pl. 5). To the southeast in Sagittarius near 18^h20^m $-16°$ is the Omega Nebula, M17 (NGC 6618), also called the Horseshoe or Swan Nebula (C.Pl. 6). The star background in this region is particularly rich. In Scutum, the nebula IC 1287, 2° south of α (alpha) Sct, requires slightly higher power.

Barnard's Star (at 18^h $+5°$), magnitude 9.5, is the second closest star to our solar system. It has the greatest apparent angular motion across the sky (proper motion) of any star (see arrows below).

Fig. 71. The path of Barnard's Star in the sky. This star appears to move across the sky with respect to more distant stars, as marked, covering a distance as great as our moon's diameter in about 180 years. (Wil Tirion)

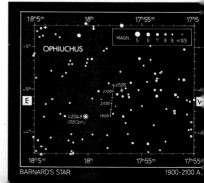

19h	18	40m	20m	18h	40m	20m	17	17h

HERCULES

OPHIUCHUS

AQUILA

SCUTUM

SERPENS CAUDA

...TARIUS

MAGNITUDES	OPEN CLUSTERS	QUASAR	GREEK ALPHABET

MAGNITUDES

- >-0.4
- -0.4 - +0.5
- 0.6 - 1.5
- 1.6 - 2.5
- 2.6 - 3.5
- 3.6 - 4.5
- 4.6 - 5.5
- 5.6 - 6.5
- 6.6 - 7.5

DOUBLE or MULTIPLE

VARIABLE

OPEN CLUSTERS
>10' TO SCALE <10'

GLOBULAR CLUSTERS
>10'   5'-10'   <5'

PLANETARY NEBULAE
>1'   0.5'-1'   <0.5'

BRIGHT DIFFUSE NEB.
>10' TO SCALE <10'

GALAXIES
>30' 20'-30' 10'-20' <10'

QUASAR

PULSAR

BLACK HOLE

MILKY WAY

GALACTIC EQUATOR
70°

ECLIPTIC
100°

CONSTELLATION BOUNDARIES

GREEK ALPHABET

α	Alpha	ν	Nu
β	Beta	ξ	Xi
γ	Gamma	o	Omicron
δ	Delta	π	Pi
ε	Epsilon	ϱ	Rho
ζ	Zeta	σ	Sigma
η	Eta	τ	Tau
θ ϑ	Theta	υ	Upsilon
ι	Iota	φ ϕ	Phi
κ ϰ	Kappa	χ	Chi
λ	Lambda	ψ	Psi
μ	Mu	ω	Omega

ATLAS CHART 31. Altair From Cygnus into Aquila, the Milky Way appears to be divided lengthwise by enormous dust clouds. These clouds form a band called the Great Rift, which lies along the central plane of our galaxy in a lane like the equatorial lanes visible in photographs of edge-on galaxies. The dark nebulae and rich star fields provide beautiful views during the summer months.

Near the top of the chart in Sagitta (the Arrow), about halfway between γ (gamma) and δ (delta) Sge, is the rich globular cluster M71 (NGC 6838), easy to locate at magnitude 7.

Above the center of the chart is α (alpha) Aquilae, Altair, the 12th brightest star in the sky. About 2° north of Altair is γ (gamma) Aql. Slightly northwest of γ Aql is B143, a dark nebula about 30 arc min in diameter that is visible in medium-sized telescopes. About 5° to the west, the planetary nebula NGC 6803 shows a small disk, as does the fainter planetary, NGC 6804, nearby. The planetary nebula NGC 6891, 7° northeast of Altair, appears as a bright disk surrounded by a fainter ring. Five degrees southwest of Altair is the planetary NGC 6781 (C.Pl. 50); it can be photographed only with large apertures.

At lower center, close to the Milky Way in Sagittarius, is NGC 6822, an irregular dwarf galaxy and one of the few members of the Local Group that is readily visible. Use a rich-field telescope to view it. Although it is one of the nearer and apparently larger galaxies, an intervening dust cloud reduces its brightness.

Slightly northwest of NGC 6822 and visible in the same field is NGC 6818, the Little Gem, one of the closest planetary nebulae. Use a large telescope to observe its 15th-magnitude central star; smaller telescopes show its pale blue-green disk.

Double stars in this region include: π (pi) Aql, a 6th-magnitude double about 3° north of Altair; γ (gamma) Del (at $20^h47^m +16°$), a wide double located at the northeast corner of the lozenge-shaped asterism formed by α, β, δ, and γ (alpha, beta, delta, and gamma) Del; and 15 Aql, a wide double lying northeast of the Scutum star cloud, about 1 arc min north and slightly west of λ (lambda) Aql (at $19^h05^m -05°$).

Among the variable stars on the chart are: η (eta) Aql, a Cepheid variable that ranges from magnitude 3.7 to 4.5 during a period of slightly more than 7 days; σ (sigma) Aql, an eclipsing variable that drops about 0.2 magnitudes within a period of 47 hours; R Aql, a reddish, long-period variable often visible to the naked eye at maximum brightness; and U Sagittae, an Algol-type eclipsing binary. The brightness of U Sge ranges from 6.3 to 9.2 with a period of 3 days, 9 hours.

One of the most unusual objects in the sky, SS433 (at $19^h15^m +05°$) is too faint (14th magnitude) for small telescopes. Its spectrum shows that it is emitting jets of gas in opposite directions at 25% of the speed of light, a speed otherwise unknown in our galaxy. It is probably a binary system containing a neutron star.

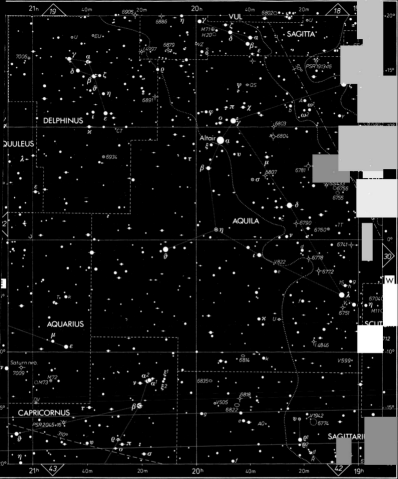

MAGNITUDES

-1	● >-0.4
0	● -0.4 -+0.5
1	● 0.6 -1.5
2	● 1.6 -2.5
3	● 2.6 -3.5
4	● 3.6 -4.5
5	● 4.6 -5.5
6	● 5.6 -6.5
7	● 6.6 -7.5

DOUBLE or MULTIPLE ●−
VARIABLE ◉ ○

OPEN CLUSTERS	◯ ◯ ○ ⬚
	>10' TO SCALE <10'
GLOBULAR CLUSTERS	⊕ ⊕ ⊙
	>10' 5'-10' <5'
PLANETARY NEBULAE	◇ ◇ ◇
	>1' 0.5'-1' <0.5'
BRIGHT DIFFUSE NEB.	⬭ □
	>10' TO SCALE <10'
GALAXIES	◯ ◯ ○ ○
	>30' 20'-30' 10'-20' <10'

QUASAR	△
PULSAR	⭤
BLACK HOLE	⋎
MILKY WAY	⌇
GALACTIC EQUATOR	70° ─┼─
ECLIPTIC	100° ─┼─
CONSTELLATION BOUNDARIES	⌐ ⌐

GREEK ALPHABET

α	Alpha	ν	Nu
β	Beta	ξ	Xi
γ	Gamma	o	O
δ	Delta	π	P
ε	Epsilon	ϱ	R
ζ	Zeta	σ	Sigma
η	Eta	τ	Tau
θ ϑ	Theta	υ	Upsilon
ι	Iota	φ φ	Phi
κ ϰ	Kappa	χ	Chi
λ	Lambda	ψ	Psi
μ	Mu	ω	Omega

ADAPTED FROM "SKY ATLAS 2000.0" BY WIL TIRION

ATLAS CHART 32. Saturn Nebula   NGC 7009, one of the brightest planetary nebulae, is located near $21^h -11°$, about 1° west of ν (nu) Aquarii (Aqr). It consists of a bright inner ring surrounded by an outer disk with faint extensions. It is sometimes called the Saturn Nebula. Small telescopes show it only as a star-like object.

Slightly southwest of NGC 7009 is M72 (NGC 6981), a small globular cluster difficult to resolve even with large telescopes and high power. The open cluster M73 (NGC 6994) is nearby.

Two globular clusters on this chart, M15 (NGC 7078) and M2 (NGC 7089), are among the brightest in the sky. Both are clearly visible to the naked eye, and even the smallest optical aid shows them as hazy patches that are condensed toward the centers. M15 (at $21^h30^m +12°$) is located about 4° northwest of ε (epsilon) Pegasi. An unusually rich and compact cluster, M15 has been identified as an x-ray source. M2 is located just below the celestial equator, about 5° north of β (beta) Aqr. Since M2 has such a highly condensed center, even large telescopes resolve only its outer edges.

The globular cluster NGC 7006 in Delphinus is very remote; it lies about 185,000 light-years from the solar system, about the same distance as the Magellanic Clouds. The cluster is very difficult to resolve; in a medium-sized telescope, it appears as a hazy spot about 1 arc min in diameter.

East of α (alpha) Aqr, the four stars γ, ζ, η, and π Aqr (gamma, zeta, eta, and pi Aqr) form a small Y-shaped asterism called the Water Jar, a characteristic feature of this constellation. The star ζ Aqr at the center of the Y is one of the finest doubles in the sky, with components of magnitudes 4.3 and 4.5.

This ends the set of equatorial charts; we now turn to charts of intermediate southern declinations.

Fig. 72. NGC 7009, the Saturn Nebula, a planetary nebula in Aquarius. (The Kitt Peak National Observatory)

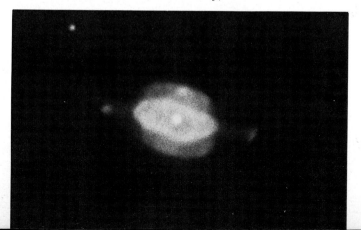

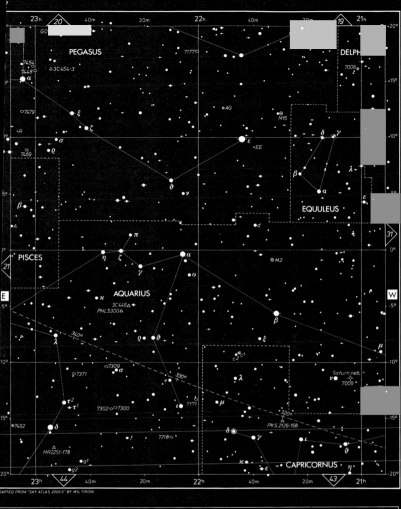

MAGNITUDES		OPEN CLUSTERS				QUASAR		GREEK ALPHABET			
-1			>10', TO SCALE	<10'				α Alpha		ν Nu	
0						PULSAR		β Beta		ξ Xi	
1		GLOBULAR CLUSTERS				BLACK HOLE		γ Gamma		o Omicron	
2			>10'	5'-10'	<5'	MILKY WAY		δ Delta		π Pi	
3		PLANETARY NEBULAE						ε Epsilon		ϱ Rho	
4			>1'	0.5'-1'	<0.5'	GALACTIC EQUATOR		ζ Zeta		σ Sigma	
5								η Eta		τ Tau	
6		BRIGHT DIFFUSE NEB.				ECLIPTIC		θ ϑ Theta		υ Upsilon	
7			>10', TO SCALE	<10'				ι Iota		ϕ φ Phi	
DOUBLE or MULTIPLE						CONSTELLATION BOUNDARIES		κ $\varkappa$ Kappa		χ Chi	
VARIABLE		GALAXIES	>30'	20'-30'	10'-20' <10'			λ Lambda		ψ Psi	
								μ Mu		ω Omega	

ATLAS CHART 33. South Galactic Pole The stellar background in this region is thin, and even galaxies are rare. The galaxy
NGC 253, near the south galactic pole in Sculptor, is a large and
bright spiral (C.Pl. 33), often considered the finest in the sky except
for the Andromeda Galaxy. Because NGC 253 is at such a southern
declination, observers in midnorthern latitudes need a medium-
sized telescope and good viewing conditions to see it well.

The spiral galaxy NGC 247, about 4° north of NGC 253, is
nearly as large as NGC 253 but about four magnitudes fainter.
NGC 247 may be found most easily by looking with low power
about 3° south of β (beta) Ceti and then waiting for the galaxy to
drift into view. NGC 247 and NGC 253 are very close to us (for
galaxies) — only about 12 million light-years away.

About 4° northwest of α (alpha) Phe, near the border between
Sculptor and Phoenix, NGC 55 is another prominent galaxy in the
southern sky, though difficult for northern observers. It is a large,
nearly edge-on spiral that shows in small telescopes as an oval
patch of light. All three galaxies belong to the Sculptor group,
most of which are loose, large spirals.

The globular cluster NGC 288 near the south galactic pole is
relatively uninteresting.

SX Phe, about 7° west of α Phe, is one of the best-known dwarf
Cepheid variables. During its 79-minute period, it varies only
slightly, from magnitude 7.1 to about 7.5.

Fig. 73. NGC 253, a large bright spiral galaxy (type Sc) in Sculptor.
(© 1979 Anglo-Australian Telescope Board)

ADAPTED FROM "SKY ATLAS 2000.0" BY WIL TIRION

MAGNITUDES		OPEN CLUSTERS				QUASAR	△	GREEK ALPHABET			
-1	>-0·4		>10' TO SCALE <10'			PULSAR	☆	α Alpha		ν Nu	
0	-0·4 - +0·5					BLACK HOLE		β Beta		ξ Xi	
1	0·6 - 1·5	GLOBULAR CLUSTERS				MILKY WAY		γ Gamma		ο Omicron	
2	1·6 - 2·5		⊕ >10'	⊕ 5'-10'	⊕ <5'			δ Delta		π Pi	
3	2·6 - 3·5	PLANETARY NEBULAE				GALACTIC EQUATOR		ε Epsilon		ϱ Rho	
4	3·6 - 4·5		>1'	0·5'-1'	<0·5'		70°	ζ Zeta		σ Sigma	
5	4·6 - 5·5	BRIGHT DIFFUSE NEB.				ECLIPTIC		η Eta		τ Tau	
6	5·6 - 6·5		>10' TO SCALE		<10'		100°	θ ϑ Theta		υ Upsilon	
7	6·6 - 7·5							ι Iota		φ φ Phi	
DOUBLE or MULTIPLE		GALAXIES				CONSTELLATION BOUNDARIES		κ κ Kappa		χ Chi	
			>30' 20'-30'	10'-20'	<10'			λ Lambda		ψ Psi	
VARIABLE	◎ ○							μ Mu		ω Omega	

ATLAS CHART 34. NGC 1300   Left of the center of the chart, near the border between Fornax and Eridanus, is the Fornax Cluster of galaxies. The cluster consists of a number of faint galaxies as well as 18 bright ones. Nine of these bright galaxies can be seen in the same 1° field of view.

NGC 1316, an elliptical or S0 galaxy at a distance of about 55 million light-years, is the brightest member of the group. It has been identified as the radio source Fornax A and may be undergoing some type of nuclear outburst. Its companion galaxy, NGC 1317, is a spiral with a bright nucleus.

NGC 1365, the second brightest galaxy of the Fornax Cluster, is an excellent example of a barred spiral galaxy. Its central bar extends about 45,000 light-years and appears to be about 3° long.

Near $3^h20^m -20°$ is NGC 1300, the prototype of a barred spiral galaxy with moderately wide arms (called SBb).

Just to the right of the chart's center, southwest of β (beta) Fornacis, is the globular cluster NGC 1049. It is part of the Fornax System, a faint dwarf galaxy that is a member of our Local Group of galaxies. The system contains five groups of stars that somewhat resemble the globular clusters in our galaxy. The clusters in Fornax, though, differ in density, brightness, and size. For example, NGC 1049 (at $2^h40^m -34°$) is 50 times larger than the largest globular cluster in our galaxy. It is 630,000 light-years away.

θ (theta) Eri (at $2^h58^m -40°$) is a wide double that can be resolved with binoculars. Its components are bright, of magnitudes 3.2 and 4.4. The double star f Eri (at $3^h48^m -38°$) has widely spaced components of magnitudes 5.9 and 5.4.

Fig. 74. NGC 1300, a typical barred spiral galaxy with moderately wide arms (type SBb) in Eridanus. (North is at left.) (Palomar Observatory photo)

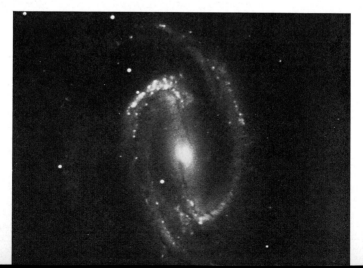

MAGNITUDES		OPEN CLUSTERS				QUASAR	△	GREEK ALPHABET			
-1	>-0.4		>10′: TO SCALE <10′			PULSAR		α Alpha		ν Nu	
0	-0.4--0.5	GLOBULAR CLUSTERS				BLACK HOLE		β Beta		ξ Xi	
1	0.6-1.5		>10′ 5′-10′ <5′			MILKY WAY		γ Gamma		o Omicron	
2	1.6-2.5							δ Delta		π Pi	
3	2.6-3.5	PLANETARY NEBULAE				GALACTIC EQUATOR		ε Epsilon		ϱ Rho	
4	3.6-4.5		>1′ 0.5′-1′ <0.5′				70°	ζ Zeta		σ Sigma	
5	4.6-5.5	BRIGHT DIFFUSE NEB.				ECLIPTIC		η Eta		τ Tau	
6	5.6-6.5		>10′: TO SCALE <10′				100°	θ ϑ Theta		υ Upsilon	
7	6.6-7.5							ι Iota		ϕ φ Phi	
DOUBLE or MULTIPLE		GALAXIES				CONSTELLATION BOUNDARIES		κ $\varkappa$ Kappa		χ Chi	
			>30′ 20′-30′ 10′-20′ <10′					λ Lambda		ψ Psi	
VARIABLE	◉ ○							μ Mu		ω Omega	

ATLAS CHART 35. M79, Lepus, Columba, Caelum, Eridanus In the constellation Lepus, the Hare, we find M79 (NGC 1904), one of the few globular clusters visible during the winter. Under good conditions it can be detected even with binoculars. M79 is not very impressive with small or medium-sized telescopes; only larger instruments begin to resolve its edges into their component stars. Located near $5^h24^m -25°$, M79 forms a triangle with β (beta) Lep and ε (epsilon) Lep. M79 is an 8th-magnitude cluster about 3 arc min in diameter and lies about 50,000 light-years away.

In the constellation Columba, the Dove, about 1° east of α (alpha) Col, is the 12th-magnitude spiral galaxy NGC 2090. Because of its extreme southern declination, NGC 2090 is difficult for northern-hemisphere observers to find; it can be glimpsed only when the sky above the southern horizon is exceptionally clear. If you can find NGC 2090, try to find the brighter spiral NGC 1792, which lies even farther south. NGC 1792 lies on the border of Columba and Caelum, about 3° south and slightly east of the 4.6-magnitude star γ (gamma) Caeli.

Further south of NGC 1792, near $5^h14^m -40°$ in Columba, the globular cluster NGC 1851 is barely visible to the eye but is an interesting object through a telescope.

Among the double stars visible in this region, γ (gamma) Lep is an easily resolved double that offers a nice color contrast even in small telescopes. Its components are magnitudes 3.6 and 6.2, separated by 95 arc sec.

About ½° to the southwest of M79 is ADS 3954, a double star with components of magnitudes 5.5 and 6.7, separated by 3.1 arc sec. This double can be resolved with small telescopes.

μ (mu) Columbae is a well-known "runaway star" that seems to be moving at high speeds — for a star in our galaxy — away from the Orion Molecular Cloud, the star nursery that lies behind the Orion Nebula. Two similar hot stars are known. All three lie at the same distance from the Orion Molecular Cloud, where infrared and radio studies have found stars in formation (see p. 118). The hot stars may have been ejected from that region during the last few million years.

Fig. 75. Lepus, the Hare, from the atlas of Bayer (1603). (Smithsonian Institution Libraries)

FROM "SKY ATLAS 2000.0" BY WIL TIRION

MAGNITUDES	OPEN CLUSTERS				QUASAR	△	GREEK ALPHABET			
● >-0·4		○ >10'	○ TO SCALE	○ <10'	PULSAR	✴	α Alpha		ν N
● -0·4-+0·5					BLACK HOLE	⅄	β Beta		ξ X
● 0·6-1·5	GLOBULAR CLUSTERS	⊕ >10'	⊕ 5-10'	⊕ <5'	MILKY WAY		γ Gamma		ο Omicron
● 1·6-2·5							δ Delta		π Pi
● 2·6-3·5	PLANETARY NEBULAE	◇ >1'	◇ 0·5-1'	◇ <0·5'	GALACTIC EQUATOR		ε Epsilon		ϱ Rho
● 3·6-4·5						70°	ζ Zeta		σ Sigma
● 4·6-5·5							η Eta		τ Tau
● 5·6-6·5	BRIGHT DIFFUSE NEB.		□ >10'.TO SCALE	□ <10'	ECLIPTIC	100°	θ ϑ Theta		υ Upsilon
● 6·6-7·5							ι Iota		φ ϕ Phi
DOUBLE or MULTIPLE ●—●					CONSTELLATION BOUNDARIES		κ ϰ Kappa		χ Chi
							λ Lambda		ψ Psi
VARIABLE ◉ ○	GALAXIES	⬭ >30'	○ 20'-30'	○ 10'-20' ○ <10'			μ Mu		ω Omega

ATLAS CHART 36. Sirius, M93, M41, Canis Major, Puppis

In the winter Milky Way, many beautiful star fields are visible low in the southern sky. In particular, Canis Major and Puppis are worth sweeping with a telescope or binoculars.

The open cluster M41 (NGC 2287) is one of the finest in the sky, located near $6^h47^m -21°$, about 4° south of Sirius in Canis Major. At magnitude 6 it is sometimes visible with the naked eye. M41 is a good object for viewing with binoculars and is easy to resolve with medium power. It is a fairly rich cluster, with about 25 bright stars and many fainter ones in a field about the size of the full moon.

A second open cluster suitable for viewing with a low-power telescope is M93 (NGC 2447), about 2° northwest of 3rd-magnitude ξ (xi) Pup, near the galactic equator at $7^h45^m -24°$. M93 lies in a dense star field at a distance of about 3300 light-years. It is a very rich cluster, with about 300 component stars. About 2° on the other side of ξ Pup is the diffuse nebula NGC 2467.

A third outstanding open cluster in this region is NGC 2477, located 15° south of M93 in Puppis, near $7^h52^m -39°$. It is rich and compact, with about 300 stars in a field 20 arc min across.

ε (epsilon) Canis Majoris lies above the center of the chart, near $7^h -29°$. Its 8th-magnitude companion is about 7 arc sec away.

Northeast, near τ (tau) CMa, is the variable star UW CMa, near $7^h19^m -25°$. About 5° to the east, the double star n Pup has two widely separated components of almost equal magnitude.

k Pup, near $7^h39^m -27°$, is a bright double easily resolved with small telescopes. Its bluish components are about magnitudes 4.5 and 4.6, separated by 9.8 arc sec. Near $7^h29^m -43°$, σ (sigma) Pup is a double with orange and white components of magnitudes 3.3 and 8.5, separated by 22.4 arc sec.

About 3° southwest of σ Pup, near $7^h14^m -45°$, L^2 Pup is one of the brightest red variables. Since its minimum brightness is about 5th magnitude, L^2 Pup is visible to the naked eye during most of its cycle. Another variable star visible to the naked eye is V Pup, about 3° southwest of γ (gamma) Velorum. V Pup is an eclipsing variable, ranging in magnitude from 4.4 to 5.3.

Fig. 76. Canis Major, the Big Dog (drawn backwards), from the star atlas of Hevelius (1690).

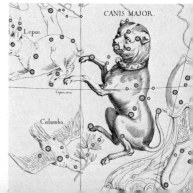

MAGNITUDES		OPEN CLUSTERS	○	◌	○	∘	QUASAR	△	GREEK ALPHABET			
●	>-0.4		>10'	TO SCALE		<10'	PULSAR	⨁	α Alpha		ν Nu	
●	-0.4–0.5	GLOBULAR CLUSTERS	⊕	⊕	⊛		BLACK HOLE		β Beta		ξ Xi	
●	0.6–1.5		>10'	5'–10'	<5'		MILKY WAY		γ Gamma		o Omicron	
●	1.6–2.5								δ Delta		π Pi	
●	2.6–3.5	PLANETARY NEBULAE	✧	✧	✧		GALACTIC EQUATOR	70°	ε Epsilon		ϱ Rho	
●	3.6–4.5		>1'	0.5'–1'	<0.5'				ζ Zeta		σ Sigma	
●	4.6–5.5								η Eta		τ Tau	
●	5.5–6.5	BRIGHT DIFFUSE NEB.					ECLIPTIC	100°	θ ϑ Theta		υ Upsilon	
●	6.6–7.5		>10'. TO SCALE		<10'				ι Iota	φ φ Phi	
DOUBLE or MULTIPLE	●—●								κ ϰ Kappa		χ Chi	
VARIABLE	◉ ○	GALAXIES	⬭	⬮	○	∘	CONSTELLATION BOUNDARIES		λ Lambda		ψ Psi	
			>30'	20'–30'	10'–20'	<10'			μ Mu		ω Omega	

ATLAS CHART 37. Gum Nebula In this region of the southern sky, we are looking toward the Milky Way along the galactic equator. Unlike other regions of the Milky Way, though, this region includes few interesting objects.

The Gum Nebula is too large to mark on the chart; it is a huge emission nebula that extends through the constellations Puppis and Vela over an area at least 35° in diameter. It is named after its discoverer, Colin Gum. The object that provides the energy for the central part of the nebula is probably PSR 0833–45 (the pulsar that also provides the energy for the x-ray source Vela X), near $8^h30^m -45°$. The pulsar emits both radio and optical pulses with a period of .09 seconds, and lies about 1,500 light-years distant.

ζ (zeta) Pup, left of 8^h at $-40°$, is a star of spectral type O, one of the hottest stars in the sky. About 5° north-northeast of ζ Pup, near $8^h12^m -35°$, is the site where Nova Puppis 1942 erupted. One of the brightest novae in modern times, it rapidly reached a maximum of about 0.3 magnitude.

About 4° east-northeast of α (alpha) Pyxidis is T Pyx, near $9^h05^m -32°$. T Pyx is one of the few recurrent novae. It is normally about 14th magnitude but may rise to magnitude 6.5 or 7 at maximum. More eruptions have been observed in this source than in any other. Its five maxima were recorded in 1890, 1902, 1920, 1944, and 1966. Given this average interval of 19 years, we might expect another eruption in about 1985. Check T Pyx regularly.

NGC 2997, a spiral galaxy in Antlia, is shown in C.Pl. 41.

Fig. 77. A part of the Gum Nebula in Vela. (Bart Bok)

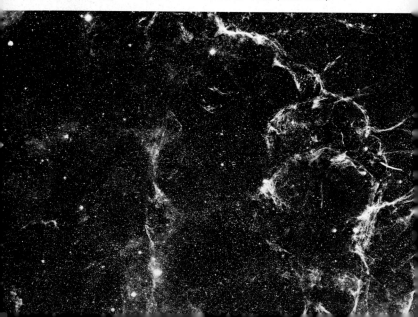

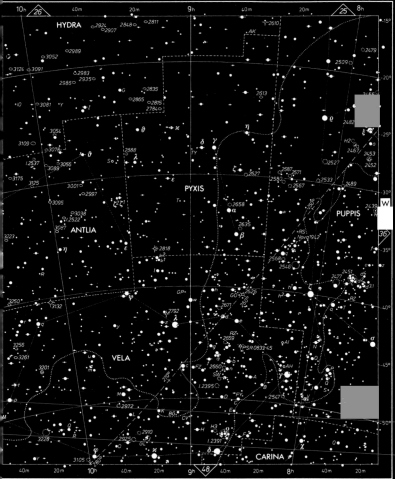

MAGNITUDES

●	>-0.4
●	-0.4–+0.5
●	0.6–1.5
●	1.6–2.5
●	2.6–3.5
●	3.6–4.5
●	4.6–5.5
●	5.6–6.5
●	6.6–7.5

OUBLE or MULTIPLE ●

ARIABLE ◎ ○

OPEN CLUSTERS	◯ ○ ○ ○
	>10' TO SCALE <10'
GLOBULAR CLUSTERS	⊕ ⊕ ⊛
	>10' 5–10' <5'
PLANETARY NEBULAE	◇ ◇ ◇
	>1' 0.5'–1' <0.5'
BRIGHT DIFFUSE NEB.	▱ □
	>10' TO SCALE <10'
GALAXIES	◯ ◯ ○ ○
	> 30' 20'–30' 10'–20' <10'

QUASAR	△
PULSAR	⚡
BLACK HOLE	⅄
MILKY WAY	
GALACTIC EQUATOR	– – – – – ✝ 70°
ECLIPTIC	– – – – – ✝ 100°
CONSTELLATION BOUNDARIES	

GREEK ALPHABET

α	Alpha	ν	Nu
β	Beta	ξ	Xi
γ	Gamma	ο	Omicron
δ	Delta	π	Pi
ε	Epsilon	ϱ	Rho
ζ	Zeta	σ	Sigma
η	Eta	τ	Tau
θ ϑ	Theta	υ	Upsilon
ι	Iota	φ φ	Phi
κ ϰ	Kappa	χ	Chi
λ	Lambda	ψ	Psi
μ	Mu	ω	Omega

ATLAS CHART 38. NGC 3132 The number of stars in this area diminishes with increasing distance from the Milky Way. Antlia, the Water Pump, is one of the least interesting regions of the sky, both to the telescope and the naked eye. Hydra and the northern part of Centaurus also contain few noteworthy objects.

One notable object is NGC 3132, one of the few planetary nebulae visible in the spring, located right on the Vela/Antlia border, near $10^h07^m -40°$. This planetary is about 8th magnitude in total brightness and has a bright central star of about 9th or 10th magnitude. It is nearly the same size as the Ring Nebula in Lyra. Though brighter than the Ring Nebula, NGC 3132 is harder to find and not as well known because of its southern declination and the lack of nearby stars to aid in finding it. One method for locating NGC 3132 is to wait for Regulus in Leo to reach your meridian and then point your telescope about 52° south.

NGC 3132 shows more detail than the Ring Nebula. On photographs, its elliptical disk looks as though several oval rings have been superimposed and tilted at different angles. Because of this appearance, it is sometimes called the Eight-Burst Nebula.

N Hydrae (ADS 8202), at $11^h32^m -29°$, is a wide pair of closely matched yellow stars of magnitudes 5.8 and 5.9. Another interesting double is β (beta) Hya (at $11^h53^m -34°$). Its components of magnitudes 5.0 and 5.4 can be resolved in a medium-sized telescope.

Fig. 78. Centaurus, the Centaur (drawn backwards), from the star atlas of Hevelius (1690).

ROM "SKY ATLAS 2000.0" BY WIL TIRION

GNITUDES	OPEN CLUSTERS	QUASAR	GREEK ALPHABET	
>-0.4	>10', TO SCALE <10'	PULSAR	α Alpha	ν Nu
-0.4−+0.5		BLACK HOLE	β Beta	ξ Xi
0.6−1.5	GLOBULAR CLUSTERS	MILKY WAY	γ Gamma	o Omicron
1.6−2.5	>10' 5'-10' <5'		δ Delta	π Pi
2.6−3.5	PLANETARY NEBULAE	GALACTIC EQUATOR	ε Epsilon	ϱ Rho
3.6−4.5	>1' 0.5'-1' <0.5'		ζ Zeta	σ Sigma
4.6−5.5			η Eta	τ Tau
5.5−6.5	BRIGHT DIFFUSE NEB.	ECLIPTIC	θ ϑ Theta	υ Upsilon
6.6−7.5	>10', TO SCALE <10'		ι Iota	φ φ Phi
JBLE or LTIPLE		CONSTELLATION BOUNDARIES	κ ϰ Kappa	χ Chi
RIABLE	GALAXIES		λ Lambda	ψ Psi
	>30' 20'-30' 10'-20' <10'		μ Mu	ω Omega

ATLAS CHART 39. Omega Centauri, M83, NGC 5128
Here we find one of the most spectacular objects within range of
amateur telescopes: NGC 5139, ω (omega) Centauri, one of the
brightest and largest globular clusters. Its apparent diameter is
about the size of the full moon. Because it is so far south of the
celestial equator, ω Cen is visible in the U.S. only from the southernmost states. Observers there can sometimes detect it low on the
horizon on clear spring and summer evenings, about 36° almost
due south of Spica, near 13^{h}27^m −47°. Binoculars may help you
find it. To the naked eye, ω Centauri looks like a fuzzy 4th- or
5th-magnitude star. Through small telescopes or binoculars, it
looks like a hazy patch (Fig. 79, p. 286). The cluster contains over
one million stars.

An outstanding galaxy in the summer sky is the spiral M83
(NGC 5236), near 13^{h}37^m −30° on the Hydra/Centaurus border,
about 18° south of Spica and nearly halfway between γ (gamma)
Hya and θ (theta) Cen. A spiral seen face-on (C.Pl. 23), it is one of
the 10 largest galaxies and one of the 25 brightest. M83 is hard for
northern observers to see because it lies very low in the southern
sky, in a field rich with star clouds and dust.

About 3° south of M83 is NGC 5253, another bright galaxy. A
medium-sized telescope shows it as an oval with a bright center.
Two of the brightest extragalactic supernovae ever recorded erupted in this galaxy. The first, in 1895, reached magnitude 7.2; another, in 1972, reached magnitude 7.9.

Perhaps the most famous galaxy in this field is NGC 5128, located at 13^{h}25^m −43°, about 4½° north of ω Cen. NGC 5128 is a
giant peculiar galaxy, one of the largest and most luminous galaxies known. It looks like a glowing sphere of 7th magnitude crossed
by a thick dust lane (C.Pl. 42) and shows well in medium-sized
telescopes and sometimes in good binoculars. A jet of gas extends
from the galaxy's center. NGC 5128 has been identified as the powerful radio source Centaurus A. Both the jet of gas and the radio
waves and x-rays may result from matter falling into a giant black
hole at the galaxy's center.

The globular cluster M68 (NGC 4590) is relatively bright even
though it is small and highly concentrated. It is located at
12^{h}40^m −27° near ADS 8612, a 6th-magnitude double star slightly
to the southwest — the only nearby star visible to the naked eye.
M68 appears as a fuzzy glowing ball of 8th magnitude, about 3 arc
min across. To locate it, sweep with binoculars from δ (delta) Crv
through β (beta) Crv and continue about 4° farther south. NGC
5367 is a nebula with an embedded open cluster.

R Hya is a notable long-period variable about 3° east of γ Hya,
at 13^{h}30^m −23°. The third long-period variable discovered, R Hya
is one of the easiest to observe. It ranges from magnitude 3.6 to 10.9
with a period of 387 days and is noted for its deep red color.

ATLAS CHART 40. Centaurus, Lupus, Libra Few interesting objects lie in this area of the southern sky, which is dominated by sparse regions in Centaurus, Lupus, and Libra.

The globular cluster NGC 5694 is located in the upper right-hand part of the chart, at the eastern tip of the tail of Hydra, almost halfway between σ (sigma) Lup and π (pi) Hya, near $14^h40^m -27°$. NGC 5694 is difficult to resolve because of its small apparent diameter. Its total magnitude is 10.9, though none of its 10 brightest stars is brighter than about 16th magnitude. Lying about 100,000 light-years away, on the far side of our galaxy, NGC 5694 is one of the most distant globular clusters.

Toward the bottom of the chart, near $14^h08^m -48°$, is the open cluster NGC 5460. Its total magnitude is 6.3, and none of its members is brighter than 8th magnitude. Though sparse, this cluster covers an area about the size of the full moon. NGC 5460 is best observed with a rich-field telescope or binoculars.

Though it is very hard to observe visually, an extended region of nebulosity can be detected in Scorpius, between δ (delta) and π (pi) Sco. δ Sco is an extremely hot star, spectral type B0. It lies about 540 light-years away and is one of the brighter members of the Scorpio-Centaurus Association. The Scorpio-Centaurus Association is a scattered group of stars that extends over 90° of the southern sky, and includes many of the brighter stars in the constellations Scorpius, Lupus, Centaurus, and Crux. This group of stars is part of a larger group of about 100 stars often called the Local Star Cloud, located in one of the spiral arms in our galaxy. Typically, the stars in the group are spectral type B, about 20 million years old. The reddish Antares, α (alpha) Sco, about 7° southeast of δ Sco (Atlas Chart 41), is the brightest member of this group.

Many double stars lie in the constellation Lupus. One of these is π (pi) Lupi (near $15^h05^m -47°$), with components of magnitudes 4.1 and 6.0. ε (epsilon) Lup, east of λ (lambda) Lup, near $15^h23^m -45°$, can be resolved with a medium-sized telescope. κ (kappa) Lup, southeast of π Lup, near $15^h12^m -49°$, is a very wide pair, of magnitudes 4.1 and 6.0. ξ (xi) Lup, northeast in Lupus, near $15^h57^m -34°$, is a wide double, with stars of magnitudes 5.2 and 5.6.

Fig. 79. ω (omega) Centauri, one of the largest, brightest globular clusters. (Chart 39) (© 1974 Anglo-Australian Telescope Board)

GNITUDES	OPEN CLUSTERS				QUASAR	△		GREEK ALPHABET		
>-0.4		>10', TO SCALE	<10'		PULSAR			α Alpha	ν Nu	
-0.4 - -0.5					BLACK HOLE			β Beta	ξ Xi	
0.6-1.5	GLOBULAR CLUSTERS	⊕ >10'	⊕ 5'-10'	⊛ <5'	MILKY WAY			γ Gamma	ο Omicron	
1.6-2.5								δ Delta	π Pi	
2.6-3.5	PLANETARY NEBULAE	◇ >1'	◇ 0.5'-1'	◇ <0.5'	GALACTIC EQUATOR			ε Epsilon	ρ Rho	
3.6-4.5						70°		ζ Zeta	σ Sigma	
4.6-5.5								η Eta	τ Tau	
5.6-6.5	BRIGHT DIFFUSE NEB.				ECLIPTIC			θ ϑ Theta	υ Upsilon	
6.6-7.5		>10', TO SCALE		<10'		100°		ι Iota	φ φ Phi	
UBLE or LTIPLE					CONSTELLATION BOUNDARIES			κ ϰ Kappa	χ Chi	
								λ Lambda	ψ Psi	
RIABLE	GALAXIES	>30'	20'-30'	10'-20' <10'				μ Mu	ω Omega	

ATLAS CHART 41. Antares, Southern Milky Way   One of the most exciting regions of the entire sky, this area offers outstanding sights for summer observers. Brilliant star fields lie among the dark clouds and winding dust lanes of the Milky Way, and many open and globular clusters are present. First sweep the area with binoculars or a telescope at low power and then switch to higher magnifications for individual objects.

A good starting point is the most prominent object in this region — Antares, α (alpha) Scorpii, near $16^h29^m -26°$. At magnitude 0.96, the reddish Antares (its name means "compared with Mars," the reddish planet) is the 15th brightest star. It becomes visible in midspring or early summer in the southeastern sky. IC 4606, a reddish nebula about 5 light-years in diameter, surrounds Antares (C.Pl. 55). IC 4604, a reflection nebula around ρ (rho) Ophiuchi, about 3° north-northwest of Antares, is too faint for visual observations but shows up as a rich bluish-purple cloud in long-exposure photographs through large telescopes (see front cover, C.Pl. 55). In between, IC 4603 is a reflection nebula around a fainter star.

Less than 2° due west of Antares is M4 (NGC 6121), one of the largest and nearest globular clusters. At 6th magnitude, M4 is visible with binoculars and can sometimes be detected with the naked eye. M4 is one of the easiest globular clusters to locate. Because of its very loose structure, small telescopes can resolve its edges and medium-sized telescopes under high power can completely resolve its stars.

Less than 1° northeast of M4 in the same field is NGC 6144, a smaller and fainter globular cluster. To observe this cluster, use high power.

A little over 4° northwest of Antares, sweep about halfway between Antares and β (beta) Sco to find M80 (NGC 6093), a globular cluster that looks smaller and brighter than M4. It lies on the western border of a dark cloud. Because M80 is very compact, it is much harder to resolve than M4.

About 8° east of Antares is M19 (NGC 6273), a compact globular cluster. To find M19, point at Antares, then sweep east and slightly north. Alternatively, point your telescope at Antares and wait about 33 minutes for M19 to drift into view. It is one of the most oval globular clusters, with an apparent diameter of 5 arc min and a magnitude of 6.9. Its outer edges can be easily resolved.

About 7° southeast of Antares and about 5° south of M19 is another pretty globular, M62 (NGC 6266). M62 is about the same size and brightness as M19, though it looks round in a small telescope. However, M62 is one of the most asymmetrical clusters, as large telescopes show. Its nucleus lies southeast of the center of the cluster, making it look somewhat like a comet.

Slightly east and about $1\frac{1}{2}°$ north of M19 is NGC 6284, a small, bright globular cluster that looks like a 10th-magnitude star in a

MAGNITUDES					
-1	⬤	>-0·4			
0	⬤	-0·4-0·5			
1	●	0·6-1·5			
2	●	1·6-2·5			
3	●	2·6-3·5			
4	•	3·6-4·5			
5	·	4·6-5·5			
6	·	5·6-6·5			
7	·	6·6-7·5			

DOUBLE or MULTIPLE

VARIABLE ◎ ○

OPEN CLUSTERS	◌◌◌ >10'	◦◦ TO SCALE	◦ <10'
GLOBULAR CLUSTERS	⊕ >10'	⊕ 5'-10'	⊕ <5'
PLANETARY NEBULAE	✧ >1'	✧ 0·5'-1'	◦ <0·5'
BRIGHT DIFFUSE NEB.	⬭ >10' TO SCALE		□ <10'
GALAXIES	⬭ >30'	○ 20'-30'	◦ 10'-20'	○ <10'

QUASAR	△
PULSAR	⌖
BLACK HOLE	Υ
MILKY WAY	∿
GALACTIC EQUATOR	⊢ 70°
ECLIPTIC	100°
CONSTELLATION BOUNDARIES	

GREEK ALPHABET

α	Alpha	ν	Nu
β	Beta	ξ	Xi
γ	Gamma	o	Omicron
δ	Delta	π	Pi
ε	Epsilon	ϱ	Rho
ζ	Zeta	σ	Sigma
η	Eta	τ	Tau
θ ϑ	Theta	υ	Upsilon
ι	Iota	φ φ	Phi
κ ϰ	Kappa	χ	Chi
λ	Lambda	ψ	Psi
μ	Mu	ω	Omega

medium telescope. The 8th-magnitude globular cluster NGC 6293 can be found about $1\frac{3}{4}°$ east-southeast of M19.

East of Antares and north of M19, the southern part of Ophiuchus is one of the most striking regions of the Milky Way, with a vast expanse of dark nebulae lit by bright stars. Here we are looking through the spiral arms of our galaxy toward the galactic center. A good star to help you orient yourself in this area is θ (theta) Ophiuchi, a 3rd-magnitude star near $17^h22^m -25°$. About 2° east and below θ Oph, the huge Pipe Nebula extends about 7°. It is readily visible to the naked eye and is one of the largest dark clouds in the Milky Way. Absorption from the Pipe Nebula obscures the stars in an area several degrees across.

For northern-hemisphere observers, the finest portions of the Milky Way lie in Sagittarius. Moonless nights in the summer are ideal for sweeping the region with binoculars or viewing with medium-sized telescopes. The brightest part is the Great Sagittarius Star Cloud, just north of γ (gamma) Sgr. Farther west, star clouds are obscured by the Great Rift, which marks the equatorial plane of our galaxy. In this region, we are looking directly toward the center of our galaxy (the *galactic center*), about 30,000 light-years from earth. The galactic center is hidden from our optical view by interstellar dust and dark nebulae. It lies about $1\frac{1}{2}°$ southwest of the 4th-magnitude Cepheid variable X Sgr, a foreground object. The galactic center can be detected as part of the radio source Sgr A. Infrared, x-rays, and gamma rays from the galactic center also penetrate the absorbing dust, providing ways to study this part of our galaxy, which may contain a giant black hole.

About $2\frac{1}{2}°$ north and slightly west of γ (gamma) Sgr is one of the few dark nebulae that can be seen in amateur telescopes.

Farther southwest along the galactic equator in Scorpius, M6 (NGC 6405) and M7 (NGC 6475) are two large bright open clusters, both beautiful in small telescopes. Looking somewhat like the Pleiades, M6 (at $17^h40^m -32°$) has a total magnitude brighter than 6, and contains many stars between 8th and 12th magnitude. Even a small telescope resolves M6 into its member stars.

Easily located about 3° southeast of M6, near $17^h50^m -35°$, M7 is a brilliant cluster of 5th magnitude, sometimes visible to the naked eye on a dark night. To the naked eye, M7 looks like a hazy glow with a rich star cloud in the background. It is one of the few

Fig. 80. Two globular clusters, NGC 6522 and NGC 6528, close to each other in Sagittarius at 18^h02^m and 18^h03^m, respectively, and $-30°$. (The Kitt Peak National Observatory)

Fig. 81. The Milky Way in Sagittarius and Scorpius. (Harvard College Observatory)

open clusters that look impressive in binoculars. Medium-sized telescopes with higher power completely resolve the cluster. M7 is much larger than M6, with many stars from 6th to 8th magnitude. On the northwest edge of M7 is the globular cluster NGC 6453, visible in the same field of view under low power.

Still farther southwest — near the galactic equator at $16^h55^m -41°$, about $\frac{1}{2}°$ north of ζ (zeta) Sco — is NGC 6231 (C.Pl. 47), another fine open cluster (though it is unfavorably located for northern-hemisphere observers). Under good conditions, it can be seen with the naked eye by observers who are far enough south; it looks like a miniature version of the Pleiades because of its central group of seven or so brilliant white stars. NGC 6231 lies about 5700 light-years away; if it were as close to us as the Pleiades, it would look almost the same size but would be about 50 times brighter.

NGC 6231 appears to be the center of a large group of bright stars of spectral types O and B. About 1° north and slightly east of NGC 6231 lies H12, the richest part of this group, with about 200 stars. A small telescope shows the stars of H12 as a trail extending northeast from NGC 6231, ending in a bright nebulous region. This extended group of stars marks one of the spiral arms of our galaxy, which lies over 5000 light-years away from us — closer to the galactic center than the arm that contains our sun. IC 4628 is a faint nebula about $1\frac{1}{2}°$ north of NGC 6231. A much larger loop of nebulosity over 300 light-years in diameter circles this entire association of stars in a giant irregular ring about 4° wide. The nebula is one of the giant regions of glowing hydrogen that outline the spiral arms in our own galaxy and other spirals.

Other open clusters in this region include: NGC 6242, a compact cluster about 1° north of H12; NGC 6281, about $2\frac{1}{2}°$ further to the northeast, a compact and elongated cluster; NGC 6124, a rich cluster about 6° west of H12 near $16^h25^m -41°$; and NGC 6193, a large cluster located about 7° south-southwest of ζ (zeta) Sco.

The remnant of the supernova that Kepler saw as a nova in 1604 is marked at upper left. However, it is faint even in photographs.

ATLAS CHART 42. Lagoon Nebula, Trifid Nebula   In this region we are still looking toward the center of our galaxy from our vantage point more than halfway out on an outer spiral arm. Moonless nights in July are best for viewing the bright star clouds, open and globular clusters, and curious dark nebulae (C.Pl. 45) in this area of the Milky Way, visible low above the southern horizon.

γ (gamma) Sgr marks the top of the Archer's arrow in the figure of Sagittarius, on the right-hand side of the chart near 18h06m −30°. North and west of γ Sgr, just off the mainstream of the Milky Way near 18h02m −23°, are two outstanding emission nebulae — the Lagoon Nebula and the Trifid Nebula. Both are about 2500 light-years away and, like M42 (the Orion Nebula), get their energy from ultraviolet radiation from young, hot stars in their midst.

M8 (NGC 6523), the Lagoon Nebula, is the more spectacular of the two, visible to the naked eye as a cometlike glow. It is an irregular nebula about the size of the full moon, crossed by a broad absorbing band of dust that gives the nebula its name (C.Pl. 2). Smaller dust clouds can also be seen. Because of the nebula's great extent, it is best viewed with a wide-field eyepiece. The Lagoon Nebula contains dust and gas clouds where new stars form and a region of glowing hydrogen resulting from the presence of the hot stars that can be seen in its midst. The eastern half of the nebula contains the prominent star cluster NGC 6530, a scattered open cluster about 10 arc min in diameter. It is one of the youngest clusters known, not more than a few million years old, and includes several T Tauri stars (variable stars with irregular fluctuations, indicating that they have not yet settled down to a regular existence).

About 1½° north-northwest of the Lagoon Nebula is M20 (NGC 6514), the Trifid Nebula. The lower part of the Trifid is a diffuse emission nebula. In photographs we see the reddish radiation from the cloud of glowing hydrogen, which is divided into three parts by dust lanes (C.Pl. 7). Where the three dark lanes overlap, we find a central triple star. These lanes are fairly easy to detect with a medium-sized telescope at medium power. The bluish upper part of M20, physically unconnected to the lower part, is a reflection nebula.

M21 is a small, rich open cluster of magnitude 6.5, located in the same low-power field as M20 (the Trifid). About 1° southeast of M8 (the Lagoon) is NGC 6544 — a small, very remote globular cluster that is difficult to find. One degree farther southeast is NGC 6553, another globular cluster; it is very difficult to resolve in part because thick obscuring dust dims it by about 6 magnitudes. About 1½° east-northeast of M8 are several irregular nebulae, possibly connected to the Lagoon Nebula by a faint gaseous haze.

λ (lambda) Sgr, a few degrees to the east of M8 near 18h28m −25°, also lies in a rich field. M22 (NGC 6656) is an impressive

MAGNITUDES	
-1	>-0.4
0	-0.4 - +0.5
1	0.6 - 1.5
2	1.6 - 2.5
3	2.6 - 3.5
4	3.6 - 4.5
5	4.6 - 5.5
6	5.6 - 6.5
7	6.6 - 7.5

DOUBLE or MULTIPLE

VARIABLE

OPEN CLUSTERS			
	>10'	TO SCALE	<10'

GLOBULAR CLUSTERS			
	>10'	5'-10'	<5'

PLANETARY NEBULAE			
	>1'	0.5'-1'	<0.5'

BRIGHT DIFFUSE NEB.		
	>10', TO SCALE	<10'

GALAXIES			
	> 30'	20'-30'	10'-20'	<10'

QUASAR

PULSAR

BLACK HOLE

MILKY WAY

GALACTIC EQUATOR
70°

ECLIPTIC
100°

CONSTELLATION BOUNDARIES

GREEK ALPHABET

α	Alpha	ν	Nu
β	Beta	ξ	Xi
γ	Gamma	ο	Omicron
δ	Delta	π	Pi
ε	Epsilon	ϱ	Rho
ζ	Zeta	σ	Sigma
η	Eta	τ	Tau
θ ϑ	Theta	υ	Upsilon
ι	Iota	φ φ	Phi
κ χ	Kappa	χ	Chi
λ	Lambda	ψ	Psi
μ	Mu	ω	Omega

Fig. 82. (*left*) The Lagoon Nebula, M8 (NGC 6523), in Sagittarius. The shorter exposure, **Fig. 83** (*right*), shows the "hourglass" structure near its center. (Lick Observatory photos)

globular cluster, 2° northeast of λ Sgr, near $18^h36^m -24°$. At 5th magnitude, M22 is the third brightest and most easily resolved globular cluster in the sky. Its only equal in the northern sky is M13, the Hercules Cluster (see Chart 17). M22's brightest stars are 11th magnitude, and it has a total population of at least half a million stars. It is one of the nearest globular clusters, only about 9600 light-years away. Since it is less than 1° from the ecliptic, we can sometimes see a planet in the same field.

About 1° west-northwest of M22 is NGC 6642, a small, faint globular cluster that looks like a hazy spot.

Slightly northwest of λ (lambda) Sgr, at $18^h24^m -25°$ in the same low-power field, is M28 (NGC 6626), a bright globular cluster that looks like a star in a rich star background. It is one of the more compact globulars, not very striking in a medium-sized telescope because it is hard to resolve. Slightly closer on the east side of λ Sgr is NGC 6638, a small, remote globular cluster that is difficult to resolve.

About 5° southwest of λ Sgr is δ (delta) Sgr, with the faint globular cluster NGC 6624 located slightly to the southeast. About 3° southeast of δ and about $2\frac{1}{2}°$ northeast of ε (epsilon) Sgr, near $18^h30^m -32°$, is M69 (NGC 6637), a small, 8th-magnitude globular cluster resolved only with large telescopes. About 2° east of M69 and almost halfway between ζ (zeta) and ε Sgr is the globular cluster M70 (NGC 6681), similar in size and brightness to M69. A third, smaller globular cluster is NGC 6652, about 1° southeast of M69. All three globular clusters look like bright stars in a uniformly rich star field and lie on the far side of the center of the galaxy.

Northeast of M70 and about 2° southwest of ζ (zeta) Sgr, near 18h55m −31°, is M54 (NGC 6715), a small, bright globular cluster of 7th magnitude. Because M54 is highly concentrated, though, only large telescopes begin to resolve it.

About 7½° south of ζ Sgr is a field of bright and dark nebulae and obscuring dust in Corona Australis. NGC 6726–27 is the brightest part of the field. Slightly southeast is the reflection nebula NGC 6729, whose variations in brightness follow the variations of its illuminating star, R CrA.

Further east in Sagittarius the richness of the star fields diminishes rapidly. Sweeping through this region does reveal one noteworthy globular cluster, M55 (NGC 6809), near 19h40m −31°, about 7° east from ζ Sgr and slightly to the south. Readily visible to the naked eye, M55 is a large, loose globular cluster of 6th magnitude that appears very similar to M13. To the naked eye and in binoculars, M55 looks like a hazy star; in a small telescope it looks like a circular glow about 10 arc min in diameter. Since M55 appears so low in the sky to northern observers, it is difficult to resolve when viewed from northern latitudes.

Fig. 84. Sagittarius, the Archer, from the star atlas of Hevelius (1690). It is drawn backwards from its appearance in the sky because Hevelius drew the celestial sphere as it would appear on a star globe seen from the outside.

ATLAS CHART 43. Capricornus   In the rather barren area near the borders of Capricornus and Sagittarius, it is easy to locate M75 (NGC 6864), near 20^h06^m $-22°$ and less than 7° southwest of β (beta) Cap (drawn above the top border of the chart). M75 is a small, rich globular cluster of 8th magnitude, bright enough to be readily visible in good binoculars when it is low in the southern sky on clear August nights. Since it is one of the most compact globular clusters, it can be resolved only in telescopes of fairly large aperture. M75 lies on the far side of the galaxy, about 95,000 light-years away, and is one of the most remote of the Messier objects.

The most prominent globular cluster in Capricorn is M30 (NGC 7099), on the top left side of this chart. It is easily found with binoculars or small telescopes, about 4° southeast of ζ (zeta) Cap and slightly northwest of the 5th-magnitude star 41 Cap, near 21^h40^m $-23°$. Like M75, though, M30 is difficult to resolve even with larger telescopes because of its extremely compact center. It lies about 40,000 light-years away.

The farthest object known in the universe, the quasar PKS 2000–330, is at the right middle of the chart. Because of the expansion of the universe, it is receding from us so rapidly that its spectral lines are redshifted by 3.7 times (370% of) their original wavelengths. That puts this quasar over 12 billion light-years away, which means that we are seeing it as it was 12 billion years ago.

Fig. 85. Capricornus, the Goat (drawn backwards), from the star atlas of Hevelius (1690).

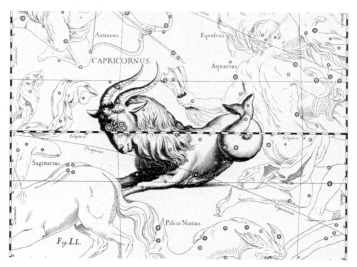

3

ADAPTED FROM 'SKY ATLAS 2000.0' BY WIL TIRION

MAGNITUDES		OPEN CLUSTERS		QUASAR		GREEK ALPHABET			
-1	>-0.4		>10', TO SCALE <10'	PULSAR		α Alpha		ν	Nu
0	-0.4-+0.5	GLOBULAR CLUSTERS		BLACK HOLE		β Beta		ξ	Xi
1	0.6-1.5		>10' 5'-10' <5'			γ Gamma		ο	Omicron
2	1.6-2.5			MILKY WAY		δ Delta		π	Pi
3	2.6-3.5	PLANETARY NEBULAE				ε Epsilon		ϱ	Rho
4	3.6-4.5		>1' 0.5'-1' <0.5'	GALACTIC EQUATOR		ζ Zeta		σ	Sigma
5	4.6-5.5					η Eta		τ	Tau
6	5.6-6.5	BRIGHT DIFFUSE NEB.				θ ϑ Theta		υ	Upsilon
7	6.6-7.5		>10', TO SCALE <10'	ECLIPTIC		ι Iota		φ φ	Phi
						κ ϰ Kappa		χ	Chi
DOUBLE or MULTIPLE				CONSTELLATION BOUNDARIES		λ Lambda		ψ	Psi
VARIABLE		GALAXIES	>30' 20'-30' 10'-20' <10'			μ Mu		ω	Omega

ATLAS CHART 44. Fomalhaut, Helix Nebula This region is uninteresting except for α (alpha) Piscis Austrini, a bright white star whose brilliance is intensified by the comparative darkness of the starless background. α PsA is also called Fomalhaut, a name derived from the Arabic for "mouth of the fish." At magnitude 1.16, Fomalhaut is the 18th brightest star in the sky. To northern-hemisphere observers it is visible in autumn, low above the horizon in an empty region of the southern sky.

Though not readily visible to observers at northern latitudes, an interesting object in this region is the planetary nebula NGC 7293, near $22^h30^m -21°$ in Aquarius; find it by looking 21° south from ζ (zeta) Aqr. NGC 7293 is the Helix Nebula, a huge bubble of gas about $1\frac{1}{2}$ light-years in diameter surrounding a 13th-magnitude central star (C.Pl. 54). A planetary results from the death of a star containing about as much mass as the sun. The Helix is the largest planetary, about $\frac{1}{2}$ the apparent diameter of the full moon. Although its total brightness is magnitude 6.5, its surface brightness is low because of its large size, making it difficult to detect except with low power on very dark nights. Even on the best nights, the nebula shows only as a faint, circular haze; little structure is visible except in larger telescopes and on long-exposure photographs (Fig. 86). To observe it, try the technique of averted vision: look not directly at the nebula but toward one side, which brings its image onto a more sensitive area of your retina.

The double star 41 Aqr lies several degrees west and slightly south of the Helix Nebula. Its contrasting components are magnitudes 5.5 and 7.5. Separated by 5.2 arc sec, the components can be resolved in small telescopes.

Southern constellations like Grus, the Crane, are relatively unfamiliar to most northern-hemisphere observers. θ (theta) Gru (at $23^h07^m -44°$) is another double star that can be separated in a small telescope, with components of magnitudes 4.5 and 7.0.

In the constellation Phoenix, SX Phe, a dwarf Cepheid variable, lies about $7\frac{1}{2}°$ west of α (alpha) Phe. SX was originally discovered as a star with large proper motion in 1938; it is about 400 light-years away. Its period is about 79 minutes, ranging in magnitude from about 7.1 to 7.5.

Fig. 86. The Helix Nebula (NGC 7293), a planetary nebula in Aquarius. See also C.Pl. 54. (The Cerro Tololo Inter-American Observatory)

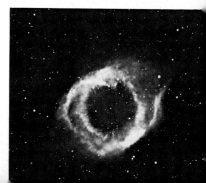

MAGNITUDES		OPEN CLUSTERS			QUASAR		GREEK ALPHABET			
-1	>-0.4		>10', TO SCALE <10'		PULSAR		α Alpha		ν Nu	
0	-0.4–+0.5	GLOBULAR CLUSTERS			BLACK HOLE		β Beta		ξ Xi	
1	0.6–1.5		>10' 5'–10' <5'		MILKY WAY		γ Gamma		o Omicron	
2	1.6–2.5						δ Delta		π Pi	
3	2.6–3.5	PLANETARY NEBULAE			GALACTIC EQUATOR		ε Epsilon		ϱ Rho	
4	3.6–4.5		>1' 0.5'–1' <0.5'			70°	ζ Zeta		σ Sigma	
5	4.6–5.5						η Eta		τ Tau	
6	5.6–6.5	BRIGHT DIFFUSE NEB.			ECLIPTIC	100°	θ ϑ Theta		υ Upsilon	
7	6.6–7.5		>10', TO SCALE <10'				ι Iota		ϕ φ Phi	
DOUBLE or MULTIPLE							κ $\varkappa$ Kappa		χ Chi	
VARIABLE		GALAXIES			CONSTELLATION BOUNDARIES		λ Lambda		ψ Psi	
			>30' 20'–30' 10'–20' <10'				μ Mu		ω Omega	

ATLAS CHART 45. Small Magellanic Cloud, 47 Tucanae
We begin here a series of eight polar charts showing the region surrounding the south celestial pole. The most interesting objects in this chart are the Small Magellanic Cloud and the bright globular cluster 47 Tucanae (NGC 104). Both can be found slightly below the center of the chart, near 0^h30^m $-72°$.

The Small Magellanic Cloud and its companion galaxy, the Large Magellanic Cloud (see Chart 47), are two of the closest galaxies to the Milky Way Galaxy. The SMC and the LMC are separated by about 80,000 light-years and are linked by a scattered group of stars and star clusters.

The Small Magellanic Cloud can be observed with all ranges of telescopic power by observers in the southern hemisphere. To the naked eye it looks like a cloud about $3\frac{1}{2}°$ across, with its hazy outline contrasting with the dark background (Fig. 95, p. 316).

Many open and globular clusters have been found in the Small Magellanic Cloud. Both the Small and the Large Magellanic Cloud also contain many Cepheid variables. In 1917 Harlow Shapley, using the relationship between the periods and luminosities of the Cepheid variables, calculated that the Magellanic Clouds were about 150,000 light-years distant. This calculation sparked a tremendous debate about whether the Magellanic Clouds were within our galaxy or were far outside its bounds and therefore were separate galaxies themselves. We now know that they are two of the approximately 24 members of the "Local Group" of galaxies to which our Milky Way Galaxy belongs.

Slightly to the west of the Small Magellanic Cloud, though not connected to it, is 47 Tucanae, a large, highly concentrated globular cluster that is bright enough (total magnitude 5) to be easily visible to the naked eye. 47 Tuc is one of the nearest globular clusters, only 20,000 light-years away.

In a similar direction is a second globular cluster, NGC 362, about twice the distance of 47 Tuc from the Milky Way. It consists of stars of magnitudes 13 to 14 and also is highly compact. NGC 362 is barely visible to the naked eye at magnitude 6.

The noteworthy double stars in this region include β (beta) Tuc (at 0^h32^m $-63°$), a wide double that can be resolved with low power into a pair of 4.4- and 4.5-magnitude stars. This system is actually a sextuple, though the other components are relatively faint. κ (kappa) Tuc (near 1^h16^m $-69°$) is a blue-white pair, of magnitudes 5.1 and 7.3, that can be resolved in a small telescope. At least two fainter components are also present.

Fig. 87. The globular cluster 47 Tucanae, NGC 104 (Chart 45). (The Cerro Tololo Inter-American Observatory)

7690○

σ

ϱ

π

7689○

o

ε

ζ

ϰ

ο7796

η

PHOENIX

○7702

○S

GRUS

7796

δ

δ●

ϱ●

○585
·q² ● ●q¹

ERIDANUS

ξ

ω●

η

υ

γ

22h
-55°

Achernar α

○434

○7205 ε

δ

π

782

BS● ι

○80

β³ ● β1.2

τ

χ

α

7126
7125

W

-60°

α

λ

ϱ

ζ

η

ε

TUCANA

ξ

δ

○7192

52

HOR

HYDRUS

ϰ

λ

π

406

ϑ

⊕ 47 Tuc

ϱ

○7329

7096○
7083 ○

INDUS

7020

γ

362⊕

Small Magellanic
Cloud

ο

SX
○Y

β

υ

602 □

λ

ν

σ

L.5052

KZ

7 ○6943

●●643

ϰ

β ●●

ϑ

PAVO

6876

20h
-70°

ϑ

θ

ν

ψ

ν

○6808

γ

σ τ¹

μ

τ2

μ¹
μ²

α

ε

τ

γ¹
γ³ γ²

ξ

ϑ ε

OCTANS

19h

MENSA

δ

ν

ξ

1841⊕

υ

φ

57

γ

ι

TZ ●

○R

τ

B
● σ

χ

○S

APUS

18h

ϰ π

-75°   -80°   -85°   -90° 49   -85°   -80°   -75°

FROM "SKY ATLAS 2000.0" BY WIL TIRION

MAGNITUDES					
● >-0.4					
● -0.4--0.5					
● 0.6-1.5					
● 1.6-2.5					
● 2.6-3.5					
● 3.6-4.5					
● 4.6-5.5					
· 5.6-6.5					
· 6.6-7.5					
DOUBLE or MULTIPLE ●—●					
VARIABLE ◎○					

OPEN CLUSTERS	○ ○ ○ ◇
	>10' TO SCALE <10'
GLOBULAR CLUSTERS	⊕ ⊕ ⊛
	>10' 5'-10' <5'
PLANETARY NEBULAE	◇ ◇ ◇
	>1' 0.5'-1' <0.5'
BRIGHT DIFFUSE NEB.	☁ □
	>10' TO SCALE <10'
GALAXIES	⬭ ○ ○ ○
	>30' 20'-30' 10'-20' <10'

QUASAR	△
PULSAR	☆
BLACK HOLE	⅄
MILKY WAY	～
GALACTIC EQUATOR	— 70° —+
ECLIPTIC	— 100° —+
CONSTELLATION BOUNDARIES	┈┈

GREEK ALPHABET			
α	Alpha	ν	Nu
β	Beta	ξ	Xi
γ	Gamma	o	Omicron
δ	Delta	π	Pi
ε	Epsilon	ϱ	Rho
ζ	Zeta	σ	Sigma
η	Eta	τ	Tau
θ ϑ	Theta	υ	Upsilon
ι	Iota	φ ϕ	Phi
κ ϰ	Kappa	χ	Chi
λ	Lambda	ψ	Psi
μ	Mu	ω	Omega

ATLAS CHART 46. Achernar   This chart shows an area with few bright stars, except for Achernar, α (alpha) Eridani, near 1ʰ38ᵐ −57°. Achernar, whose name means "the star at the end of the river," lies at the southernmost point of Eridanus. At magnitude 0.46, it is the ninth brightest star in the sky. Though it is never visible from most of the continental U.S., on autumn evenings it can sometimes be seen just above the southern horizon from southern Texas and Florida, and appears higher in the sky from Hawaii. Achernar is a hot blue giant, about 650 times as bright as the sun and about 120 light-years away. Just north of Achernar is p Eri, a wide orange pair of 6th-magnitude stars.

Below α (alpha) Reticuli (the Net), which is located at 4ʰ14ᵐ −62°, θ (theta) Ret is a double star with components of magnitudes 6.2 and 8.3, separated by 3.9 arc sec. A couple of degrees northeast of θ Ret, R Ret is a long-period variable that ranges in magnitude from 6.8 to 14.0. Its period is 278 days.

Fig. 88. Eridanus, the river into which Cygnus dived in search of a friend. Cygnus was then changed by Apollo into a swan and placed in the sky. The sea-serpent Hydrus is nearby. The drawing of the constellation is backwards, as are all drawings from the star atlas of Hevelius.

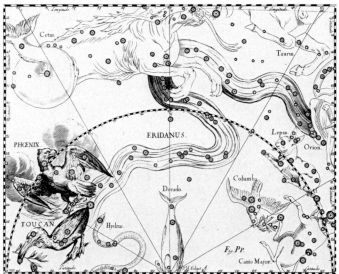

MAGNITUDES		OPEN CLUSTERS				QUASAR		GREEK ALPHABET			
●	>-0·4		>10'	TO SCALE	<10'		△	α	Alpha	ν	Nu
●	-0·4 -·0·5					PULSAR		β	Beta	ξ	Xi
●	0·6-1·5	GLOBULAR CLUSTERS	⊕	⊕	⊕	BLACK HOLE		γ	Gamma	o	Omicron
●	1·6-2·5		>10'	5'-10'	<5'			δ	Delta	π	Pi
●	2·6-3·5					MILKY WAY		ε	Epsilon	ϱ	Rho
●	3·6-4·5	PLANETARY NEBULAE	◇	◇	◇			ζ	Zeta	σ	Sigma
●	4·6-5·5		>1'	0·5'-1'	<0·5'	GALACTIC EQUATOR		η	Eta	τ	Tau
●	5·6-6·5						70°	θ ϑ	Theta	υ	Upsilon
	6·6-7·5	BRIGHT DIFFUSE NEB.			□	ECLIPTIC		ι	Iota	φ φ	Phi
			>10'	TO SCALE	<10'		100°	κ ϰ	Kappa	χ	Chi
DOUBLE or MULTIPLE						CONSTELLATION BOUNDARIES		λ	Lambda	ψ	Psi
VARIABLE	◉ ◎	GALAXIES	◯	◯	◯			μ	Mu	ω	Omega
			>30'	20'-30'	10'-20'	<10'					

ATLAS CHART 47. Large Magellanic Cloud, Tarantula Nebula, Canopus In the center of this region is the Large Magellanic Cloud, the huge, irregular companion galaxy to the Small Magellanic Cloud (Chart 45). Covering a vast area of the sky, the Large Magellanic Cloud (C.Pl. 34) can be resolved even with relatively small telescopes. It looks like a hazy cloud to the naked eye and is bright enough to be visible even under a full moon. (See Fig. 95 opposite Chart 52.)

The central part of the LMC extends over 20,000 light-years in length, while the whole galaxy covers at least 50,000 light-years. It contains at least 30 billion stars, including some supergiants larger than any in our galaxy. At least 400 planetary nebulae and more than 700 open clusters lie within the Large Magellanic Cloud, along with about 60 globular clusters, most of which are similar to those found in our galaxy.

The most striking areas of the LMC are the huge bright regions of glowing hydrogen where supergiant stars provide the energy for emission nebulae. Over 50 diffuse nebulae are visible with medium-sized telescopes. The most outstanding of these is the Tarantula Nebula (NGC 2070), a looped nebula surrounding the star 30 Doradus (in Dorado, the Swordfish). Visible to the naked eye, the Tarantula Nebula is the largest diffuse nebula known in the universe. Its complex structure shows much detail (Fig. 89). More than 100 supergiant stars cluster at its center, a region of nebulosity where stars are forming. The Tarantula Nebula lies about 190,000 light-years away. If it were as close to us as the Orion Nebula is, it would cover about 30° of the sky and would shine three times brighter than Venus.

The brightest star in the Large Magellanic Cloud is S Doradus in the open cluster NGC 1910. S Dor is an irregular variable star that ranges from magnitude 8.4 to 9.5. Its average luminosity is over 500,000 times that of the sun.

Near 6^h24^m $-53°$ is α (alpha) Carinae, also known as Canopus. At magnitude -0.72, Canopus is the second brightest star in the sky, surpassed in brightness only by Sirius. From the southern half of the U.S. it can be seen during the winter months low on the southern horizon, passing south of us about 20 minutes before Sirius does. Because it is so bright and so isolated from other bright stars in the sky, Canopus is often used for navigation by spacecraft going to the outer planets.

Fig. 89. The Tarantula Nebula (NGC 2070). (The Cerro Tololo Inter-American Observatory)

MAGNITUDES		OPEN CLUSTERS				QUASAR	△	GREEK ALPHABET			
-1	>-0.4		>10' TO SCALE <10'			PULSAR	⚡	α	Alpha	ν	Nu
0	-0.4 -0.5					BLACK HOLE		β	Beta	ξ	Xi
1	0.6-1.5	GLOBULAR CLUSTERS	⊕ ⊕ ⊛			MILKY WAY		γ	Gamma	ο	Omicron
2	1.6-2.5		>10' 5'-10' <5'					δ	Delta	π	Pi
3	2.6-3.5	PLANETARY NEBULAE	✧ ✦ ✧			GALACTIC EQUATOR		ε	Epsilon	ϱ	Rho
4	3.6-4.5		>1' 0.5'-1' <0.5'				70°	ζ	Zeta	σ	Sigma
5	4.6-5.5							η	Eta	τ	Tau
6	5.6-6.5	BRIGHT DIFFUSE NEB.			□	ECLIPTIC	100°	θ ϑ	Theta	υ	Upsilon
7	6.6-7.5		>10' TO SCALE <10'					ι	Iota	φ φ	Phi
DOUBLE or MULTIPLE								κ ϰ	Kappa	χ	Chi
		GALAXIES	◯ ◯ ○ ∘			CONSTELLATION BOUNDARIES		λ	Lambda	ψ	Psi
VARIABLE	◉ ○		>30' 20'-30' 10'-20' <10'					μ	Mu	ω	Omega

ATLAS CHART 48. Southern Milky Way In Carina, Vela, and Centaurus, the southern Milky Way again comes into view. Here, along the galactic equator, the Milky Way's brilliant star fields are extremely rich, rewarding amateurs for hours of random scanning with binoculars or small telescopes.

The numerous open clusters all merit observation. Among these are NGC 3114 (at 10^h03^m $-60°$), NGC 3293 (at 10^h35^m $-58°$), and NGC 2516 (at 8^h $-61°$). One of the finest open clusters visible to observers in the lower latitudes of the northern hemisphere, NGC 2516 should be viewed with a wide-field telescope because of its great size (about 1° or 15 light-years). It lies about 1300 light-years away and is older than the double cluster in Perseus (an extremely young cluster, only 10 million years old), and younger than the Pleiades (still fairly young at 100 million years old, less than 1% of the age of the universe).

One noteworthy globular cluster in this region is NGC 2808, above the center of the chart, near 9^h12^m $-65°$. It is a large, rich cluster of 13th- to 15th-magnitude stars, covering an area about 5 arc min across. At its center is a brilliant group of loosely packed stars.

Of the double stars, H Vel can be resolved in a small telescope. It is a blue-orange pair with magnitudes 4.9 and 7.7. v (upsilon) Car (at 9^h47^m $-65°$), at magnitude 3.0, is another double that can be resolved in small telescopes; its 6th-magnitude secondary component is 5 arc sec away from the 3rd-magnitude primary. t^2 Car, near 10^h40^m $-59°$, is a wide orange-green pair.

l ("ell") Car (at 9^h45^m $-63°$) is one of the brightest Cepheid variables. It reaches magnitude 4.1 at maximum, and is visible to the naked eye throughout its cycle. It is too far south for most U.S. observers, though. This star is one of the largest Cepheids known, with a diameter about 200 times that of the sun. l Car has a period of about $35\frac{1}{2}$ days and is over 3000 light-years away.

Fig. 90. Star trails around the south celestial pole, photographed from Australia. At lower left, the bright star Acrux (alpha Crucis) in the foot of the Southern Cross has risen. The Large and Small Magellanic Clouds are also visible, as hazy blurs. This is an unguided 24-minute exposure with a wide-angle (24 mm) lens on ASA 250 film. (Dennis di Cicco/*Sky & Telescope*)

VELA   PUPPIS
                                                          2427
2547   2427

-50°

CARINA   W
                                                         -60°

PICTOR

VOLANS   47
                              6h
                        DORADO   -65°

Tarantula
Neb.
2070
1910
Large
Magellanic
Cloud

CHAMAELEON   5h
                                          -70°

MUSCA

MENSA

APUS   OCTANS   HYDRUS   4h

                                         46
                                         3h

-75°   -80°   -85°   -90° 52 -85°   -80°   -75°

MAGNITUDES		OPEN CLUSTERS	>10' TO SCALE <10'	QUASAR	△	GREEK ALPHABET			
●	>-0.4			PULSAR		α	Alpha	ν	Nu
●	-0.4 -+0.5	GLOBULAR CLUSTERS	⊕ ⊕ ✳	BLACK HOLE		β	Beta	ξ	Xi
●	0.6-1.5		>10' 5-10' <5'	MILKY WAY		γ	Gamma	o	Omicron
●	1.6-2.5	PLANETARY NEBULAE	✧ ✧ ✧			δ	Delta	π	Pi
●	2.6-3.5		>1' 0.5'-1' <0.5'			ε	Epsilon	ϱ	Rho
●	3.6-4.5			GALACTIC EQUATOR		ζ	Zeta	σ	Sigma
·	4.6-5.5				70°	η	Eta	τ	Tau
·	5.6-6.5	BRIGHT DIFFUSE NEB.	▭	ECLIPTIC		θ ϑ	Theta	υ	Upsilon
·	6.6-7.5		>10' TO SCALE <10'		100°	ι	Iota	φ φ	Phi
DOUBLE or MULTIPLE	●-			CONSTELLATION BOUNDARIES		κ ϰ	Kappa	χ	Chi
VARIABLE	◉ ○	GALAXIES	◯ ⬭ ○ · · ·			λ	Lambda	ψ	Psi
			>30' 20'-30' 10'-20' <10'			μ	Mu	ω	Omega

ATLAS CHART 49. Eta Carinae Nebula, Coalsack Nebula, Southern Cross The Milky Way continues here through Carina, Crux, and Centaurus, offering outstanding sights for observers in southern latitudes. In the upper right-hand corner of the chart, near $10^h45^m -59°$, the Eta Carinae Nebula (NGC 3372) warrants special attention (C.Pl. 40). Visually the brightest part of the Milky Way, this extended nebula shows a highly complex structure with intermingling dark and light lanes and patches. The dark cloud superimposed on the brightest area at the center of the nebula is the Keyhole Nebula (NGC 3324). The brightest star to the left of the Keyhole is the variable star η (eta) Carinae, lying about 4,000 light-years from us.

η (eta) Carinae itself is a 6th-magnitude star. On the chart, it is located at the end of a short line that extends to the right and downward from the Greek letter η. η Car is surrounded by a small nebulous shell that is expanding at the rate of about 4 arc sec per century. Although this variable star shows some similarities to most novae, it also shows some peculiar differences: It remained bright for more than a century and was unusually luminous at maximum. First noted by Halley in 1677, it varied in brightness during the 19th century, finally reaching its maximum of magnitude -1 in 1843. The star faded to 7th magnitude and has since brightened to magnitude 6.2. With a mass approximately 100 times that of the sun, it may explode as a supernova; the presence of some of the heavy elements in unusual amounts, discovered by analysis of its spectrum, shows that this star is in an advanced stage of life.

Slightly north of the Eta Carinae Nebula is the open cluster NGC 3293, surrounded by reflection nebulae. The cluster is only 1,500 light-years away from us and is not physically connected to the Eta Carinae Nebula. About 3° northeast of Eta Carinae is another open cluster, NGC 3532. Since it is very large and elongated, it is best viewed with a wide-field telescope. This superb cluster contains at least 400 members, of which over 150 stars are 12th magnitude or brighter. About 5° south of the nebula is the scattered open cluster IC 2602. θ (theta) Car is the bright central star of this cluster, which contains 30 stars that are brighter than 9th magnitude and many fainter ones. The cluster is over 1° across and is best viewed with a rich-field telescope or at low power. With its central star θ Car at a distance of only 750 light-years from earth, IC 2602 may be one of the nearest open clusters to us.

Further east (left) along the Milky Way is IC 2944 (at $11^h35^m -63°$), a faint shell around the star λ (lambda) Cen. IC 2944, along with many other regions of nebulosity visible here, lies in a region of ionized hydrogen that marks one of the spiral arms of our galaxy. Slightly east (left) of IC 2944 is the loose open cluster IC 2948, lying behind the densest part of IC 2944. To the north, on the galactic equator (near $11^h35^m -62°$), NGC 3766 is a very rich, concentrated open cluster that contains at least 200 stars from 8th

MAGNITUDES		OPEN CLUSTERS				QUASAR	△	GREEK ALPHABET			
-1	⬤ >-0.4		>10' TO SCALE		<10'	PULSAR	⊠	α Alpha		ν Nu	
0	⬤ -0.4–+0.5					BLACK HOLE	⅄	β Beta		ξ Xi	
1	● 0.6–1.5	GLOBULAR CLUSTERS	⊕	⊕	⊕	MILKY WAY		γ Gamma		ο Omicron	
2	● 1.6–2.5		>10'	5–10'	<5'			δ Delta		π Pi	
3	● 2.6–3.5	PLANETARY NEBULAE	✧	✧	✧	GALACTIC EQUATOR		ε Epsilon		ϱ Rho	
4	● 3.6–4.5		>1'	0.5'–1'	<0.5'		70°	ζ Zeta		σ Sigma	
5	• 4.6–5.5						+	η Eta		τ Tau	
6	• 5.6–6.5	BRIGHT DIFFUSE NEB.			▫	ECLIPTIC	100°	θ ϑ Theta		υ Upsilon	
7	· 6.6–7.5		>10' TO SCALE		<10'		+	ι Iota		φ φ Phi	
DOUBLE or MULTIPLE	⬤○					CONSTELLATION BOUNDARIES		κ κ Kappa		χ Chi	
		GALAXIES	⬭	◯	○	∘		λ Lambda		ψ Psi	
VARIABLE	◉ ○		>30'	20'–30'	10'–20'	<10'		μ Mu		ω Omega	

to 15th magnitude; it covers an area about 10 arc min across. Beautiful even through binoculars, it is similar in appearance to M37 in Auriga (see Chart 11). About 5° to the north is the 8th-magnitude planetary nebula NGC 3918. It appears as a rich blue, featureless disk and has been compared to the planet Uranus.

Continuing eastward (to the left) along the Milky Way, we come to Crux, a small constellation nearly surrounded by Centaurus. Crux contains the famous Southern Cross, a diamond-shaped constellation. α (alpha) Cru — Acrux — marks the foot of the cross. β (beta) Cru — Mimosa — to the northeast marks the eastern end of the crosspiece. γ (gamma) Cru — Gacrux — to the northwest (upper right on the chart) marks the head of the cross and is nearly the same brightness as β. δ (delta) Cru, to the southwest of γ, marks the western end of the crosspiece. Three of these stars are among the 30 brightest stars in the sky. Because the cross has no central star, it actually looks more like a kite than a cross.

At 1st magnitude, α Cru (Acrux) lies nearly 400 light-years away. It is an easily resolved double, with two blue stars of magnitudes 1.4 and 1.9, separated by 4½ arc sec. Each of these components is also a spectroscopic double.

β Cru (Mimosa), 20% farther away than α Cru, is the 20th brightest star at magnitude 1.25. Nearly as bright is γ Cru, only about 90 light-years away. Like Antares, it is a red giant, though not so large and bright. An optical double, its components show a striking contrast in both color and magnitude.

Fig. 91. The Jewel Box (NGC 4755), an open cluster surrounding the star κ (kappa) Crucis. The 8 brightest stars range from about 6th to 10th magnitude; the cluster is about 10 arc min across. (Gabriel Martin, AURA, Inc.; The Cerro Tololo Inter-American Observatory)

Fig. 92. The Milky Way. Acrux (alpha Crucis) is at center, just at the right edge of the dark Coalsack. Hadar is the next bright star to the left. The complex of nebulosity around η (eta) Carinae is at right center. (Royal Astronomical Society)

Just to the south of β Cru is a magnificent open cluster called the Jewel Box (NGC 4755). It is one of the finest objects in the southern Milky Way. This cluster contains more than 100 stars scattered over an area 50 light-years across. About 50 of its brightest stars, most notably the 6th-magnitude star κ (kappa) Cru, are concentrated at its center. Many of the cluster's stars are supergiants, including some of the brightest ones known in our galaxy. Most of its members are bluish-white or white, contrasting with the scattered red supergiants. Not more than a few million years old, the Jewel Box is one of the youngest open clusters known. Visible with the naked eye, it should be viewed with low power because of the richness of the surrounding star field.

South of κ Cru and east of α Cru is an almost starless area, caused by a vast dust cloud about 7° by 5° across (60–70 light-years in diameter) that obscures the Milky Way in the background. This is the Coalsack, the most famous of the naked-eye dark nebulae. The few stars visible through telescopes are foreground objects lying between us and the obscuring dust. The Coalsack is one of the nearest dark nebulae, only 500–600 light-years away. It is easily apparent to the naked eye.

ATLAS CHART 50. Alpha Centauri Through Centaurus and Norma, the brightness of the Milky Way is divided by clouds of obscuring dust. The most distinctive features in this region are the two 1st-magnitude stars α (alpha) Centauri and β (beta) Centauri, located above the center of the chart. The former is known in navigational circles as Rigil Kent (short for Kentaurus), while the latter is sometimes called Hadar or Wazn. A bright pair of stars in the southern sky separated by $4\frac{1}{2}°$ (two fingers' width) often turns out to be α and β Centauri (Cen).

The fourth brightest star in the sky at magnitude 0.00, α Cen A is in the nearest star system to the earth. Alpha is 4.34 light-years away from us. It is a multiple system; its primary star is almost three times as bright as its secondary. A third, fainter companion lies further away from the primary and secondary stars, in a direction closer to the sun. This third component is Proxima Centauri (see Fig. 93, below). Proxima ("the nearest") is a red dwarf with a diameter about $\frac{1}{20}$ that of the sun. One of the smallest stars known, it is $\frac{1}{3}$ the size of Jupiter, though its mass is much greater.

β Centauri is the 11th brightest star at magnitude 0.61. It also has a companion star only 1.3 arc sec away; the pair is difficult to resolve because of the primary's brightness. β Cen is a blue-giant star, about 10,000 times as bright as the sun.

Circinus X-1, an x-ray source, is an interesting object on this chart, though it is not visible to the eye. It probably contains a neutron star, a compact remnant left over by the death of a massive star.

Fig. 93. A chart showing the region near Proxima Centauri, the closest star to the sun. Because it is so close, over a period of years we can see it move with respect to more distant stars, as marked. (Wil Tirion)

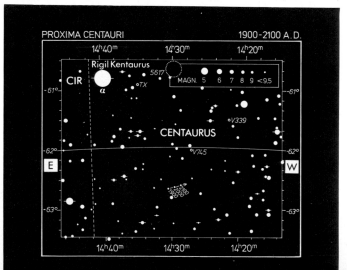

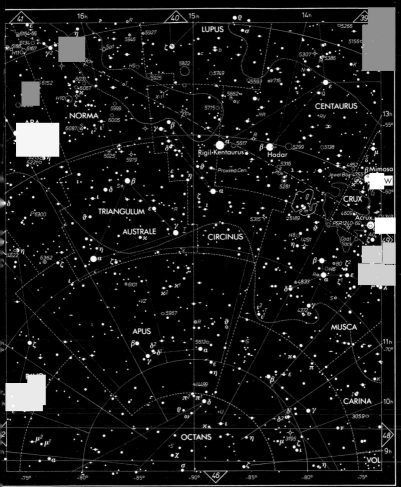

MAGNITUDES

-1	>-0.4
0	-0.4–+0.5
1	0.6–1.5
2	1.6–2.5
3	2.6–3.5
4	3.6–4.5
5	4.6–5.5
6	5.6–6.5
7	6.6–7.5

DOUBLE or MULTIPLE

VARIABLE

OPEN CLUSTERS	○ ○ ○ ○
	>10' TO SCALE <10'
GLOBULAR CLUSTERS	⊕ ⊕ ⊕
	>10' 5'–10' <5'
PLANETARY NEBULAE	◇ ◇ ◇
	>1' 0.5'–1' <0.5'
BRIGHT DIFFUSE NEB.	>10' TO SCALE <10'
GALAXIES	⬭ ○ ○ ○
	>30' 20'–30' 10'–20' <10'

QUASAR	△
PULSAR	☆
BLACK HOLE	Y
MILKY WAY	～
GALACTIC EQUATOR	70°
ECLIPTIC	100°
CONSTELLATION BOUNDARIES	

GREEK ALPHABET

α	Alpha	ν	Nu
β	Beta	ξ	Xi
γ	Gamma	ο	Omicron
δ	Delta	π	Pi
ε	Epsilon	ϱ	Rho
ζ	Zeta	σ	Sigma
η	Eta	τ	Tau
θ ϑ	Theta	υ	Upsilon
ι	Iota	φ φ	Phi
κ ϰ	Kappa	χ	Chi
λ	Lambda	ψ	Psi
μ	Mu	ω	Omega

ADAPTED FROM "SKY ATLAS 2000.0" BY WIL TIRION

ATLAS CHART 51. Octans, Pavo, Apus, Ara As we move southeast of the Milky Way in this chart, the star fields rapidly diminish in richness, though globular clusters become more apparent. NGC 6397 (at 17ʰ40ᵐ −54°) in Ara, the Altar, is a naked-eye globular cluster of exceptional brilliance. It can be found by looking toward the eastern (left) edge of the Milky Way, about 10½° south of θ (theta) Sco, though it can't be seen from most of the U.S. It is a loose scattered group with a total magnitude of 7.3, easily resolved in small telescopes. About 8200 light-years distant, NGC 6397 may be the nearest globular cluster to our solar system.

Another exceptionally bright cluster, NGC 6752 (at 19ʰ10ᵐ −60°), has a total magnitude of 7.2, making it the seventh brightest globular cluster. The third largest in apparent size, exceeded only by ω (omega) Cen and 47 Tuc, NGC 6752 is also one of the nearer globular clusters, about 14,000 light-years away.

A noteworthy double star in this region is ξ (xi) Pav (at 18ʰ23ᵐ −62°), an orange-green pair with a primary component of magnitude 4.4 and an 8th-magnitude secondary.

The spiral galaxy NGC 6744 in Pavo is shown in C.Pl. 49.

Fig. 94. The constellations near the south celestial pole, from the star atlas of Hevelius (1690). They are reversed from the way they appear in the sky (and on the chart) because this is a view looking in on a star globe. Note the beautiful peacock (Pavo).

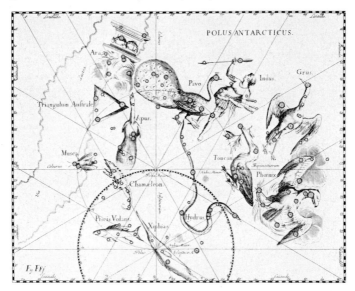

ATLAS CHART 52. The constellations in this area contain relatively few stars. Here we find only two bright stars, α (alpha) Pavonis, the Peacock Star, and α (alpha) Gruis, Al Na'ir, in Grus, the Crane. Indus, the Indian, is the other constellation that takes up most of this chart.

The region near the south celestial pole is almost barren. Although the number of background stars here does not differ greatly from the number near the north celestial pole, that pole has enough fairly bright stars to at least make the surrounding constellations look interesting. The 5th-magnitude star σ (sigma) Octantis, lying within 1° of the south celestial pole, is the southern pole star. Though it is very inconspicuous compared with Polaris and lacks the Pointers and Little Dipper that make Polaris easy to find, σ Oct is easily visible to the naked eye in a dark sky.

λ (lambda) Oct (at 21ʰ51ᵐ −83°) is a double star easily resolved in a small telescope. It shows contrasting orange and green components of magnitudes 5.5 and 7.7.

Fig. 95. The Large and Small Magellanic Clouds are 20° and 40°, respectively, above the horizon in this photograph from Australia, a 45-second exposure on 400 ASA film with a 35-mm lens at f/1.5. The south celestial pole is near the right edge of the picture. Canopus is at lower left and Achernar is at top left. (Shigetsugu Fujinami)

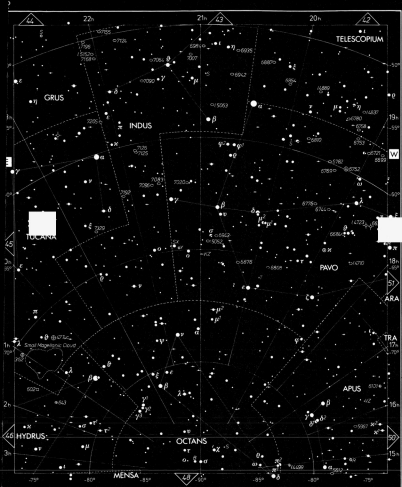

MAGNITUDES

-1	-0.4
0	-0.5
1	0.6-15
2	16-25
3	2.6-3.5
4	3.6-4.5
5	4.6-5.5
6	5.6-6.5
7	6.6-7.5

DOUBLE or MULTIPLE

VARIABLE

OPEN CLUSTERS		>10' TO SCALE	<10'
GLOBULAR CLUSTERS		>10' 5'-10'	<5'
PLANETARY NEBULAE		>1' 0.5'-1'	<0.5'
BRIGHT DIFFUSE NEB.		>10' TO SCALE	<10'
GALAXIES	>30'	20'-30' 10'-20'	<10'

QUASAR

PULSAR

BLACK HOLE

MILKY WAY

GALACTIC EQUATOR 70°

ECLIPTIC 100°

CONSTELLATION BOUNDARIES

GREEK ALPHABET

α	Alpha	ν	Nu
β	Beta	ξ	Xi
γ	Gamma	ο	Omicron
δ	Delta	π	Pi
ε	Epsilon	ϱ	Rho
ζ	Zeta	σ	Sigma
η	Eta	τ	Tau
θ ϑ	Theta	υ	Upsilon
ι	Iota	φ φ	Phi
κ ϰ	Kappa	χ	Chi
λ	Lambda	ψ	Psi
μ	Mu	ω	Omega

ADAPTED FROM "SKY ATLAS 2000.0" BY WIL TIRION

8

The Moon

For much of each month, the moon is the most prominent object in the nighttime sky. The moon is somewhat more than $\frac{1}{4}$ the diameter of the earth. This makes it the largest moon in the solar system in comparison to its parent planet. Thus the earth and moon essentially form a double-planet system. (Note that the moon is largest only in comparison to the earth; three moons of Jupiter and one each of Neptune and Saturn are physically larger than our moon.)

The moon orbits the earth every $27\frac{1}{3}$ days with respect to the stars. But during that time, the earth and moon have moved as a system about $\frac{1}{12}$ of the way in their yearly orbit around the sun. So if the moon at a certain point in its orbit is directly between the earth and the sun, $27\frac{1}{3}$ days later it has not quite returned to that point directly between the earth and sun. The moon must orbit the earth a bit further to get back to the same place with respect to the line between the earth and the sun. The moon reaches this point in a couple of days, making the *synodic period* of the moon equal to $29\frac{1}{2}$ days. (The synodic period is the interval between two successive conjunctions — coming to the same celestial longitude — of two celestial bodies, in this case conjunctions of the moon and sun as observed from the earth.) It is the synodic months that are taken into account in lunar calendars.

The Phases of the Moon

The phases of the moon repeat with this $29\frac{1}{2}$-day period, since the phases simply depend on the angle between the earth, sun, and moon. One-half of the moon is always illuminated by the sun. To us on earth, the moon appears to go through phases because in the course of the month we see different fractions of its lighted half. When the sun and moon are on opposite sides of the earth, the moon is *full* (Fig. 96). Everyone on earth who can see the moon above the horizon sees the same phase of the moon at the same time.

When the sun and moon are on the same side of the earth, we are looking past the moon at the sun. The far side of the moon is illuminated by the sun, but the side that faces us receives no sunlight. We say that the moon is *new*. Actually, we can often see the dark face of the moon by *earthshine* — sunlight that bounces off the earth and back up to the moon.

Fig. 96. A full moon. North on the moon is up, as it would be for a northern-hemisphere observer viewing the moon with the naked eye or with binoculars. West is at left. Mare Crisium is the small mare on the northeast limb (just above center at the right). Tycho is the crater at lower center with prominent rays emanating from it. (Lunar and Planetary Laboratory, University of Arizona)

In between the new moon and the full moon, the moon is a *crescent* (less than half-illuminated, Fig. 97), a *half moon* (half the face we see illuminated), or is *gibbous* (more than half-illuminated, Fig. 98). (Gibbous comes from the Latin word that means "a hump," or "bulging.") The first half moon after the new moon is called the first quarter, since we are $\frac{1}{4}$ of the way through the monthly cycle of phases. The half moon after the moon is full is called the third quarter. While the phases are changing from new to full, with more of the illuminated side of the moon becoming visible, we say the moon is *waxing*. Between full and new phases the moon is *waning*.

Table 15 lists dates for the phases of the moon. The fact that the moon's orbit is elliptical means that the moon travels at different speeds around the earth at different locations in its orbit, making the period between its phases vary slightly.

Since the angle between the earth, the moon, and the sun determines which phase will appear, the moon rises at a specific time of night, depending on the phase it is in. By its nature, a full moon must rise when the sun sets. A new moon must rise when the sun rises and must set when the sun sets. The moon rises about 50 minutes later each night, on the average. A first-quarter moon — the half moon that occurs about one week after the new moon —

(*text continues on p. 322*)

Table 15. Phases of the Moon

New Moon	First Quarter	Full Moon	Last Quarter
d h m	d h m	d h m	d h m
(U.T.)	(U.T.)	(U.T.)	(U.T.)

1984

New Moon	First Quarter	Full Moon	Last Quarter
Jan. 3 05 16	Jan. 11 09 49	Jan. 18 14 06	Jan. 25 04 48
Feb. 1 23 47	Feb. 10 04 00	Feb. 17 00 42	Feb. 23 17 13
Mar. 2 18 32	Mar. 10 18 28	Mar 17 10 10	Mar. 24 08 00
Apr. 1 12 11	Apr. 9 04 53	Apr. 15 19 11	Apr. 23 00 27
May 1 03 46	May 8 11 50	May 15 04 29	May 22 17 45
May 30 16 49	June 6 16 42	June 13 14 42	June 21 11 11
June 29 03 20	July 5 21 05	July 13 02 21	July 21 04 02
July 28 11 52	Aug. 4 02 34	Aug. 11 15 45	Aug. 19 19 42
Aug. 26 19 26	Sept. 2 10 30	Sept. 10 07 02	Sept. 18 09 32
Sept. 25 03 12	Oct. 1 21 53	Oct. 9 23 59	Oct. 17 21 14
Oct. 24 12 09	Oct. 31 13 09	Nov. 8 17 43	Nov. 16 07 00
Nov. 22 22 58	Nov. 30 08 01	Dec. 8 10 54	Dec. 15 15 26
Dec. 22 11 48	Dec. 30 05 28		

1985

New Moon	First Quarter	Full Moon	Last Quarter
		Jan. 7 02 18	Jan. 13 23 27
Jan. 21 02 30	Jan. 29 03 31	Feb. 5 15 19	Feb. 12 07 57
Feb. 19 18 44	Feb. 27 23 42	Mar. 7 02 14	Mar. 13 17 35
Mar. 21 11 59	Mar. 29 16 12	Apr. 5 11 33	Apr. 12 04 42
Apr. 20 05 22	Apr. 28 04 26	May 4 19 54	May 11 17 35
May 19 21 41	May 27 12 56	June 3 03 51	June 10 08 20
June 18 11 59	June 25 18 54	July 2 12 09	July 10 00 50
July 17 23 58	July 24 23 40	July 31 21 42	Aug. 8 18 29
Aug. 16 10 07	Aug. 23 04 38	Aug. 30 09 28	Sept. 7 12 17
Sept. 14 19 21	Sept. 21 11 04	Sept. 29 00 09	Oct. 7 05 05
Oct. 14 04 34	Oct. 20 20 13	Oct. 28 17 38	Nov. 5 20 08
Nov. 12 14 21	Nov. 19 09 04	Nov. 27 12 43	Dec. 5 09 02
Dec. 12 00 55	Dec. 19 01 59	Dec. 27 07 31	

1986

New Moon	First Quarter	Full Moon	Last Quarter
			Jan. 3 19 49
Jan. 10 12 23	Jan. 17 22 14	Jan. 26 00 31	Feb. 2 04 42
Feb. 9 00 57	Feb. 16 19 56	Feb. 24 15 03	Mar. 3 12 18
Mar. 10 14 52	Mar. 18 16 39	Mar. 26 03 03	Apr. 1 19 31
Apr. 9 06 09	Apr. 17 10 35	Apr. 24 12 48	May 1 03 23
May 8 22 10	May 17 01 01	May 23 20 46	May 30 12 56
June 7 14 01	June 15 12 01	June 22 03 42	June 29 00 54
July 7 04 56	July 14 20 11	July 21 10 41	July 28 15 35
Aug. 5 18 36	Aug. 13 02 22	Aug. 19 18 55	Aug. 27 08 39
Sept. 4 07 11	Sept. 11 07 42	Sept. 18 05 35	Sept. 26 03 18
Oct. 3 18 55	Oct. 10 13 30	Oct. 17 19 23	Oct. 25 22 27
Nov. 2 06 03	Nov. 8 21 12	Nov. 16 12 13	Nov. 24 16 51
Dec. 1 16 44	Dec. 8 08 03	Dec. 16 07 06	Dec. 24 09 18
Dec. 31 03 11			

1987

New Moon	First Quarter	Full Moon	Last Quarter
	Jan. 6 22 36	Jan. 15 02 31	Jan. 22 22 47
Jan. 29 13 46	Feb. 5 16 21	Feb. 13 20 59	Feb. 21 08 57
Feb. 28 00 51	Mar. 7 11 58	Mar. 15 13 13	Mar. 22 16 22
Mar. 29 12 47	Apr. 6 07 49	Apr. 14 02 32	Apr. 20 22 16

320

Table 15 (contd.). Phases of the Moon

New Moon			First Quarter			Full Moon			Last Quarter		
d	h	m	d	h	m	d	h	m	d	h	m
(U.T.)			(U.T.)			(U.T.)			(U.T.)		

1987 (contd.)

New Moon			First Quarter			Full Moon			Last Quarter		
Apr.	28	01 36	May	6	02 26	May	13	12 51	May	20	04 04
May	27	15 14	June	4	18 54	June	11	20 50	June	18	11 03
June	26	05 37	July	4	08 35	July	11	03 33	July	17	20 18
July	25	20 38	Aug.	2	19 25	Aug.	9	10 18	Aug.	16	08 26
Aug.	24	12 00	Sept.	1	03 49	Sept.	7	18 14	Sept.	14	23 45
Sept.	23	03 09	Sept.	30	10 40	Oct.	7	04 13	Oct.	14	18 06
Oct.	22	17 29	Oct.	29	17 11	Nov.	5	16 47	Nov.	13	14 39
Nov.	21	06 34	Nov.	28	00 38	Dec.	5	08 02	Dec.	13	11 42
Dec.	20	18 26	Dec.	27	10 01						

1988

New Moon			First Quarter			Full Moon			Last Quarter		
						Jan.	4	01 41	Jan.	12	07 04
Jan.	19	05 27	Jan.	25	21 55	Feb.	2	20 53	Feb.	10	23 01
Feb.	17	15 56	Feb.	24	12 16	Mar.	3	16 02	Mar.	11	10 57
Mar.	18	02 03	Mar.	25	04 42	Apr.	2	09 22	Apr.	9	19 22
Apr.	16	12 01	Apr.	23	22 32	May	1	23 42	May	9	01 24
May	15	22 11	May	23	16 50	May	31	10 54	June	7	06 23
June	14	09 15	June	22	10 24	June	29	19 47	July	6	11 38
July	13	21 53	July	22	02 16	July	29	03 26	Aug.	4	18 23
Aug.	12	12 32	Aug.	20	15 52	Aug.	27	10 57	Sept.	3	03 52
Sept.	11	04 51	Sept.	19	03 19	Sept.	25	19 08	Oct.	2	16 59
Oct.	10	21 50	Oct.	18	13 02	Oct.	25	04 36	Nov.	1	10 12
Nov.	9	14 21	Nov.	16	21 36	Nov.	23	15 54	Dec.	1	06 50
Dec.	9	05 37	Dec.	16	05 41	Dec.	23	05 29	Dec.	31	04 58

1989

New Moon			First Quarter			Full Moon			Last Quarter		
Jan.	7	19 23	Jan.	14	14 00	Jan.	21	21 34	Jan.	30	02 04
Feb.	6	07 38	Feb.	12	23 15	Feb.	20	15 33	Feb.	28	20 08
Mar.	7	18 20	Mar.	14	10 11	Mar.	22	09 59	Mar.	30	10 23
Apr.	6	03 34	Apr.	12	23 14	Apr.	21	03 14	Apr.	28	20 46
May	5	11 48	May	12	14 21	May	20	18 17	May	28	04 01
June	3	19 54	June	11	07 00	June	19	06 58	June	26	09 10
July	3	05 00	July	11	00 20	July	18	17 43	July	25	13 33
Aug.	1	16 06	Aug.	9	17 29	Aug.	17	03 07	Aug.	23	18 41
Aug.	31	05 45	Sept.	8	09 50	Sept.	15	11 51	Sept.	22	02 11
Sept.	29	21 48	Oct.	8	00 53	Oct.	14	20 33	Oct.	21	13 19
Oct.	29	15 29	Nov.	6	14 12	Nov.	13	05 52	Nov.	20	04 45
Nov.	28	09 42	Dec.	6	01 26	Dec.	12	16 31	Dec.	19	23 55
Dec.	28	03 21									

d = day, h = hour, m = minute.

Note: To convert Universal Time (U.T.) to time in U.S. time zones,
  subtract: 5 hours to get E.S.T., 4 hours to get E.D.T.
  6 hours to get C.S.T., 5 hours to get C.D.T.
  7 hours to get M.S.T., 6 hours to get M.D.T.
  8 hours to get P.S.T., 7 hours to get P.D.T.
  9 hours to get Alaska S.T.
  10 hours to get Hawaii S.T.

Fig. 97. (*left*) A crescent moon, 5 days old (that is, 5 days past new). Mare Crisium is still visible. (Lunar and Planetary Laboratory, University of Arizona)

Fig. 98. (*right*) A gibbous moon, 10 days old (that is, 10 days past new). (Lunar and Planetary Laboratory, University of Arizona)

thus rises at about noon and sets around midnight, and so is at its highest point each day at sunset. An interesting project is to observe the moon on as many days and nights of the month as possible, and to sketch each phase and its orientation with respect to the sun. You can often see the moon in the daytime, especially if you first figure out approximately where to look for it.

The Moon's Surface

Even with the naked eye, we can see that the surface of the moon is varied in structure. The fact that the moon has large flat areas called *mare* (pronounced "mar′ā"; plural — *maria,* pronounced "mar′ē a") and craters was discovered by Galileo when he first turned his small telescope to look at the moon in 1609. Maria means "seas," though there is no water in these lunar seas.

The view of the moon through even a small telescope can be breathtaking. You can see that the maria are very flat, and that there are other regions, the *highlands,* that are covered with craters. There are relatively few craters on the maria. These "seas" have, instead, been made flat by volcanic material — lava — that flowed from beneath the lunar surface over 3 billion years ago. This lava covered whatever craters existed at that time, so the only craters now visible in the maria were formed by the impact of interplanetary rocks — meteorites — that have hit the moon since then.

When observing the moon with binoculars or with a telescope, it is often most interesting to observe the *terminator*, the line separating light from dark (day from night, on the moon). At the terminator, we are seeing the region where the sun's rays are hitting the moon at the most oblique angle, and therefore shadows are longest there. This increases the contrast of the surface features. One can even calculate the height of lunar mountains by measuring the lengths of their shadows.

In this chapter, we provide detailed maps of the lunar surface. They are artist's drawings based on National Aeronautics and Space Administration (NASA) photographs, which are able to bring out the structure of features of all parts of the moon. The maps are drawn in a projection that enlarges the areas near the edges (called *limbs*) of the moon, which are otherwise too foreshortened to be seen clearly. In this projection, a dime held over any portion of the map covers the same area on the moon.

The moon keeps essentially the same face toward the earth as it orbits the earth, presumably because a slight bulge of matter exists in the side of the moon that faces the earth, allowing the earth's gravity to capture the moon's rotation. (As a result, the moon turns once on its axis with respect to the stars during each of its orbits around the earth.)

We are not, however, completely limited to seeing only 50% of the moon. The moon is sometimes turned slightly one way and sometimes slightly the other. These variations, called *librations*, enable us to see about $\frac{5}{9}$ of the surface. Because of this fact, the special projection of the lunar maps in this guide may be more useful to an observer than a photograph taken at a certain instant when the moon may have been turned at a different libration angle. The librations result from several causes, the chief of which is the fact that the moon is turning at a constant rate, while it is moving in its elliptical orbit at slightly different speeds, depending on how far away it is from the earth.

When we observe the moon with the naked eye or through a telescope, we see various maria (Fig. 99, p. 324). The largest mare is actually called an ocean, Oceanus Procellarum. Other maria have names such as Palus, signifying a marsh, and Lacus, meaning a lake, though we now know that there is certainly no water on the moon. At the edges of some of the maria we find indentations, each known as a bay — a *sinus*.

Craters of all sizes can be found, ranging up to hundreds of kilometers across. Many of the craters have *central peaks* — raised regions at their centers. Scientists have been able to duplicate these central peaks in laboratories on earth by shooting bullets at sand; the central peaks arise from a rebound of material. The cra-

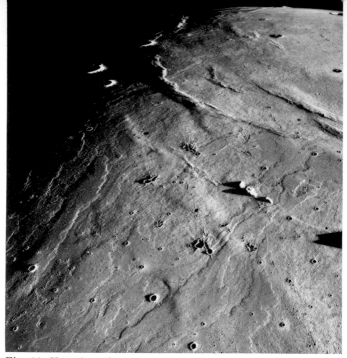

Fig. 99. Huge lava flows in southern Mare Imbrium, photographed from Apollo 15 in orbit around the moon. (NASA)

ters, for the most part, have been formed where meteorites have hit the moon. When the meteorites crash, they release their energy explosively, as though an atomic bomb or tons of TNT have gone off. It is this energy that forms each crater, which explains why the craters on the moon are relatively round, even though the meteorites may have hit at an angle.

The moon has *mountain peaks,* and *mountain ranges,* along with smaller raised *ridges.* A few *valleys* can be seen on the lunar surface. Some small valleys are known as *rilles; rilles* often seem sinuous in form, winding their ways for hundreds of kilometers across the lunar surface.

The photographs show that at certain phases, one can see *rays* that appear to radiate outward from a few craters, most prominently from Tycho. These rays consist of material ejected when the craters were formed. Over millions of years, the material in the rays darkens until it becomes invisible. The fact that the rays extending from Tycho are still clearly visible tells us that this is one of the youngest of the large craters on the moon.

For thousands of years, humans had to guess what the moon was like. But in 1969, with the first Apollo landings of U.S. astronauts, we began to acquire first-hand knowledge of the lunar surface. Even such basic knowledge as the fact that the moon has a hard surface with a thin coating of dust had to await the first landings of unmanned spacecraft on the moon and then observations that the Apollo astronauts didn't sink into the surface. Twelve astronauts on six Apollo missions landed on the moon (Table 16) and carried out many types of experiments there (C.Pl. 56). The astronauts measured the flow of heat from beneath the lunar surface, collected particles from the sun that hit the moon, and measured the hardness of rocks on the surface. Of course, the collecting of moon rocks and moon dust for detailed analysis back on earth was the most widely known of the experiments. Three unmanned Soviet landers also brought back to earth a few samples of moon rock and dust, though in far smaller quantities than the 382 kilograms (843 pounds) of material brought back by the American astronauts. Many of the lunar rocks are basalts, a common type of rock on earth formed by vulcanism.

As a result of the analysis of moon rocks on earth, scientists were able to work out an accurate chronology of the moon's history. The oldest rocks were 4.42 billion years old, which presumably represents the time when the moon's surface began to solidify, shortly after the overall formation of the moon 4.6 billion years ago. Then, from about 4.2 to 3.9 billion years ago, the surface was bombarded heavily by meteorites. A hundred thousand years later, radioactive elements inside the moon had generated so much heat that lava flowed onto the moon's surface, filling the largest basins and making the maria that we see today. This era of lunar vulcanism ended about 3.1 billion years ago, and the surface of the moon has changed little since then.

Though pictures of the moon taken from earth are often spectacular, whether taken with small telescopes or with some of the largest ones, any view of the moon from earth is always somewhat blurred by the effect of looking through the earth's atmosphere

Table 16. Manned Missions to the Moon

Mission	Year	Landing Site — see Moon Map
Apollo 11	1969	6
Apollo 12	1969	3
Apollo 14	1971	3
Apollo 15	1971	6
Apollo 16	1972	2
Apollo 17	1972	6

(Fig. 100). The photographs from the Apollo spacecraft therefore can show much finer detail. And from the earth, we see only the moon's near side. The far side was completely unknown to us until spacecraft circled the moon. In this chapter, in addition to providing eight detailed maps of the moon's near side, we present a map of the far side, on a less expanded scale. These maps show some of the names that have been assigned to craters and other features by the International Astronomical Union, which has the responsibility for naming astronomical features in the universe.

Lunar Eclipses

Though the moon revolves around the earth once each month, its orbital plane is tilted 5° with respect to the earth's. Thus when the moon passes behind the earth (in relation to the sun) during its monthly orbit, it usually passes either above or below the earth's shadow. (The earth's shadow is a cone that — at the moon's distance from earth — is about twice as broad as the moon's diameter.) But every few months, the moon passes either partially or entirely into the earth's shadow, and we have a partial or total *lunar eclipse* (Fig. 101).

Fig. 101. A drawing of a lunar eclipse, showing the umbra of the earth's shadow (the area defined by solid lines) and the penumbra (defined by dotted lines). (Susan Eder)

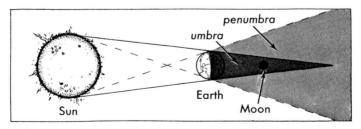

The earth's shadow has a darker inner cone, the *umbra,* from which the sun cannot be seen at all, and an outer cone, the *penumbra,* from which part of the sun can be seen. Most of the penumbra is not dark enough to have a noticeable effect on the moon's brightness, so penumbral eclipses are largely ignored.

During a partial lunar eclipse, the umbra of the earth's shadow advances onto part of the moon's surface and then moves off the other side. No other effect is seen.

During a total lunar eclipse (Table 17), the umbra hits the moon (first contact) and then gradually covers the moon until the moon is entirely covered (second contact). Then, for an hour or two, the

Table 17. Lunar Eclipses

Date	Moon enters shadow	Total-ity begins	Total-ity ends	Moon leaves shadow	Maximum magnitude	Overhead at maximum (long.)	(lat.)
5/4/85	18:17	19:22	20:30	21:35	1.24	300°	−16°
10/28/85	15:55	17:20	18:04	19:29	1.07	269°	13°
4/24/86	**11:04**	**12:11**	**13:15**	**14:22**	**1.20**	**191°**	**−13°**
10/17/86	17:30	18:41	19:55	21:06	1.24	293°	10°
8/27/88	10:08	none	none	12:00	0.29	165°	−11°
2/20/89	13:47	14:57	16:15	17:27	1.28	230°	11°
8/17/89	**1:21**	**2:20**	**3:56**	**4:55**	**1.60**	**46°**	**−14°**

Notes: Entries in **boldface** = totality visible in the U.S. and Canada.
All times are given in Universal Time (U.T.). To convert Universal Time to time in U.S. time zones, subtract:

   5 hours to get E.S.T., 4 hours to get E.D.T.
   6 hours to get C.S.T., 5 hours to get C.D.T.
   7 hours to get M.S.T., 6 hours to get M.D.T.
   8 hours to get P.S.T., 7 hours to get P.D.T.
   9 hours to get Alaska S.T.
   10 hours to get Hawaii S.T.

The April 24, 1986 eclipse will be visible in the early morning in the western U.S. and Canada. In 1989, a lunar eclipse will be visible throughout the U.S. and Canada on the night of August 16–17. You can calculate how much of each eclipse will be visible from your location by comparing the eclipse times (after changing the times above from Universal Time to your local time zone) with those for sunrise and sunset, which can be obtained from the Graphic Timetables in Chapters 1, 4, and 9. You could, alternatively, see how far your longitude is from the longitude at which the moon will be overhead at maximum eclipse. The maximum magnitude is the fraction of the moon's diameter that is in the umbra at the deepest part of the eclipse. The whole moon is barely in the umbra for maximum magnitude of 1, and is further in the umbra for maximum magnitude greater than 1.

(Based on "Canon of Lunar Eclipses" by Jean Meeus and Hermann Mucke)

moon is immersed in the earth's shadow. Though no sunlight falls directly on the totally eclipsed moon, some sunlight is bent around the earth's atmosphere. From this refracted sunlight, the blue light is removed when the light is scattered in the earth's atmosphere. This creates blue skies overhead for people on earth; only the red light gets through to the moon. Thus the totally eclipsed moon appears reddish (C.Pl. 57). Just how reddish it looks depends on whether volcanic dust is present in the earth's atmosphere; the dust makes the moon look darker and less reddish. Even giant storms or cloudy regions on earth can affect the earth's shadow, perhaps making the darkness of the shadow appear uneven on the moon. The moon will also appear less evenly illuminated if it passes closer to the sides of the umbra instead of through the umbra's center.

Observers in the western U.S. and western Canada will be able to observe the lunar eclipse of April 24, 1986. The August 16, 1989 (August 17, U.T.) lunar eclipse will be observable from a wider area across North America (see Table 17, p. 327).

To photograph a lunar eclipse, use a telephoto lens with a focal length of at least 200 mm for close-ups, though wide-angle views can also be interesting. During the partial phases, the bright part of the moon is illuminated by the sun; since it is about the same distance from the sun as the earth is, use the same exposure you would use to photograph people outdoors on a sunny day on earth. (For example, a lens setting of $f/11$ at $\frac{1}{125}$ second for ASA 64 film would be typical.) During the total part of the eclipse, however, the moon is much darker, and time exposures of a minute or more with your lens wide open may be necessary. These will blur because of the moon's motion, unless your camera or telescope is on a tracking mount. For best results, take a wide range of exposures in both black-and-white and color.

The Moon Maps

Ewen A. Whitaker

Map Projection

As seen from earth, the limb (outer) regions of the moon appear very crowded together (foreshortened) because of the viewing angle; formations situated within 10° of the map edges can be observed only with difficulty, and then only when illumination and libration conditions are favorable. The lunar maps used here (which were prepared in a cooperative effort of the National Geographic Society and the U.S. Geological Survey) are drawn in a projection that spreads out the limb regions and makes them easier to see. The projection preserves the relative areas that features cover on the lunar surface but slightly alters their true dimensions.

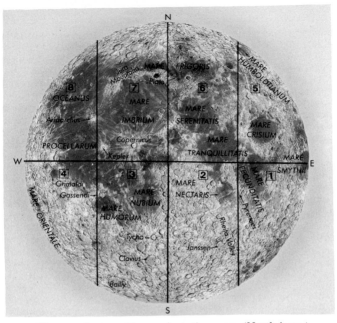

The moon's near side — orientation map. (North is up.)

Directions on the Moon

Since the moon has north and south poles, these directions are unambiguous. Prior to 1961, east and west on the moon were reckoned in the same directions as in the sky, *i.e.,* for an observer in the earth's northern hemisphere facing south (toward the moon, from U.S. and Canadian latitudes), east was to the left and west to the right. However, these directions are opposite to those used universally in terrestrial mapping, and the advent of the Space Age required that a common convention be adopted. Thus east and west in the accompanying moon charts and descriptions, as with all professional documents published since 1961, conform to these true lunar directions. Readers who may wish to consult some of the more detailed descriptive works and maps of the moon published prior to this date should be aware that "east" and "west" in these publications refer to directions in the sky. All common astronomical telescopes invert the moon's image, so the enlarged maps of the moon's near side in this guide are oriented with south at the top and lunar east to the left, to make the maps easy to use at the telescope. The orientation chart of the near side, however, is not inverted; it directly corresponds with the view seen with the naked eye or through binoculars.

MOON MAP 1. This area consists largely of the lunar high-lands, which are the highly cratered and lighter-toned areas of the moon, although most of Mare Fecunditatis lies within the boundaries of this map, as do portions of Mare Smythii and Mare Australe. The major craters are Humboldt (toward the left edge), Furnerius, Petavius, Vendelinus, and Langrenus.

Humboldt, a relatively fresh crater with a diameter of 125 miles (200 km), would be a very impressive feature if it were situated closer to the center of the moon's face (disk). It is best viewed one night after full moon, but favorable librations are essential. Vendelinus and Furnerius are better placed for good viewing but are older, as evidenced by the number of more recent craters that have overlapped them and by their generally smoother appearance. Petavius, 110 miles (177 km) in diameter, is somewhat younger, and is interesting because of a major rille that runs from its west wall to the imposing group of central peaks. Langrenus, named for (and by!) Van Langren, who produced the first lunar map, is also fairly fresh and clear-cut in appearance, with a small central peak and finely terraced walls. To the southwest (upper right) of Petavius are situated the sources of two extensive systems of bright surface streaks called rays. These are best seen when the terminator lies some distance away, *i.e.,* during the period towards full moon. The rays are remarkable in that their source craters, Furnerius A and Stevinus A, are so small.

Mare Fecunditatis consists of two contiguous, nearly round areas of dark basaltic lavas; the northern part has about three times the diameter of the southern. The larger part of this "sea" displays a typical system of fairly prominent ridges, the result of compressional forces. The small craters Messier and Messier A, situated near the lunar equator, are interesting both because of their somewhat anomalous shapes and their unusual pair of rays, resembling searchlight beams, that stretch westward from Messier A (see Fig. 102, lower right). These craters were studied extensively in the past because of suspected changes in their appearance, which are now known to be the result of changes in illumination and viewing angles.

Fig. 102. The craters Messier and Messier A, in Mare Fecunditatis, photographed from lunar orbit during the Apollo 8 mission. Messier is 8.5 miles long, 5.5 miles wide, and 7000 feet deep. (NASA)

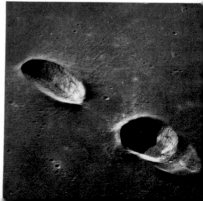

1

S

Jeans

Lyot

Brisbane

Peirescius

Hamilton

Gum

Oken

Vega

VALLIS RHEITA

MARE AUSTRALE

Marinus

Abel

Fraunhofer

Furnerius

Barnard

Adams

Furnerius A

Rheita

Legendre

Stevinus

Hase

Stevinus A

Phillips

Palitzsch

Snellius

Reichenbach

Humboldt

Petavius

2

Hecataeus

Schorr

Wrottesley

Borda

Gibbs

Biot

Santbech

Balmer

Holden

Mange

Behaim

Cook

Colombo

Montes Pyrenaeus

Lamé

Vendelinus

McClure

La Pérouse

Kapteyn

Crozier

Bellot

Magelhaens

Lohse

Goclenius

Langrenus

MARE FECUNDITATIS

Kästner

Kiess

Lubbock

Gilbert

Van Vleck

Maclaurin

Messier A

Weierstrass

Webb

Luna 16

Messier

MARE SMYTHII

E

3

MOON MAP 2. This area also consists largely of highlands, but is interesting because of the presence of Mare Nectaris (lower left) and its associated basin, and of the Rheita Valley (Vallis Rheita, upper left). This lunar valley, which is best viewed when the moon is either about 4–5 days old (during the crescent phase), or 2–3 days after the full phase, is about 300 miles long. However, there is some evidence that the valley is not a single structure but consists of two segments of different origin. The 200-mile (320-km) northern segment appears to have been caused by a subsidence of the moon's crust; the southern segment, which lies at an angle to the northern one, resembles many of the valleys that appear to radiate from the Imbrium basin (Map 7), such as those southeast of the crater Hipparchus (lower right of this map; Fig. 103).

Janssen is one of the largest craters (120 miles — 190 km — in diameter) shown on this map (upper left of center); most of its floor is very rough. This floor also contains a prominent rille, with many lesser rilles and a clifflike fault branching from it. Maurolycus (70 miles — 112 km — in diameter), at center, is one of the deepest (15,000 ft. — 4,570 m) of the larger craters; it looks very impressive when situated near the terminator. Theophilus (bottom center), with a diameter of 62 miles (100 km), looks somewhat fresher (sharper-edged). It is well situated for observations of part of its *ejecta blanket* — the blanket of material ejected from the crater — that extends in a northerly direction.

A medium-sized telescope will reveal the roughness of this area compared with the fairly smooth surface of Mare Nectaris. Mare Nectaris is a prominent, somewhat circular area of dark, basaltic lavas some 220 miles (354 km) in diameter that displays some concentric ridges on the eastern part. This mare represents a filling by lavas of a pre-existing central depression caused by the impact that formed the Nectaris basin. The outer boundary of this basin is marked, on the southwest, by the fault line named the Altai Scarp (Rupes Altai) and on the east, by the Pyrenees Mountains. These and other segments of scarps and mountain blocks form a circle that is concentric with and has twice the diameter of the mare's surface — a characteristic shared by all the larger impact structures on the moon.

Fig. 103. Lunar valleys radial to the Imbrium basin, near the crater Hipparchus (bottom center). (Lunar and Planetary Laboratory, University of Arizona)

Amundsen
Wexler
Demonax
Scott
Malapert
Neumayer
Helmholtz
Petov
Gill
Boguslawsky
Boussingault
Simpelius
Pontécoulant
Curtius
Manzinus
Mutus
Pentland
Hanno
Hagecius
Zach
Biela
Nearch
Kinau
Rosenberger
Tannerus
Asclepi
Jacobi
Reimarus
Hommel
Lilius
Watt
Vlacq
Pitiscus
Baco
Cuvier
Steinheil
Ideler
Mallet
Löckyer
Dove
Breislak
Clairaut
Heraclitus
Young
Spallanzani
Licetus
Janssen
Nicolai
Barocius
Fabricius
Maurolycus
Faraday
Metius
Wöhler
Büsching
Stöfler
Nasireddin
Rheita
Buch
Stiborius
Gemma Frisius
Fernelius
Kaiser
Rabbi Levi
Celsius
Nonius
Neander
Lindenau
Walter
Zagut
Gooddacre
Piccolomini
Rothmann
Wilkins
Paisson
Aliacensis
Weinek
Pontanus
Werner
Pons
Krusenstern
Santbech
Sacrobosco
Blanchinus
Fracastorius
Catharina
Azophi
Playfair
Beaumont
Geber
Abenezra
Faye
Almanon
Delaunay
MARE
Tacitus
Ary
Argelander
Vogel
NECTARIS
Cyrillus
Abulfeda
Parrot
Montes
Descartes
Burnham
Pyrenaeus
Kant
Andel
Klein
Gaudibert
Mädler
Albategnius
Gutenberg
Theophilus
Apollo 16
Halley
Capella
Isidorus
Zöllner
Taylor
Hind
Hipparchus
Torricelli
Alfraganus
Censorinus
Hypatia
Horrocks
Pickering
Gyldén
Delambre
Seeliger
Réaumur
Moltke
SINUS
MEDII

E
1
3
6

RUPES
ALTAI
Vallis
Rheita

MOON MAP 3. For an earth-based observer, the area covered by this map is dominated by the dark expanses of maria — Mare Nubium, Mare Cognitum, Mare Humorum, and a portion of Oceanus Procellarum (see also Map 8). The highlands area to the south also displays some interesting formations, including the 55-mile (90-km) diameter crater Tycho (left center). When situated near the terminator, this crater appears both rougher and deeper than its neighbors, with strongly terraced inner walls and a well-formed central peak. At phases near full moon, you can see that Tycho is the center of a very extensive system of bright rays, some of which extend almost a quarter of the way around the moon.

There are several large craters to the south of Tycho, but Clavius (140 miles — 225 km — in diameter) is undoubtedly the most eye-catching, with an arc of smaller craters spanning its level floor. A smaller crater to its west, Schiller (upper right of center), is unique because it is noticeably elongated, a condition that is further enhanced by the foreshortening that occurs near the moon's limbs. Nearby (farther west) are the large crater Schickard and the unique formation Wargentin. Wargentin looks like a typical crater but its interior has been filled to the rim with material. The elevated floor is smooth and flat, with a few low, curving ridges.

Nearer to the center of the moon's disk, at the lower left of this map, we find the large craters Ptolemaeus, Alphonsus, and Arzachel, which gradually decrease in diameter and increase in depth and freshness of appearance. Southwest of Arzachel, near the eastern edge of Mare Nubium, lies the Straight Wall (Rupes Recta), a 70-mile-long (113-km-long) fault in the mare surface that averages almost 1000 ft. (300 m) in height, with a slope of about 40°. When the moon is 8–9 days old, the Wall is impressive at sunrise, for it casts a wide, wedge-shaped shadow.

To the west, Mare Humorum (at right) is a reasonably circular dark area similar in size to Mare Nectaris (Map 2), but the surroundings look more degraded; as a result the edge of the basin is scarcely discernible in a telescopic view. The terrain between the two rings that surround Mare Humorum has been largely flooded with lavas (in Palus Epidemiarum, for example); the area between the rings surrounding Mare Nectaris is largely highlands material. The area lying between Mare Humorum and Palus Epidemiarum includes three parallel, curving rilles that are concentric with the Mare. These rilles are true *graben* — areas of subsidence between parallel faults in the lunar crust. The crater Gassendi, at the northern end of the Mare, is a much-observed and interesting formation. It is about 70 miles (110 km) in diameter, with a cluster of central peaks and a rough floor that displays a complicated network of rilles. Transient Lunar Phenomena (TLPs) — temporary appearances such as apparent obscurations, colorations, glows, etc. — have been reported in this crater on many occasions.

MOON MAP 4. Besides Oceanus Procellarum (lower left; see also Moon Map 8), this area contains the eastern half of Mare Orientale, the Orientale basin, and its extensive ejecta blanket (Fig. 104). However, these latter formations are too close to the limb to permit good observations from earth; lunar features that are located further from the limb command more interest in the telescope. In the highlands areas these include the crater Grimaldi (lower right), the Sirsalis-Darwin rille system, and the crater Mersenius and its environs (left center). Grimaldi is really a small basin, since it has the characteristic inner and outer concentric scarps. Its floor has also been flooded with mare lavas except toward the north, where they have not reached the crater wall.

The Sirsalis Rille (Rima Sirsalis) is one of the longest graben on the moon, and is contiguous with the Darwin system. Where the rille crosses crater walls, the walls have been cut by the parallel faults, as has the lower terrain.

The crater Mersenius, 51 miles (83 km) in diameter, is unusual in that its floor is noticeably convex — it is more convex than the moon's surface. This bulge is noticeable at sunrise, when the eastern part of the floor is well illuminated while the western part is still largely in shadow.

The western "shores" of Mare Humorum display some interesting features, such as the rough surface northeast of Mersenius, the clifflike faults southeast of that crater, and a small crater (Liebig F) that sits squarely on one of these faults. The flooded crater Doppelmayer is interesting for its concentric structure.

The dark-floored crater Billy and its companion crater Hansteen, which is of similar size but quite different appearance, are situated near the shoreline of Oceanus Procellarum. Between these craters lies the exceptionally bright mountain Mons Hansteen; special photography has shown it to be notably redder than the highlands surface nearby.

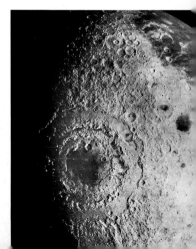

Fig. 104. This bull's-eye structure is known as the Orientale basin, from the dark central area, Mare Orientale (Eastern Sea). The basin was caused by the impact of a body that was probably about 30 miles (50 km) across. At right, we see part of the moon's front side. Grimaldi is the small dark region to the right (east) of the Orientale basin, with Oceanus Procellarum above (northeast of) it. North is up in this photo taken by Lunar Orbiter 4. (NASA)

S

4

Catalan

Inghirami

Vallis Baade

Baade

Graff

Schickard

Vallis Inghirami

Vallis Bouvard

Lehmann

Piazzi

Lagrange

Shaler

Lacroix

Wright

Fourier

Krasnov

Nicholson

Pettit

Vieta

3]

Palmieri

Byrgius

MONTES

MARE ORIENTALE

Doppelmayer

Cavendish

Eichstadt

Lamarck

MONTES ROOK

Liebig F

Lacus Veris

Liebig

Henry Freres

Henry

MARE HUMORUM

Mersenius

De Vico

Rima Sirsalis

Darwin

MONTES CORDILLERA

Gassendi

Crüger

LACUS AUTUMNI

Kopff

Zupus

Fontana

Rocca

Billy

Sirsalis

Hansteen

Mons Hansteen

Hartwig

PROCELLARUM

Grimaldi

Schlüter

Flamsteed

Damoiseau

Orbiter 5

"Surveyor 1"

OCEANUS

Hermann

Lohrmann

Riccioli

Hevelius

8

W

MOON MAP 5. The most notable feature on this map is Mare Crisium, an oval dark spot that is readily visible to the naked eye. This lunar sea was once named "The Caspian." It is comparable in size with both Mare Nectaris and Mare Humorum, but unlike those two seas, it is totally isolated from the other major maria. The outer ring of the Crisium basin, twice again the diameter of the mare, is visible mostly on the northern (lower) side when the sun is setting on this region, about two days after full moon. The area between the rings contains several patches of dark lavas, which have been named Mare Anguis (Serpent Sea), Mare Undarum (Sea of Waves), and more recently, Lacus Bonitatis (Lake of Goodness). Although it is only $17\frac{1}{2}$ miles (28 km) in diameter, the crater Proclus, to the west (right) of Mare Crisium, catches the eye during most phases because of its extreme brilliance. As the moon approaches the full phase, a striking ray system, which extends from all but the southwest sector of the crater, also becomes apparent. Similar asymmetrical ray patterns can be reproduced in the laboratory by using low-angle impacts.

At the north end of Mare Fecunditatis lies the crater Taruntius, which displays a weak ray system but has a flooded floor with concentric rilles and a small central peak.

Of the larger craters on the map, Neper (named for Napier, the inventor of logarithms), with its great terraced walls and prominent central peak, would be very impressive if it were situated nearer the central regions. However, it is on the lunar limb and can only be seen either soon after new moon or full moon, and then only if the moon's libration is favorable. Nearby are the dark areas of Mare Marginis and Mare Smythii (Smyth's Sea, Fig. 105), equally difficult to observe. The crater Gauss (110 miles — 177 km — in diameter), at lower left, is somewhat easier to observe; its floor exhibits some smaller craters and other detail. Burckhardt, north of Mare Crisium, is interesting in that it has partially obliterated two older, slightly smaller craters, which now protrude like ears on opposite sides of it.

Fig. 105. The earth is rising over the lunar limb as the Apollo 11 Lunar Module rises to rendezvous with the orbiter. In this westward-looking view, the large, dark area in the background is Smyth's Sea, which is on the moon's east limb as seen from earth. (NASA)

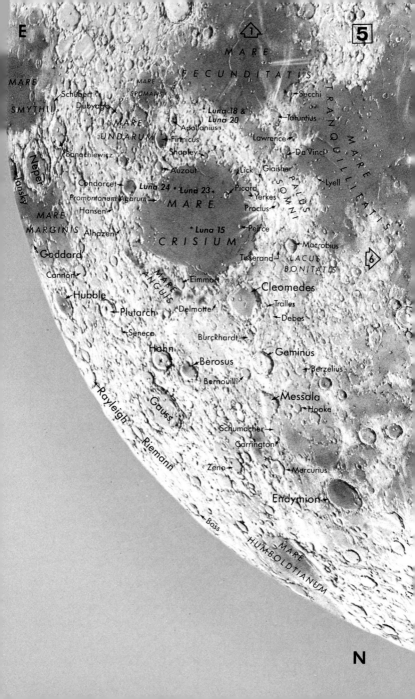

MOON MAP 6. As seen from earth, this area consists mostly of mare surfaces, including Mare Serenitatis and its adjacent Lacus Mortis and Lacus Somniorum, Mare Tranquillitatis and Mare Vaporum, Sinus Medii, and Mare Frigoris. The only reasonably pristine highlands areas on this map are those lying well north of Mare Frigoris. The area towards the center of the moon's face, which lies in Sinus Medii at upper right, abounds with interesting formations. Triesnecker marks the approximate center of a branching system of rilles that can be observed with medium-sized telescopes.

North of this system lies the more prominent Hyginus Rille, which cuts right through the small crater Hyginus. The western arm of this rille is punctuated with a string of craterlets, visible only in large instruments, attesting to some type of lunar volcanic or internal activity here in the past. Hyginus itself is a flat-floored rimless crater, which again indicates a volcanic-tectonic origin.

East of Hyginus lies the Ariadaeus Rille, a great linear graben that cuts through lowlands and mountain ridges alike. The area immediately north of the Hyginus Rille is one of the darkest spots on the moon; from lunar satellites and data acquired from other ground-based studies, we know that this darkness is largely due to a higher than average content of the mineral ilmenite (ferrous titanate) in the soil; in this area the soil consists mostly of the lunar equivalent of volcanic ash (mainly small glass beads).

The full moon appears almost silvery white in a dark sky, but its true color is mostly brownish gray. You can see color variations, however, especially if you view Mare Serenitatis and Mare Tranquillitatis in a telescope at low power when the terminator is not too close by. The rather warm, brownish tone of Serenitatis contrasts noticeably with the more steely gray tones of Tranquillitatis; the grayish color results from the unusually high ilmenite content of the basaltic lavas of Mare Tranquillitatis. This mare has a very strong system of ridges that traverse its western half.

Mare Serenitatis displays the typical concentric ridges of a flooded basin. The landing site of Apollo 17 is situated just off the easternmost corner of the mare, in a patch of noticeably darker material. Lunokhod 2, carried by the Soviet Luna 21 spacecraft, landed and operated near Le Monnier crater 100 miles (160 km) farther north. Slightly farther north is Posidonius, a prominent crater on the shore of Mare Serenitatis. Both sinuous and graben-type rilles cross the floor of Posidonius. The sinuous rilles follow tortuous paths over the lunar surface and are true lava-flow channels or collapsed lava tubes.

South of Mare Frigoris, craters Aristoteles, Eudoxus, Atlas, and Hercules are the remaining large, easily observable craters in this map area. Atlas (55 miles — 88 km — in diameter) is unusual because of the rilles and dark spots on its floor.

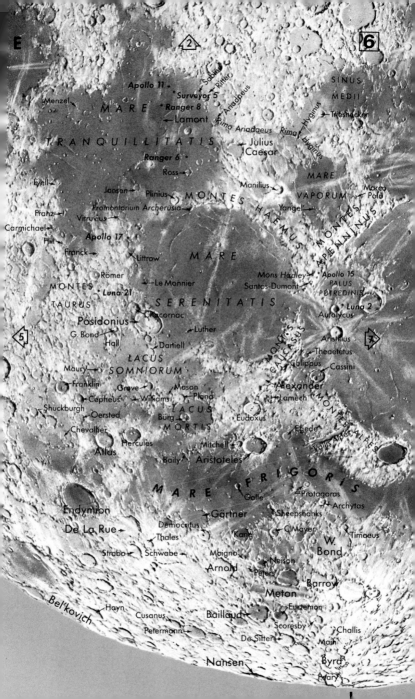

E

△2

6

MENZEL

MARE

TRANQUILLITATIS

Sabine
Ritter
* Surveyor 5
* Ranger 8
← Lamont

Amadaeus

SINUS
MEDII

→ Triesnecker

Hyginus

Rima Ariadaeus

Rima Hyginus

Ranger 6 →
Ross →

Julius
Caesar

MARE

VAPORUM

Lyell →

Jansen →
Plinius ←

Manilius

Marco
Polo

MONTES HAEMUS

Franz →
Vitruvius →

Promontorium Archerusia

MARE

Yangel ←

MONTES

APENNINUS

Carmichael →
Hill →
Franck →

Apollo 17

Littrow →

SERENITATIS

Mons Hadley
Santos-Dumont ←

Apollo 15
PALUS
PUTREDINIS

Römer →
Luna 21 →

Le Monnier ←

MONTES

TAURUS

Posidonius →
G. Bond →
Hall →

Chacornac →

Luther ←

Dartiell →

Luna 2
Autolycus →

Aristillus →
Theaetetus →
Calippus →
Cassini →

Maury →

LACUS

SOMNIORUM

Alexander →
Lamech ←

MONTES CAUCASUS

Franklin →
Cepheus →

Grove →
Williams ←

Mason →
Plana

Eudoxus

Egede →

MONTES ALPES

Shuckburgh →
Oersted →
Chevallier →

Bürg →

LACUS

MORTIS

Vallis Alpes ←

Hercules →

Atlas →

Baily →

Mitchell →

Aristoteles →

MARE

FRIGORIS

Endymion →

Golle →

Protagoras →
Archytas →

De La Rue →

Democritus →
Thales ←

Gärtner →

Sheepshanks →
C. Mayer ←

Timaeus →

Strabo →
Schwabe →

Kane →

W.
Bond

Moigno →

Neison →

Barrow →

Arnold →

Peters →

Meton →

Bel'kovich

Hayn →

Cusanus →

Baillaud →

Eudemon →

Scoresby →

Challis →

Petermann →

De Sitter →

Main →

Nansen →

Byrd
Peary

△5

⇦7

MOON MAP 7. This area is dominated by Mare Imbrium, with the mare areas of Mare Insularum, Sinus Aestuum, Mare Frigoris, and Sinus Roris occupying much of the remainder. The Imbrium basin is second only to the Orientale basin in freshness, with an age (it is believed) of about 3.9 billion years. On the moon's earthside face, only the much older Procellarum basin is larger than this huge Imbrium basin. Its formation therefore had a profound effect on the pre-existing lunar surface, and all craters situated south of Mare Frigoris (on this map) necessarily post-date that event. The mare lavas must also be more recent than the basin. There is strong evidence that all lunar vulcanism that produced dark mare lavas occurred no earlier than about 3.7 billion years ago. The surface of Mare Imbrium displays the usual concentric ridge systems, but under good conditions, lava-flow fronts — sharp demarcations — can be detected. Isolated mountain ranges and peaks, such as the Straight Range (Montes Recti), Montes Teneriffe, Mons Pico, Montes Spitzbergen, and Mons La Hire, lie on a circle that is half the diameter of the circle marked by the Carpathians (Montes Carpatus), Apennines (Montes Apenninus), and Caucasus (Montes Caucasus, Map 6). These inner ranges thus mark the ring that is equivalent to the general shoreline for Mare Crisium, Mare Nectaris, and Mare Humorum.

The craters Archimedes (on the east — to the left — of Mare Imbrium), Aristillus, and Autolycus form a very well-known landmark group. The first crater is flooded with mare lavas and thus is older than the lava flows that filled it, but the other two are clearly more recent than the lavas, displaying typical unflooded impact-crater floors and rough, radial ejecta deposits on top of the mare surface.

The Apennine Scarp (or Mountains, Montes Apenninus) is extremely impressive when viewed just after first quarter or near the last quarter phase. The highest peaks rise some 16,000 ft. (5 km) above the mare surface, but the pinnacle-like drawings of them in older books stem from a misinterpretation of the shapes of their shadows; photographs taken by the Apollo 15 astronauts revealed the generally smooth, rolling nature of the mountains that comprise this range. The Hadley Rille, which was visited by those astronauts, can be glimpsed with small telescopes under steady atmospheric conditions when the moon is about nine days old (past new moon).

The north shore of Mare Imbrium is marked by a strip of highlands that contains the lunar Alps (Montes Alpes), the crater Plato, and the spectacular Sinus Iridum (Bay of Rainbows). The location of this strip is quite anomalous with respect to the rest of the Imbrium basin; there is mounting evidence that the strip is actually a coalesced pair of crustal islands that were torn from their original locations some 100 miles (160 km) to the northeast, during the refilling of the Imbrium crater by magma flowing un-

7

3

SINUS
MEDII

• Surveyor 4
• Surveyor 6   • Schröter
• Murchison   Reinhold
• Pallas   • Surveyor 2
• Bode

MARE

INSULARUM

Kunowsky
Encke
Maestlin

Copernicus

Kepler

SINUS
AESTUUM   Stadius   MONTES CARPATUS

Milichius

OCEANUS

Eratosthenes   La Lussac

• Mons Wolff   T. Mayer
Draper

Mons Ampère
Wallace   Pytheas

Mons Huygens
Mons Bradley

MONTES   APENNINUS

Euler

MARE

Aristarchus

Apollo 15
PALUS
PUTREDINIS   Timocharis   Lambert   Diophantus   Montes   Prinz
Harbinger

Mons La Hire   Caventou   Delisle   Krieger

olycus   Luna 2   Archimedes   IMBRIUM   Angström   W

Heis   Gruithuisen

Aristillus   Montes   C. Herschel
Spitzbergen

Mons
Piton   Kirch   Luna 17   8

Piazzi Smyth   LeVerrier   Helicon   Promontorium Heraclides

Mont Blanc   Promontorium   SINUS   Mairan
Laplace
Mons   Montes   IRIDUM   Louville
Pico   Montes   Recti
Tenerife   Sharp

MONTES ALPES   Maupertuis   Bianchini

Vallis   Plato   Bouguer   Foucault
La Condamine

Harpalus   SINUS RORIS

ARE

FRIGORIS   Horrebow   Robinson   Markov

Timaeus   Fontenelle   South

Birmingham   J. Herschel   Oenopides

Epigenes   Babbage   Pythagoras

Goldschmidt   Anaximander   Cleostratus
Philolaus

Anaxagoras   Carpenter

Mouchez   Anaximenes   Desargues   Boole

Poncelet   Pascal   Cremona

Gioja

Sylvester   Brianchon

Hermite

N

Fig. 106. The northeast portion of Mare Imbrium, showing Montes Caucasus, Montes Alpes (the Alps), and Vallis Alpes (the Alpine Valley). South is up in this and other photos taken with the 61-in. telescope. (Lunar and Planetary Laboratory, University of Arizona)

derneath the moon's crust. The generally parallel-sided shape of Mare Frigoris, the "wrong" placement of the Alps, and other bits of evidence support this conclusion. The Alps, a group of isolated mountains and hills, contain the famous Alpine Valley (Vallis Alpes), a roughly parallel-sided "gash" over 100 miles long and averaging about 6 miles (10 km) in width (Fig. 106). It was once surmised that this valley was formed either by a tangentially travelling meteorite or else by a flying fragment from the Imbrium basin-forming impact, but this idea has long been discounted. It appears to be a graben-like feature, possibly caused by the crustal movements noted above.

Plato, 63 miles (101 km) in diameter, is prominent because of its smooth, level, dark floor, which contrasts strongly with the surrounding lighter highlands materials. The crater's floor contains craters 1½ miles (2½ km) in diameter and smaller; the largest one is located near the center. A medium-sized telescope is needed to show these objects as recognizable craters. Sinus Iridum looks beautiful when the sun is rising on it and its illuminated western edge protrudes far into the unlit surrounding area. This occurs when the moon is about 10 days old.

The crater Copernicus (57 miles — 92 km — in diameter), situated at the top of the map in Mare Insularum, is at least as famous and frequently observed as Plato. It is an excellent example of a large, comparatively fresh impact crater. It has a level but somewhat rough floor, with a group of small central peaks. The strongly terraced walls, a result of post-impact slumping, rise some 15,000 ft. (5 km) above the floor. Copernicus is the center of one of the most extensive ray systems anywhere on the moon. The smaller crater Kepler (20 miles — 32 km — in diameter), at the top right of the map, also has a notable ray system, the rays of which mingle with those of Copernicus.

MOON MAP 8. This chart is taken up almost entirely by the gray expanse of Oceanus Procellarum, whose western shore is never far removed from the lunar limb. In addition to the usual ridge systems and sprinkling of small craters, this area displays some unique features and interesting formations. The showpiece is undoubtedly the crater Aristarchus, plus the whole Aristarchus Plateau region to the northwest of the crater itself.

Aristarchus is one of the brightest craters of its size anywhere on the moon. It is a relatively fresh impact crater, 28 miles (45 km) in diameter, with a rim rising about 9000 ft. (2700 m) above the floor. It is the center of a major ray system, as are all relatively fresh impact craters. More Transient Lunar Phenomena (TLPs) have been reported in and near Aristarchus than at any other lunar location. In the past, many of these reports undoubtedly stemmed directly or indirectly from effects of the crater's brightness. Thus when the moon is less than a week old, the crater can be seen glowing in the earthlit part of the disk; observed rather low in the sky, the sunlit crater is bright enough against the dark mare background to produce a spectrum as the earth's atmosphere spreads out the lunar light, causing some of the reported color phenomena. However, dedicated observers are now aware of these "red herrings," and not all reports can be dismissed so easily. The whole area is unusual from several standpoints, and rare phenomena, such as emissions of gas, may occur here. Indeed, the orbiting Apollo spacecraft detected more of the gas radon here than anywhere else on the moon.

Close to the northwest lies the moon's largest volcano and associated lava-flow channel, visible in the smallest telescopes. Although it was discovered by Huygens, this channel is known as Schröter's Valley (Vallis Schröteri), after the German amateur astronomer who first observed it. The volcano itself is the rather unassuming hill immediately northwest of Aristarchus, and the Valley's northern flank starts with a depression nicknamed the "Cobra Head," because of the Valley's resemblance to the head and body of a snake. The Valley cuts through a rough area that is mostly somewhat darker than the surrounding mare, despite the presence of a concentration of rays from Aristarchus. Evidence from other investigations indicates that the whole area is covered with a blanket of lunar-type volcanic ash with a high ilmenite content. Although dark, the area is distinctly browner than the surrounding mare, or even other dark, ash-blanketed areas, which are all grayer than their surroundings. It is interesting that this volcanic soil with high ilmenite content can exist in two forms of identical composition but different color. Thus the black and orange soils collected near Shorty crater (which is not shown because of its small size) by the Apollo 17 astronauts are identical apart from color. The difference in color may be due to differing cooling rates, which determine whether the black ilmenite crystallizes out or stays dispersed as an orange coloration in the glass beads comprising the soil.

The bright marking known as Reiner Gamma (upper right; Fig. 107) is an example of a rare type of surface marking; other examples are confined almost entirely to the moon's far side. It is not connected with any visible topography, nor does it have any relief of its own. It does, however, exhibit a magnetic field; this field may shield the surface from the darkening effects of the solar wind and thus preserve the area's brightness. To the northwest lies a large field of rather small volcanic structures known familiarly as the Marius Hills, but the details of their topography only begin to become visible in large telescopes (Fig. 107).

In the highlands region at upper right, the crater Hevelius — 74 miles (119 km) in diameter — commands the most attention. Its floor is criss-crossed by a lattice of linear rilles, two of which cut through the crater wall and continue into the adjacent terrain.

The crater Lichtenberg, situated in the mare much further to the north, has a ray system that extends only to the north, thus superficially resembling the Proclus system. However, the cause in this case is quite different — the southern portion of the Lichtenberg ray system was covered up by a later lava flow.

Fig. 107. The craters Reiner (top), Marius (left), and the unique bright marking called Reiner Gamma, to the west (right) of Reiner. Some details of the Marius Hills are visible to the west-northwest of the crater Marius. South is up. (Lunar and Planetary Laboratory, University of Arizona)

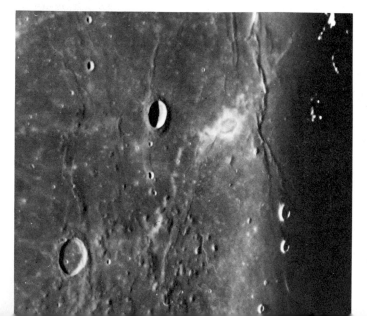

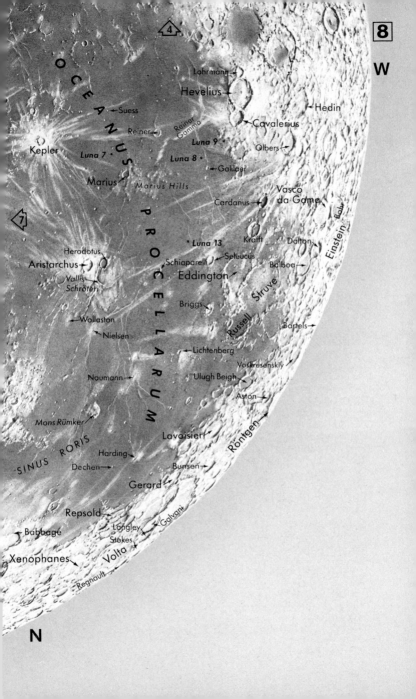

OCEANUS

Lohrmann

Hevelius →

← Hedin

Suess →

Cavalerius

Reiner
Reiner
Gamma

Luna 9 •

Olbers →

Kepler

Luna 7 •

Luna 8 •

Marius

← Galilaei

Marius Hills

Vasco
da Gama

Cardanus →

Sabi

△7

* Luna 13

Krafft

Dalton

PROCELLARUM

Herodotus

Schiaparelli →
Seleucus

Balboa →

Aristarchus →

Eddington

Struve

Vallis
Schröteri

Briggs →

Russell

Wollaston →

Bartels →

Nielsen →

← Lichtenberg

Einstein

Naumann →

Ulugh Beigh →

Voskresenskiy

Aston →

Mons Rümker

Lavoisier →

SINUS RORIS

Harding →

Röntgen

Dechen →

Bunsen →

Gerard →

Repsold →

Longley →
Galvani

← Babbage

Stokes →

Xenophanes →

Volta

Regnault →

The Moon's Far Side

The far side of the moon has been seen only by Apollo astronauts, but it has been photographed both by them and by unmanned spacecraft, so that its topography is quite well documented.

The most obvious difference from the near side is the scarcity of dark maria. Apart from the maria that straddle the limb (Mare Humboldtianum, Mare Marginis, Mare Smythii, Mare Australe, and Mare Orientale), the only named maria on the far side are Mare Moscoviense and Mare Ingenii. The Apollo basin shows signs of flooding by lava flows, as do some of the larger craters near Von Kármán. However, Tsiolkovskiy is the most prominent of the flooded craters because of its isolated location and its apparently extra-dark floor. Even though the far side lacks the degree of lava flooding that exists on the near side, there are several medium to large basins, including Hertzsprung, Korolev, Birkhoff, and so on, all of which display the double-ring structure characteristic of the near-side basins.

The far side also displays some remnants of larger basins. One is centered near the crater Buys-Ballot, where we see a small lava patch; the other is a very large structure centered between the craters Von Kármán and Bose. The presence of this structure accounts for the irregular lava patches and partially flooded crater floors in this region. The rim of this basin extends from about the crater Aitken to a point just beyond the lunar south pole.

The cratered areas of the far side resemble the heavily cratered areas towards the south on the moon's near side, with a range of diameters and degrees of apparent freshness. Among the largest of these fresh craters are Schrödinger (actually almost a basin), Compton, Antoniadi, and Tsiolkovskiy. Some smaller fresh craters include Crookes, King, Ohm, and Giordano Bruno. Ohm is remarkable in that one ray stretches from it right onto the near side, passes north of Aristarchus and ends at Timocharis! On the far-side map, these rays seem incorrectly centered on the nearby crater Robertson. The crater Giordano Bruno, only 14 miles (22 km) in diameter, also has long rays that extend hundreds of kilometers onto the near side, west and north of Mare Crisium; the presence of these rays indicates that the crater is extremely fresh. In fact, the impact that formed this crater may have occurred as recently as 1178 A.D., when Gervase of Canterbury reported that several eyewitnesses had observed some unusual phenomena in the thin crescent moon that might be explained by such an impact and the resulting ejection of powdered material along radial trajectories.

The paucity of maria (lava-flooded areas) on the far side means an almost complete absence of the volcanic-related features such as rilles, domes, vents, etc., that diversify the near-side topography.

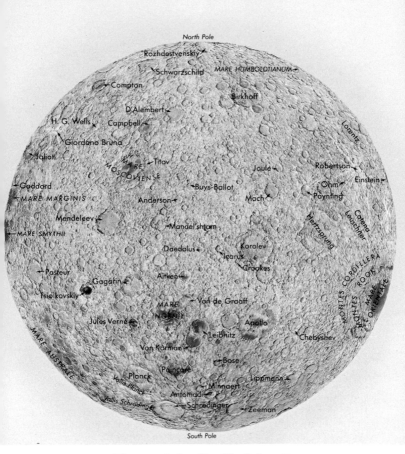

The moon's far side. (North is up.)

Two valleys — Vallis Schrödinger and Vallis Planck — are the chief features of the far side apart from craters and maria. These valleys radiate from the large crater Schrödinger (193 miles — 312 km — in diameter) and are a product of the cratering mechanism. Catena Leuschner is a similarly produced feature extending from the Orientale basin, but it looks like a chain of craters rather than a valley.

9

Finding the Planets

The brightest planets are quite easy to recognize. They often stand out because they are so bright and shine so steadily instead of twinkling. The reddish color of Mars also makes that planet distinctive. When we are observing the sky, these characteristics can help us recognize that we are observing planets instead of stars. The fact that the planets all lie close to the *ecliptic* — an imaginary line drawn across the sky to mark the yearly path of the sun among the stars — also helps us pick out the planets. The ecliptic is marked with a dotted line on the Monthly Sky Maps in Chapter 3 and on the Atlas Charts in Chapter 7.

Most of the planets can be seen with the naked eye; Venus, for example, is often conspicuous as the "morning star" or the "evening star." The planets are most enjoyable to observe, however, when you turn a telescope on them. In the next chapter, we will describe what each planet looks like in a telescope, and what some of the latest observations from space have revealed about the planets. In this chapter, we will show you how to locate the naked-eye planets in the sky, and how to know in advance which planets will be "up" — visible above the horizon — when you plan to observe. Also, if you see a bright object in the sky, you can use the information in this chapter to determine whether it is a planet.

As they orbit the sun, the planets change their positions with respect to the stars. We display the positions of the planets in the sky from night to night in a series of Graphic Timetables of the Heavens at the end of this chapter. The Graphic Timetables are a compact way of displaying the rising, setting, and transit times of the planets during the night; they show the times for sunrise, sunset, and twilight as well. (An object *transits* when it crosses your *meridian,* the line drawn from the north celestial pole through your zenith and then through the point due south on your horizon to the south celestial pole; thus a planet "transits" when it is passing due north or south of you, and is then at its highest in the sky.) Once you have become familiar with the Graphic Timetables, which takes a little practice, you will then be able to determine quickly which objects are well suited for observing at any given time of the year and at any time of night; you will also be able to identify the planets you are seeing in the sky. You can also locate the planets by using the information in Appendix Table A-8 to plot their positions along the ecliptic, which is marked (as a dotted line) on the Monthly Sky Maps in Chapter 3 and on the Atlas Charts in Chapter 7.

We have provided one Graphic Timetable for each year from 1984 through 1989. To use any of the Timetables, scan down the left-hand axis to find the date of the year. Then scan across the top or bottom axis to find the time of night in local time for your time zone. An extra line across the bottom shows the time in Daylight Saving Time.

Look along the horizontal line that corresponds to your date to see which planets can be seen. The time of a planet's transit tells you the time when it will appear due south of you. Before that time, the planet will be to the east along the ecliptic; after that time, it will be to the west along the ecliptic. If a planet sets during the night, it will generally be visible from dusk until it sets. If a planet rises during the night, it will generally remain visible until morning twilight.

The Graphic Timetables in this guide are drawn for observers at latitude 40°N; the correction for observers at different latitudes amounts to only a few minutes, so we do not give it here. The Timetables give rising and setting times for objects assuming a flat horizon; if you have mountains on your horizon, of course, planets will become visible later or will disappear from your view earlier.

Each year's planetary curves are distinct from those of other years, yet distinct patterns appear. The Timetables show the curves even before sunset and after sunrise, when the planets aren't usually visible (though the brighter ones can sometimes be seen), in order to display the patterns better.

The regular cycling of Mercury's rising and setting times as it rapidly orbits the sun is displayed as distorted sine waves oscillating on either side of the sunrise and sunset lines. Venus displays a similar pattern but over a much longer period. Notice that the rising and setting curves for Venus recur every 19.2 months, the period with which Venus returns to successive *superior conjunctions* — when Venus lines up on the far side of the sun for us on earth. (This is Venus' *synodic period,* as opposed to its *sidereal period* — the time it takes to return to the same position in the sky with respect to the stars.) Whenever Venus sets more than $\frac{3}{4}$ of an hour after sunset, you should be able to spot it easily with the naked eye. Mercury and Venus have orbits smaller than earth's, and are known as *inferior planets.*

The *superior planets* have orbits outside the earth's. They can be seen rising and setting far from the sun in the sky, and the curves showing their positions on the Graphic Timetables are more linear. Mars' pattern repeats every 25–27 months, which means that *oppositions* of Mars — when Mars is opposite the sun in our sky, transiting in the middle of the night, 12 hours after the sun has transited — occur at intervals of about 2 years, 2 months. During an opposition, Mars is on your meridian at midnight, when the sun is as far on the other side of the earth (and thus invisible) as it gets. Since Mars' orbit is elliptical, Mars is closer to earth at some oppositions than it is at others. Mars will be especially close to earth, 37 million miles (59 million km) away, at its opposition of September 1988.

Jupiter returns to opposition every 13 months and Saturn every 12.4 months; in other words, in the 12 months the earth takes to complete an orbit of the sun, Jupiter and Saturn have moved ahead a little and it takes us a few more weeks to catch up.

For practice in using the Graphic Timetable, let us pick an example. Look at the night of April 29–30, 1989. Draw a horizontal line from left to right, or follow across the Timetable with your finger. First you find the sunset line; at the bottom, you can see that it comes at about 8 p.m. Daylight Saving Time. (Since Daylight Saving Time is in common use at this time of year, we will give times in D.S.T. for this example.) About half an hour later, you see that Venus sets. Then, at around 10 p.m., Mercury sets. So they have been "evening stars." Another half-hour later, Jupiter sets, and then at midnight, Mars sets. Since the first lines you come to are the settings of these planets, you know that they must have been up since sunset (or else you would have seen the lines mark-

Fig. 108. Seven planets (including the earth) were visible at the same time when this montage of photographs was taken with a 50 mm lens. (Dennis di Cicco/*Sky & Telescope*)

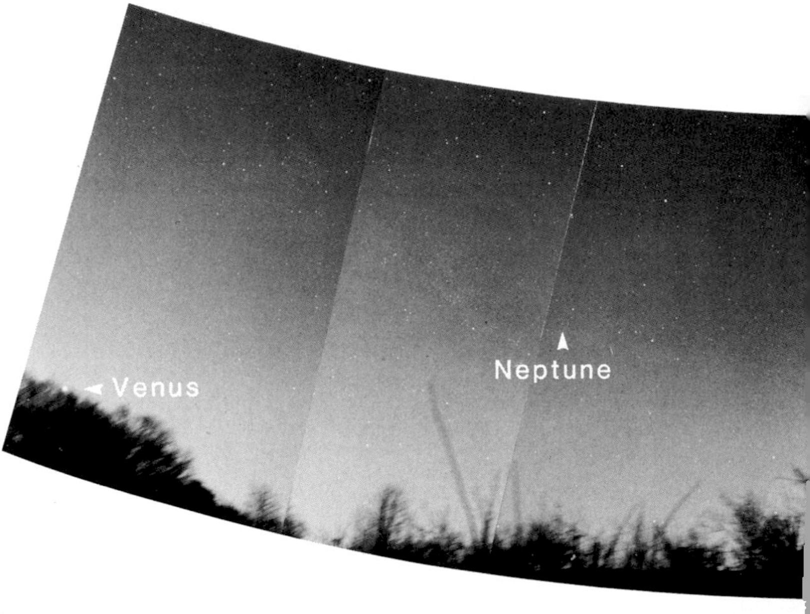

ing their risings). Thus you know that the evening in question would be a good one for observing planets.

Still on the same night, Saturn rises about an hour after midnight. You often have to wait an hour or so before it is high enough to observe. Continuing through the early morning hours of April 30th, morning twilight begins at about 4:15 a.m. D.S.T. and Saturn transits — passing due south of you — at about 5:30 a.m., just before dawn. The times of the risings of Mercury, Venus, and Jupiter are also shown, but they come after sunrise so the planets would not be visible.

You can use a Graphic Timetable in reverse to identify an object in the sky. You must check both the Graphic Timetables of the Planets to see if the object is a planet and the Graphic Timetable of the Brightest Stars (Fig. 2, p. 9) to see if it is a star. If the object is in the east, check the appropriate Timetable in this chapter for a curve representing a recent rising time of a planet. If it is to the south, then search for a transit curve and compare it with those for the brightest stars, in Fig. 2. If the object is to the west, look to see what bright planets are about to set at the time when you are observing.

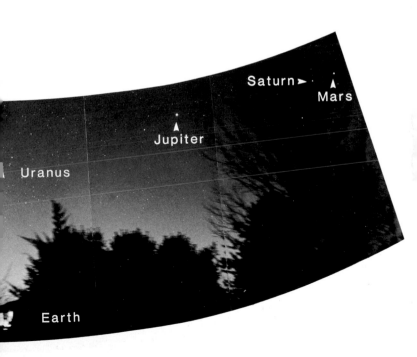

Visibility of the Planets
Robert C. Victor
1984

VENUS begins the year as a "morning star," rising in the southeast more than 3 hours before sunrise. Approaching the sun, Venus rises ever closer to dawn and becomes hard to see by late March. Watch Venus pass Antares on January 8, and very close to Jupiter on January 26 and 27.

Passing behind the sun in mid-June, Venus emerges as an "evening star" in the western sky by early August. Steadily improving in visibility, Venus sets over $3\frac{1}{2}$ hours after sundown at year's end. In the evening sky, watch Venus pass Spica on September 20, Saturn on October 7, and Jupiter on November 24.

MERCURY has three favorable morning apparitions in 1984: in January–February, in September, and in late December. In the morning sky in 1984, watch Mercury swing close to Jupiter on January 13, and close to Regulus on September 9.

In the evening sky, Mercury has an excellent apparition in March–April. A less favorable appearance takes place in July, when Mercury passes close to Regulus on the 25th.

MARS opens 1984 near Spica in Virgo in the morning sky. The red planet then rises after 1 a.m. and is high in the southern sky during morning twilight. Rising earlier each night, Mars slides close to Saturn on February 14 and stops short of the head of Scorpius in early April. On May 11 Mars stands at opposition, 180° from the sun. It appears in the southeastern sky at dusk and is visible all night. Around this date Mars is nearest to earth and reaches its greatest prominence, magnitude -1.7. Mars retrogrades toward Saturn until June 10, and soon resumes its eastward motion against the starry background. In the evening sky for the rest of the year, Mars goes past Antares on September 3 and Jupiter on October 13.

JUPITER, in early January, is just emerging from the sun's glare into the morning sky. Look for it in the twilight glow in the southeastern sky, about 26° to the lower left of Venus on January 1. In Sagittarius all year, Jupiter is soon overtaken by Venus, as mentioned above. Jupiter rises nearly 2 hours earlier each month. By June 29 it stands at opposition, rises in the southeast at sunset, remains above the horizon all night, and sets at sunrise. Jupiter is then at its brightest for the year (magnitude -2.2). Jupiter soon becomes exclusively an evening object.

SATURN is in the constellation Libra, where it passes the 3rd magnitude star alpha Librae three times in 1984: on January 12, April 8, and October 5. As the year opens, Saturn rises around 3 a.m. and is approaching the south as morning twilight begins. Rising 2 hours earlier each month, Saturn reaches opposition — all-night visibility — on May 3. The planet then shines at magnitude $+0.3$ and its rings are tilted 20° from edge-on. Saturn remains an evening object through summer into early fall.

THE GRAPHIC TIMETABLE OF THE HEAVENS

1984

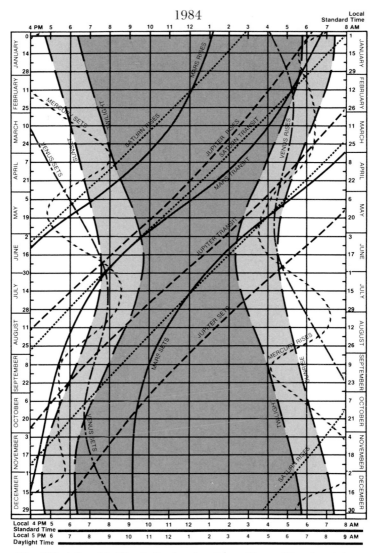

Fig. 109. A Graphic Timetable showing the positions of the planets in 1984. (© 1982 Scientia, Inc.)

1985

VENUS begins the year as a fine "evening star." It reaches 47° elongation — its greatest angular distance from the sun — on January 22 and sets nearly 4 hours after sundown. Venus and Mars remain within 5° of each other from January 12 to February 26, and the two planets appear closest on January 28. Venus rapidly gets lower in the western sky during March. Swinging between earth and sun in early April, Venus leaves the evening sky and becomes visible in the east before sunrise as the "morning star." On June 12 it reaches greatest elongation, 46° from the sun.

In the morning sky, Venus passes the Pleiades on July 6, and Aldebaran on July 15. Rising as much as 3 hours before the sun in early August, Venus goes by Pollux on the 22nd. Gradually moving toward the sun, Venus passes close to Regulus on September 21, close to Mars on October 4 and 5, and near Spica on November 4. By mid-December Venus is difficult to pick out in the sun's glare. It will pass behind the sun in January 1986.

MERCURY has three favorable morning apparitions in 1985: in January, in August–September, and in December. On September 4–6 Mercury forms a tight grouping with Mars and Regulus. On December 16 Mercury passes close to Saturn.

In the evening sky, Mercury has a fine apparition in March and a good one in June–July. On June 24–25, Mercury passes Pollux.

MARS opens the year 1985 in the western evening sky, not far from Venus. As the months pass, Mars sets earlier in the evening. Watch Mars pass the Pleiades on April 25 and Aldebaran on May 9. Later in May, Mars becomes too difficult to observe.

Mars is in conjunction — on the far side of the sun — in mid-July. By the end of August it is visible to the lower left of Mercury, very low in the eastern sky before sunrise. On September 4 Mercury is very close to Mars, and on September 9 Mars passes close to Regulus. Rising earlier each morning, Mars is within 5° of Venus from September 26 to October 12. Mars passes near Spica on December 4, and ends the year in Libra.

JUPITER is in conjunction with the sun in January and not visible. In mid-February it emerges into the southeastern morning sky and appears in Capricornus for the rest of the year. Rising earlier each month, Jupiter reaches opposition on August 4. Shining at magnitude −2.4 in the southeast at dusk, it is visible all night. Jupiter remains an evening object for the rest of the year, but sets earlier each month.

SATURN opens the year in the morning sky. Rising earlier each month, Saturn reaches opposition on May 15 at magnitude +0.2 in Libra. A telescope shows the rings tilted 23° from edge-on. Saturn sets earlier each month and remains in the evening sky until October, when it sets in the west-southwest at dusk. In conjunction with the sun in late November, Saturn emerges into the southeastern morning sky in December.

THE GRAPHIC TIMETABLE OF THE HEAVENS

1985

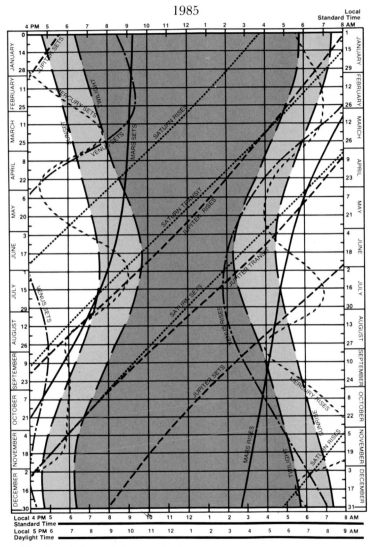

Fig. 110. A Graphic Timetable showing the positions of the planets in 1985. (© 1982 Scientia, Inc.)

1986

VENUS is at superior conjunction (beyond the sun) in January, and is not visible. By late February it sets late enough after the sun to be seen as an "evening star" low in the west. Venus passes the Pleiades on April 26, Aldebaran on May 4, Pollux on June 9, Regulus on July 11, and Spica on August 31. Venus sets as much as $2\frac{1}{2}$ hours after the sun in early June and reaches greatest elongation, 46° from the sun, on August 27. Heading back into the sun, Venus remains in the evening sky until mid-October. Venus swings through inferior conjunction on November 5 and within a week becomes a "morning star" in the southeast. By late December it rises more than $3\frac{1}{2}$ hours ahead of the sun.

MERCURY has a very favorable evening apparition in late February, and another good one in June. In the morning sky, Mercury is best seen in very early January, in August, and in November–December.

MARS opens the year in Libra in the morning sky. Mars passes between Saturn and Antares in mid-February. The red planet rises earlier each night, and stands at opposition on the night of July 9–10, when it is above the horizon virtually all night. At a distance of 38 million miles (0.4 A.U.), Mars is closer to earth at this opposition than it has been since 1971, and gleams brilliantly at magnitude −2.4. Its large disk, 23 arc seconds across, is poorly seen from midnorthern latitudes because Mars is far to the south in Sagittarius and does not rise very far above the southern horizon. Fading as it recedes from earth, Mars continues as an evening object for the rest of 1986. On December 18 Mars passes close to Jupiter. At year's end Mars is high in the southern sky at dusk, in Pisces.

JUPITER, as the year opens, is in the southwestern sky at dusk, in Capricornus. By early February Jupiter disappears into the solar glare and reaches conjunction with the sun on February 18. By late March it reappears in the east-southeast before sunrise. In Aquarius, Jupiter rises earlier each month, and by September 10 it rises at sunset. Then at opposition, Jupiter shines at magnitude −2.5 and is visible all night. Jupiter remains an evening object through the end of the year, when it is high in the south at dusk.

SATURN through this year remains within 6°–9° from Antares, the reddish star in the heart of Scorpius. In an event called a triple conjunction, Saturn passes about 6° north of this star three times in 1986: on March 6, on April 1, and on November 12. As the year begins Saturn is in the head of Scorpius, in the morning sky. Saturn rises about 2 hours earlier each month until it reaches opposition — all-night visibility — on May 27. The planet then appears at magnitude +0.2 in the southeast at dusk; a telescope shows the rings tilted 25° from edge-on. The planet remains in the evening sky until early November, when it sets in the west-southwest in twilight. Passing solar conjunction on December 4, this planet emerges into the morning sky late in December.

THE GRAPHIC TIMETABLE OF THE HEAVENS

1986

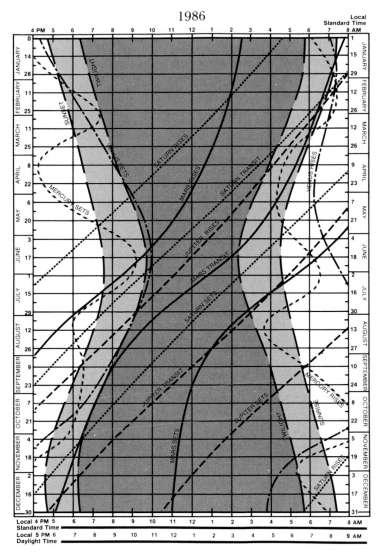

Fig. 111. A Graphic Timetable showing the positions of the planets in 1986. (© 1982 Scientia, Inc.)

1987

VENUS opens the year as a "morning star" in the southeastern sky for more than $3\frac{1}{2}$ hours before sunrise. On January 15 it reaches greatest elongation, 47° from the sun, and through telescopes appears as a tiny "half moon." Thereafter Venus approaches the sun, rises ever closer to dawn, and becomes hard to see by July. Watch Venus pass Antares on January 17, Saturn on January 25, Jupiter on May 5, and the Pleiades on June 11.

Going behind the sun in late August, Venus becomes visible as an "evening star" by late October. Venus overtakes Saturn on November 20 and by year's end sets $2\frac{1}{2}$ hours after the sun.

MERCURY has two favorable evening apparitions in 1987: in January–February and in May–June. Mercury comes within 13° of Jupiter on February 16 and within 1° of Mars during June 8–12. In morning twilight, Mercury can be found low in the east-southeast, to the lower left of Mars and Spica, during November. This apparition is the best of the year.

MARS, as the year begins, is in Pisces, high in the southern sky at dusk. Mars is still of magnitude 0.8 on January 1 and appears just 8° east of Jupiter. Drifting toward the western sky and setting earlier as the months pass, Mars remains visible in the evening sky for the first half of the year. Watch Mars pass the Pleiades on April 5, Aldebaran on April 19, and Pollux on June 25. Around that date Mars fades into the twilight glow in the west-northwest.

Mars emerges from the sun's glare in early October, when it rises due east in morning twilight. Rising ever further ahead of the sun, Mars passes Spica on November 14 and alpha Librae on December 16.

JUPITER begins the year as the most brilliant "evening star," high in the south-southwest at dusk. Jupiter crosses from Aquarius into Pisces before the end of January, and remains visible until early March, when it sets in the west in twilight.

Jupiter emerges into the morning twilight in the east in early May and is passed by Venus on May 5. As months pass, Jupiter rises progressively earlier in the night. Although it enters Aries in July, Jupiter is back in Pisces when it is at opposition on October 18. On that date Jupiter is in the east at dusk and remains above the horizon all night. Closer to earth than it will be again in this century, Jupiter gleams at magnitude -2.5 and shows a disk 50 arc seconds across.

SATURN is in Ophiuchus all year. On January 1 it is low in the southeastern sky at dawn and appears to the lower left of Venus, which overtakes it on January 25. Saturn rises nearly 2 hours earlier with each passing month, until it reaches opposition on June 9. The 0.2-magnitude planet then rises in the southeast at sunset and is up all night. The rings are tilted more than 26° to our line of sight, nearly the maximum possible. Saturn remains in the evening sky until it fades into the southwestern twilight in late November.

THE GRAPHIC TIMETABLE OF THE HEAVENS

1987

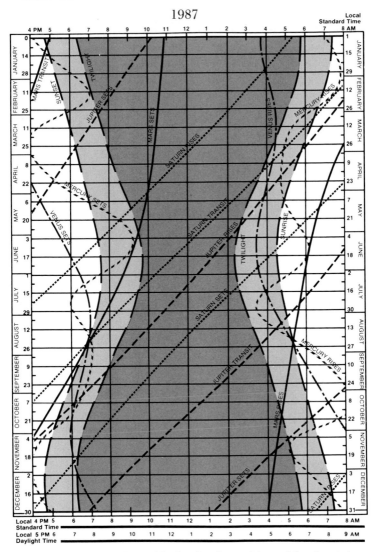

Fig. 112. A Graphic Timetable showing the positions of the planets in 1987. (© 1982 Scientia, Inc.)

1988

VENUS this year is an "evening star" in the western sky from January 1 through early June, and sets as much as 4 hours after the sun during April. Watch Venus go past Jupiter on March 5 and pass within 1° of the Pleiades on April 2 and 3. On April 3 Venus stands at greatest elongation, 46° from the sun. Continuing eastward among the stars, Venus passes Aldebaran on April 13 and stops near Gemini's feet on May 22. Venus appears in the morning sky from late June on. On August 22 Venus reaches greatest elongation again and rises more than $3\frac{1}{2}$ hours before the sun. Venus goes past Pollux on August 31, Regulus on October 4, Spica on November 18, and Antares on December 25.

MERCURY has two favorable evening apparitions in 1988: in January, when it is low in the west-southwest at dusk, and in May, low in the west-northwest.

In morning twilight, Mercury can be seen very low in the east-southeast in February–March, very low in the east-northeast in July, and low in the east-southeast in October and November. In July, October, and November Mercury appears to the lower left of Venus. The October–November apparition is the best.

MARS, as the year begins, is a 1.7-magnitude "morning star" in Libra, in the southeast at dawn. With each passing month it rises earlier and becomes brighter. Mars passes Antares on January 22, then goes past Saturn in Sagittarius on February 23. Mars reaches opposition on September 27, when it is low in the east at dusk and visible all night. Only 37 million miles from earth, the planet gleams at magnitude −2.5 and shows a red disk nearly 24 arc seconds across. Mars won't be as bright or as close to earth again until the year 2003.

JUPITER in January is in Pisces, high in the southern sky at dusk. In the following months Jupiter drifts toward the western sky and sets earlier. It is overtaken by Venus on March 5. Crossing into Aries, Jupiter remains an evening object until mid-April.

Jupiter reappears in the east-northeast during morning twilight by early June. Shortly afterward, Jupiter crosses into Taurus, where it remains for the rest of the year. Rising earlier each month, Jupiter passes the Pleiades on July 20 and stops short of Aldebaran on September 24.

Retrograding for the next 4 months, Jupiter reaches opposition on November 22, when it shines at magnitude −2.4 and is up all night. Jupiter passes the Pleiades again in early December. At year's end it is high in the east at dusk.

SATURN emerges from the sun's glare into the southeastern morning sky at the beginning of January. Moving into Sagittarius, Saturn rises 2 hours earlier with each passing month. Saturn is overtaken by Mars on February 23. Reaching opposition on June 20, 0.2-magnitude Saturn is in the southeast at dusk and is visible all night. This year the rings will be open to their maximum tilt, nearly 27° from edge-on.

THE GRAPHIC TIMETABLE OF THE HEAVENS

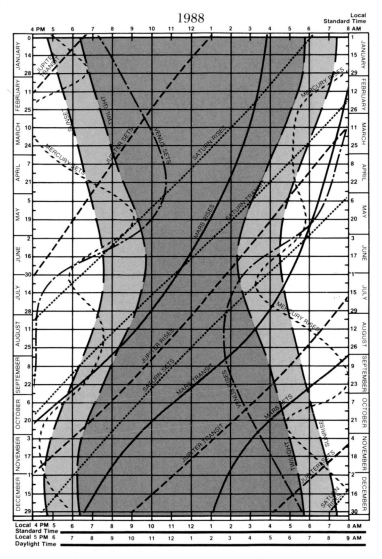

Fig. 113. A Graphic Timetable showing the positions of the planets in 1988. (© 1982 Scientia, Inc.)

1989

VENUS is a "morning star" low in the southeast at dawn for the first month of 1989. Venus passes Saturn on January 16; with each passing day, Venus appears lower in the sky and closer to the sun, while Saturn gets higher.

Venus goes beyond the sun on April 4 and by mid-May emerges into the west-northwest at dusk. Now an "evening star," it remains quite low for several months. Watch Venus pass Jupiter on May 22, Pollux on June 23, Mars on July 12, Regulus on July 23, Spica on September 6, and Antares on October 16. On November 8, Venus reaches greatest elongation, 47° from the sun. At this point Venus sets farther south than it will again until 1997. Venus goes past Saturn on November 15. Setting over 3 hours after sundown in early December, Venus reaches its best position in the sky of the year. Binoculars and telescopes show Venus as a crescent getting larger and thinner as weeks pass.

MERCURY has three good evening apparitions this year: in early January; in April–May (very favorable — look to lower right of Aldebaran); and in December. In the morning sky you can see Mercury in February, June, and October. The last of these apparitions is the best.

MARS opens this year as a zero-magnitude "evening star" in Pisces. High in the southern sky at dusk in January, Mars shifts to the southwest and fades to 1st magnitude by mid-February. Mars passes between Jupiter and the Pleiades on March 11 and on past Aldebaran on March 26. Continuing eastward through the zodiac, Mars goes past Pollux on June 5. By July Mars is very low in the west-northwest at dusk and fades to magnitude 2.0.

By early November Mars emerges as a faint 2nd-magnitude object during morning twilight in the east-southeast, to the lower left of Spica. At year's end Mars is just 5° north of Antares in the southeast at dawn.

JUPITER opens 1989 in Taurus, high in the eastern sky at dusk. On March 9 Jupiter passes the Pleiades, and 2 days later is overtaken by Mars. Jupiter passes Aldebaran on May 1 and fades into the twilight glow in the west-northwest later that month.

Jupiter emerges as a "morning star" in the east-northeast during morning twilight by early July. Jupiter rises progressively earlier until it reaches opposition on December 27. Shining at magnitude −2.3, Jupiter is then above the horizon all night.

SATURN emerges in mid-January as a "morning star" low in the southeast at dawn. In Sagittarius all year, Saturn rises about 2 hours earlier with each passing month until it reaches opposition on July 2. Then at magnitude +0.2, Saturn is in the southeast at dusk and is up all night long. Its rings are tilted 25° from edge-on.

Saturn remains an evening object until mid-December, when it disappears into the twilight glow in the southwest.

THE GRAPHIC TIMETABLE OF THE HEAVENS

1989

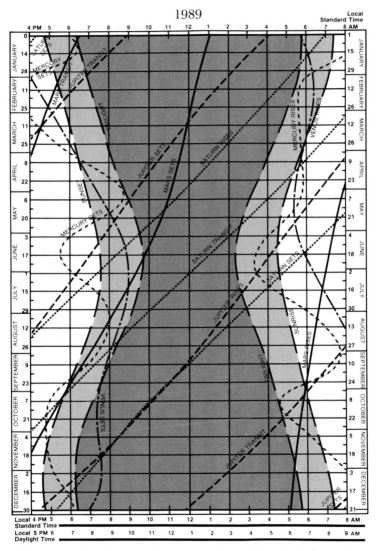

Fig. 114. A Graphic Timetable showing the positions of the planets in 1989. (© 1982 Scientia, Inc.)

10

Observing the Planets

In the previous chapter, we explained how to find the planets in the sky. This chapter deals with observing the planets themselves. With even a small telescope, you can see that the planets have shapes and that the apparent shapes of some of the planets change over time. Since the easiest and most impressive planets to observe (in order) are Jupiter, Saturn, Venus, and Mars, we describe the planets in that order here, then discuss the planets that are harder to see — Mercury, Uranus, Neptune, and Pluto.

JUPITER

Jupiter, the largest planet in the solar system, is 11 times the diameter and over a thousand times the volume of the earth. It is about 5.2 times as far away from the sun as the earth is; since the average distance from the earth to the sun is called 1 Astronomical Unit (1 A.U.), Jupiter averages 5.2 A.U. from the sun. When Jupiter is on the same side of the sun as the earth, it is high in the nighttime sky, and is only about 4 A.U. from us (5.2 A.U. minus the earth's orbital radius of 1 A.U. equals 4.2 A.U.). At this distance, the planet covers almost 50 seconds of arc in the sky — $\frac{1}{40}$ the diameter of the full moon — and can be about magnitude -2.5, three times brighter than the brightest star, Sirius.

When a planet is on a line with the earth and sun, the planet and sun appear close together in the sky (C.Pl. 74) and we say that the planet is in *conjunction*. (The actual time of conjunction is when the planet and the sun appear at the same longitude along the ecliptic (see Appendix Table A-8). Planets can also be in conjunction with each other or with the moon. Jupiter is in conjunction with the sun only when it is on the far side of the sun and consequently appears smaller than it does when it is at opposition.

A planet is at *opposition* when it is on the extension of the line of sight from the sun through the earth. Jupiter and the other outer planets are best observed at opposition, when they are above the horizon all night. Oppositions of Jupiter occur in June 1984, August 1985, September 1986, October 1987, November 1988, and December 1989.

Fig. 115. Jupiter, photographed from the earth, showing its bright horizontal zones and its dark horizontal belts. South is at top, since this is a view through a telescope. The four Galilean satellites are also visible. (Lunar and Planetary Laboratory/University of Arizona)

Jupiter's Disk

With a small telescope, you can see light and dark bands across Jupiter's disk. The number of bands you see depends not only on how close Jupiter is to earth at the time but also on how steady the atmosphere is for your observing location. The bands represent clouds drawn out into long streaks as Jupiter rotates; the planet, which is made entirely of gas, rotates at different speeds at different latitudes. (Solid bodies like the earth don't do this.) Gas near the pole rotates about 5 minutes faster than gas near the equator does during each 10-hour rotation period. (Planets rotate — spin — on their axes but revolve — orbit — around the sun.)

Figure 115 is an outstanding ground-based photograph of Jupiter; it represents the most detail that is available, even through a large telescope, because of the obscuring effects of the earth's atmosphere. The form of the bands changes slowly over long periods of time. The major bands have been given names, as shown in Fig. 116. The bright bands are called *zones* and the dark bands are called *belts* (C.Pl. 65). The belts and zones show subtle colors.

Jupiter's rotation causes it to bulge at the equator — to be

Fig. 116. A schematic diagram showing the major features of Jupiter.

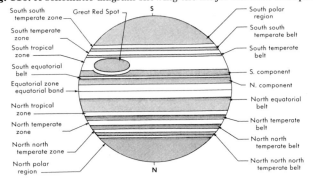

South south temperate zone

Great Red Spot

S

South polar region

South temperate zone

South south temperate belt

South tropical zone

South temperate belt

South equatorial belt

S. component

Equatorial zone equatorial band

N. component

North tropical zone

North equatorial belt

North temperate zone

North temperate belt

North north temperate zone

North north temperate belt

North polar region

North north north temperate belt

N

oblate — by 7%. It is obvious in even a small telescope that Jupiter is out-of-round.

A large reddish region known as the Great Red Spot has been observed on Jupiter from time to time for hundreds of years. The Great Red Spot is about 14,000 by 30,000 km, much larger than the earth. The Great Red Spot is a giant storm in Jupiter's clouds, which is more visible at some times than others. Sometimes its shape changes. Close-up views taken from the Voyager spacecraft revealed the rotation of the Great Red Spot (Fig. 117, C.Pl. 63).

Though the Voyagers discovered a thin ring around Jupiter (Fig. 118), it is too faint to be detected by ordinary means from earth.

A NASA mission called Galileo is planned for the end of the 1980s. Its goal is to launch a probe that will parachute through Jupiter's clouds and send back data for about an hour. It will also put a spacecraft in orbit around Jupiter for a couple of years.

Jupiter's Moons

The four brightest moons (satellites) of Jupiter can be seen with even a small telescope. Named Io, Europa, Ganymede, and Callisto, after mythological lovers of Jupiter (as are all of Jupiter's more than 16 moons), they range from 3100 to 5300 km in diameter. All but one of the four brightest moons are larger than our own moon. The largest of Jupiter's moons, Ganymede, is even larger than the planet Mercury.

Nonetheless, these four "Galilean satellites" were merely points of light in the sky until NASA's two Voyager spacecraft flew nearby in 1979 and 1980, radioing back close-up pictures of the moons' surfaces (C.Pl. 64). Io is covered with volcanoes, of which about 10 are now erupting (Fig. 119); it is the most active body in the solar system. Europa is apparently covered with ice. Ganymede — which has the largest surface diameter of any moon in the solar system — may contain water and ice surrounding a core of rock; its surface shows craters and strange grooves (Fig. 120). Callisto is covered with craters, including a huge bull's-eye called Valhalla.

Fig. 117. A close-up of Jupiter's Great Red Spot and the turbulence surrounding it. North is up on this and other photos taken from space. (JPL/NASA)

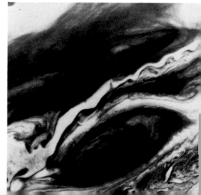

Fig. 118. Jupiter's ring, photographed from Voyager 2. (JPL/NASA)

These four bright moons orbit in the plane of Jupiter's equator, so they always seem stretched out close to a line drawn through Jupiter. The period of Io's orbit is 1 day, 18 hours; Europa's period is 3 days, 13 hours; Ganymede's is 7 days, 3 hours; and Callisto's is 16 days, 16 hours. If you observe these moons in a telescope over a night or from night to night, you can see their relative positions change. We sometimes see the moons move in front of or behind Jupiter; when the moons move in front of Jupiter, their shadows fall on the planet. When the moons are at their brightest, Ganymede can reach magnitude 4.6, and the others can be no more than a magnitude fainter.

Each year, the positions of the Galilean moons in relation to the sides of Jupiter are graphed in *The Astronomical Almanac,* and the graphs are reprinted in the magazines *Sky & Telescope* and *Astronomy.* Since separate graphs are needed for each month, we will not take space in this *Field Guide* to print the maps here.

(*Text continues on p. 371.*)

Fig. 119. (*left*) Volcanoes erupting hundreds of kilometers high above the surface of Io, a satellite of Jupiter. (JPL/NASA)
Fig. 120. (*right*) A computer-enhanced photo of Ganymede's surface, showing some of its craters (the dark areas) and grooves. (JPL/NASA)

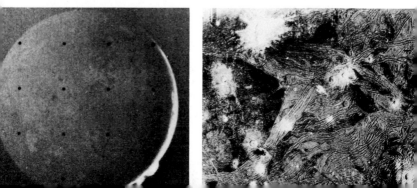

Table 18. Identifying the Moons of Jupiter

Jupiter's moons move in orbits that appear to take them from side to side across Jupiter's image. You can use a calculator that has a sine function to figure out which moon is which at any time by using the following formula:

$x = A \sin (2\pi D/B + C)$, where D is the number of days since January 0, 1984 at 0.00 U.T., and A, B, and C are tabulated below. (January 0, 1984, is the same as December 31, 1983; it is convenient to use January 0 instead of January 1 since then it is not necessary to subtract 1 from the date to get the day number: $D = 24$ for January 24. $D = 365$ for December 31 for most years but $D = 366$ for December 31 for 1984 and 1988, which are leap years. Don't forget to convert U.T. to local standard time by subtracting 5 hours for E.S.T., -6 for C.S.T., -7 for M.S.T., -8 for P.S.T., -9 for Alaska S.T., and -10 for Hawaii S.T. Many calculators have a key for $\pi = 3.1415926536. \ldots$

Note that the parenthesis in the formula is expressed in radians, so that your calculator must be set to the radian mode if you have it. If not, and your calculator takes the sine of arc measure, π radians $= 180°$, so that you must multiply the parenthesis by $180°/\pi$ before taking the sine. Thus $x = A \sin 180°/\pi \, (2\pi D/B + C_r) = A \sin (360 D/B + C_d)$, if you are calculating in degrees of arc. (The subscript d is for degrees and r is for radians.)

To the number of the day in the year, add 366 for 1985, 731 for 1986, 1,096 for 1987, 1,461 for 1988 (a leap year), and 1,827 for 1989. You may use fractions of a day; for example, 18:00 U.T. on January 5 would be day 5.75 (5 for January 5th plus 0.75 for the fraction of a day).

You can plot the x values you calculate along a horizontal line, from -100 to $+100$. They give the lineup of satellites relative to each other. Jupiter's diameter is 4 on this scale.

(A is the size of the orbit relative to Callisto's orbit, which is set to 100 units; B is the period in days; and C is the phase of the orbit, that is, when its curve crosses a zero point set at Jupiter's center.)

Object	A	B	C_r	C_d
Io	22	1.76986049	0.8533	48.89
Europa	36	3.55409417	-1.4482	82.98
Ganymede	57	7.16638722	5.2447	300.5
Callisto	100	16.75355227	1.3642	78.16

Example: for 0:00 U.T. on January 8, 1984, $D = 8.0$. Io $= -18$; Europa $= +4$; Ganymede $= -17$; Callisto $= -94$. Thus Io and Ganymede would be slightly to one side of Jupiter with Callisto (which has the largest orbit of the four Galilean satellites) much farther out on that side. Europa would be very close to Jupiter on its other side. Before you calculate the configuration for your date, check to see that you agree with this calculation for January 8, 1984, in order to test the adaptation of the method for your particular calculator. (Don't worry about the decimals or round-off error. Values rounded off to the nearest 5 or 10 should be sufficient for identification in most cases.)

Table 18, however, contains a formula that you can easily use with a pocket calculator to calculate how Jupiter's moons will appear on any specific time and date.

Observing Jupiter

Count, sketch, and note the colors of belts and zones on the disk. Observe and sketch the Great Red Spot and perhaps other disturbances in Jupiter's atmosphere; notice the oblateness of Jupiter's disk. Observe and identify the Galilean satellites and how they move; observe their occultations by Jupiter (when they pass behind Jupiter's disk), their eclipses, and the transits of the moons and their shadows across Jupiter's disk.

SATURN

Galileo, with his tiny telescope, could see that Saturn wasn't quite round. Some decades later, astronomers realized that Saturn was surrounded by a ring (Fig. 121, C.Pl. 66). As the decades and centuries passed, we realized that Saturn had more and more rings. In the 1970s, we knew of half a dozen rings; now, with the results from the Voyager flybys, we know of hundreds of thousands of rings (Fig. 122).

Saturn is 9.4 times the diameter of earth, almost the size of Jupiter. Its ring extends out much farther, up to 135,000 km from Saturn's center.

Saturn is 9.5 A.U. from the sun, which means that at best it is twice as far from earth as Jupiter is. Its maximum size is thus

Fig. 121. Saturn, photographed from the earth through a 24-in. telescope. The largest dark gap in the rings is Cassini's division, which separates the outer ring (the A-ring) from the bright middle ring (the B-ring). The C-ring, still further in, is too faint to show in this photo. South is up. (New Mexico State University Observatory)

about 20 seconds of arc, and it never gets brighter than about zero magnitude. Oppositions of Saturn occur in May 1984, May 1985, May 1986, June 1987, June 1988, and July 1989.

Saturn's Rings

Even small telescopes reveal Saturn's rings. They are often the first objects an amateur looks at, and they are certainly among the first things shown to novices. The rings are so glorious that professionals like to look at them too. When you are observing Saturn (or any of the other planets), you will find that waiting for steady air (good "seeing") will give you better images.

The main division between Saturn's bright middle ring (the *B-ring*) and fainter outer ring (the *A-ring*) is called *Cassini's division,* after its 17th-century discoverer. A faint inner *crepe ring* (the *C-ring*) is hard to see. Though Cassini's division looks dark when seen from earth, and therefore resembles a gap in the rings, the Voyagers revealed that it really contains bits of opaque material. A gap called *Encke's division* runs through the A-ring, though it is extremely hard to detect even with large telescopes.

Color Plates 67 and 68 show Voyager views of Saturn and its rings. On close examination, Saturn's rings break up into thousands of tiny "ringlets" (Fig. 122). The rings and ringlets consist of chunks of rock, ranging in size from pebbles to huge boulders, each independently orbiting Saturn.

Saturn's rings and equator are inclined by 27° to Saturn's orbit, so we see them from different aspects at different times (Fig. 123). The angle at which we see the rings varies over a 30-year period. The earth last passed through the plane of the rings, when the rings become invisible, in 1979–80, and so they will be widening to our view until 1987–88. They will again appear edge-on in 1995–96.

Fig. 122. A contrast-enhanced photo showing about 95 of Saturn's rings, taken from Voyager 1. Voyager 2 revealed hundreds of thousands of such "ringlets." (JPL/NASA)

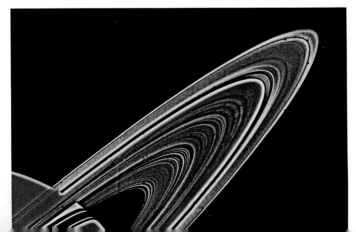

Fig. 123. The rings of Saturn as seen from earth over a period of time. (Lowell Observatory)

Saturn's Disk

Saturn's disk, out-of-round by 10%, shows less contrast in its bands than Jupiter's disk does. The effect may be because Saturn is farther from the sun and thus colder; the cooler temperatures may slow down the chemical reactions that create the colors in the clouds.

Saturn's Moons

Saturn boasts of the second largest moon in the solar system, Titan. With its clouds included, Titan is 5150 km (3625 miles) in diameter, nearly half the size of earth, and is surrounded by an atmosphere that is even thicker than earth's. But Titan is so far away that sunlight has to travel much farther to reach it than to reach Jupiter's moons and then must travel much farther back to earth. As a result, Titan never becomes brighter than about 8th magnitude; it can, however, be seen as a small, starlike point of light with even a small telescope.

Saturn has another dozen moons over 100 km across, and many smaller ones (Table A-7). Voyager close-ups of some of them appear in C.Pl. 68. A mnemonic for the major moons and their order of distance from Saturn — Mimas, Enceladus, Tethys, Dione, Rhea, Titan, Hyperion, Iapetus, and Phoebe — is "Met Dr. Thip," remembering that Titan is the second T.

Observing Saturn

Observe the rings and their orientation; observe and sketch Cassini's division; try to observe the bands on Saturn's disk.

Observe Titan, Saturn's largest moon, and plot its changes in position during its 16-day orbital period; see how many other satellites you can observe (such as Rhea, Tethys, and Dione) with, say, a 6-inch telescope.

VENUS

Venus can be the brightest object in the sky besides the sun and moon. Since its orbit is inside that of the earth, we see Venus only when we are looking in the general direction of the sun. This bright planet is visible only during the first few hours after sunset, when it is known as "the evening star," or before sunrise, when it is known as "the morning star." Venus can be brighter than magnitude −4, and can even cast shadows.

Venus is covered with thick layers of clouds through which we cannot see. From earth, we see no structure, though we do see Venus go through phases as it orbits the sun. (Only planets that have orbits smaller than the earth's — Venus and Mercury — can go through a crescent phase. The fact that Venus goes through a complete cycle of phases, including the crescent phase, was discovered by Galileo and was a major proof of the validity of Copernicus' idea that the sun rather than the earth is at the center of the solar system.)

Notice in Fig. 124 how, when Venus is just about to pass between the sun and us, it appears as a crescent and is at its largest. We may even see sunlight bent around toward us through its thick atmosphere. When Venus appears at the same longitude along the ecliptic as the sun, it is at *conjunction*. When Venus lies nearly on a line of sight between us and the sun, it is at *inferior conjunction*. When we can see all of Venus' lighted side (its "full" phase), the planet is on the far side of the sun, at its farthest point from us and therefore at its smallest. It is then at *superior conjunction*.

From spacecraft looking through ultraviolet filters from the vicinity of Venus, we have been able to study the circulation of the planet's clouds. The Pioneer Venus Orbiter in 1979 even compiled a map of Venus' surface, using radar (C.Pl. 58).

A series of Soviet spacecraft have landed on Venus and sent back photographs of its surface (Fig. 125). From studies of the composi-

Fig. 124. The phases of Venus. Note the different size of Venus at its various phases. Whenever we see a crescent, Venus must be on the near side of the sun; as a result, it appears relatively large. (New Mexico State University Observatory)

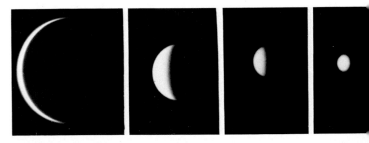

Fig. 125. The surface of Venus, photographed in 1982 by a Soviet lander, showed flat rocks and soil. The curved swath photographed shows horizons at upper left and upper right, and looks down at the base of the spacecraft in the center.

tion of the surface, we know that the same kind of geological processes that formed the earth's surface also were at work on Venus. The earth, though, has several continents and is mostly covered with deep ocean basins. Venus, on the other hand, has only a few small continents and few deep basins; it is mostly covered with a broad rolling plain.

Venus' clouds consist primarily of sulfuric acid droplets. Its atmosphere is mostly carbon dioxide, and the surface pressure is 90 times that on earth. The earth contains about the same amount of carbon dioxide, though on earth it is locked up in carbonate rocks formed under the ocean. So the presence of water on earth apparently saved us from undergoing Venus' fate.

The carbon dioxide in Venus' atmosphere traps sunlight, which enters mostly as visible light but is changed to infrared radiation when it heats Venus' surface. The infrared can't escape, mostly because of the carbon dioxide but also somewhat because of other gases and particles, so the atmosphere heats up to a temperature of $500°C$ ($900°F$) on Venus' surface. This effect is known as the "greenhouse effect." If we unbalance our earth's atmosphere in some way, perhaps by burning too much fossil fuel and thus putting too much carbon dioxide into the atmosphere, our atmosphere could become as unlivable as Venus'.

Venus' atmosphere also teaches another lesson about air pollution. If we introduce too many fluorocarbons into the earth's atmosphere by using aerosol cans that contain them or by leakage of air conditioner or refrigerator coolant, we could destroy a lot of the earth's ozone layer, which protects us from the sun's ultraviolet light. The effects of ozone and other gases are now better understood through comparative studies of the atmospheres of the earth, Venus, and other planets.

Transits of Venus

Only rarely does Venus *transit* — go directly in front of — the sun. It is then visible as a black dot projected on the solar disk; light bent forward to us by Venus' atmosphere makes a bright ring around the planet. Historically, transits of Venus have been important for setting the distance scale in the solar system; we can now find the distance scale more accurately with radar and by tracking spacecraft, so transits of Venus are now merely an observational curiosity.

Transits of Venus come in pairs separated by 8 years; the interval between successive pairs is more than 100 years. The last transits were in 1874 and 1882. The next will be on June 8, 2004, and June 5–6, 2012.

Observing Venus

Observe the phases, especially during the 10 weeks before and after inferior conjunction, when Venus switches from an "evening star" to a "morning star." An ultraviolet filter sometimes adds contrast to photographs of the disk.

MARS

Mars, "the red planet," has long been an object of interest to humanity because of its relatively rapid motion among the stars and because of its reddish color. Over a period of months, Mars' path among the stars apparently reverses itself in a giant loop (Fig. 126). In 1543, Copernicus explained this loop in Mars' orbit, called *retrograde* (backwards) *motion,* by showing how it is an effect of perspective. It occurs when the earth passes Mars, as both planets orbit the sun. The orbits of the other planets have similar retrograde loops.

Mars' surface, when viewed from the earth, shows semiperma-

Fig. 126. A planetarium simulation of the path of Mars through the regions of Libra and Scorpius from approximately January 1984 through September 1984, showing the retrograde loop in its orbit. (Allen Seltzer, American Museum-Hayden Planetarium)

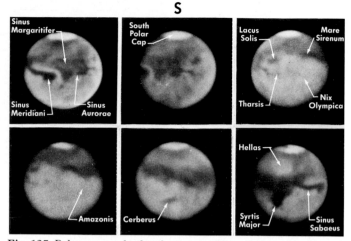

Fig. 127. Polar caps and other features on Mars that can be observed from earth. South is at top. (International Planetary Patrol/Lowell Observatory)

nent features (C.Pl. 60), which can be followed over the 25-hour period of rotation of Mars, a Martian "day." These features change with the seasons of the year on Mars; the cycle of Martian seasons takes 23 earth months, the period with which Mars orbits the sun. Mars' reddish tinge is obvious even to the naked eye and certainly with a small telescope, but little detail on Mars' surface can be seen from earth (Fig. 127).

Because of Mars' elliptical orbit, at different oppositions Mars is at different distances from us. The angular diameter of its disk at opposition ranges from about 14 arc sec to 25 arc sec, which is $\frac{1}{130}$ to $\frac{1}{75}$ the diameter of the full moon and about the same size as Saturn's disk. The opposition of September 1988, when Mars will be only 59 million kilometers from earth, will be especially favorable for viewing (see Table 19, p. 379), since the closer the planet is to us, the larger it appears. Mars can appear much smaller when not at opposition, sometimes as small as 4 arc seconds across.

Because of the seasonal changes on Mars, and because of the thin lines that appear to cross Mars' surface when it is observed visually from earth, Percival Lowell and others at the turn of the last century suggested that there was life on Mars. The seasonal changes, they thought, were caused by vegetation, and the lines were "canals" dug by Martians to carry water.

But observations from the ground and later from spacecraft have shown that the seasonal changes are the result of blowing dust, during huge seasonal storms generated by the effect of solar heating. When dust is blown off certain surfaces, the dark underly-

Fig. 128. (*right*) Mars' giant volcano, Olympus Mons, is over 600 km in diameter and is about 25 km high, larger than any mountain on earth. The volcano's crater is so large that Manhattan Island could be dropped inside it. (JPL/NASA)

Fig. 129. (*below*) Sand dunes and rocks on Mars, photographed by the Viking 1 Lander. (JPL/NASA)

ing material is revealed. Close-ups have proved that the "canals" don't really exist; they were presumably an effect of the eye and brain, which tend to imagine connections even when they don't exist. In fact, the features that show up in telescopes as bright and dark to earthbound observers don't necessarily correspond to any physical features on Mars.

The polar caps on Mars, which wax and wane with the seasons of each Martian year, turn out to be largely made of frozen carbon dioxide that condenses over a core of frozen water. The existence of large quantities of water is thought to be essential for life as we know it, so the discovery of signs of water on Mars is encouraging for those who hope to find that life has started there independently of life on earth.

The American Viking missions in 1976 carried out the closest reconnaissance of Mars to date. Photographs taken during their approach (C.Pl. 61) show much of the surface of Mars and the giant volcanoes on it (Fig. 128). Each of the two Viking spacecraft had an orbiter that took photographs of Mars' surface over a period of years. The Viking orbiters' photographs revealed giant volcanoes, a huge canyon larger in diameter than the continental United States, many craters in certain regions, and many other

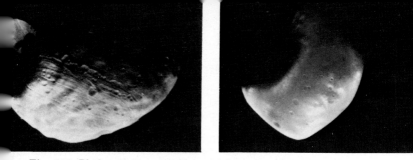

Fig. 130. Phobos (*left*) and Deimos (*right*). (JPL/NASA)

geological features. Branching channels appear like streambeds on earth, indicating that water may have flowed on Mars in the past.

Each Viking carried a Lander, which took photographs of the surface (Fig. 129, C.Pl. 59), and a small biology laboratory that searched for signs of life. The signs, after much investigation, seem to be negative, though the possibility that life exists cannot be completely ruled out on the basis of these studies.

Mars has two moons (Fig. 130), Phobos ("Fear," in Greek) and Deimos ("Terror," in Greek), named after the mythological companions of the Roman war god, Mars. Photographs from the Voyagers and the earlier Mariner 9 spacecraft revealed that both moons are elliptical, with Phobos having a longest diameter of only 27 km and Deimos a longest diameter of only 15 km. These moons are not large and are not made round by gravity like our moon, which is really a small planet. The moons of Mars are really only small, orbiting chunks of rock, comparable to some of the smaller asteroids. They do not become brighter than 11th and 12th magnitudes, and so are not easily observed from earth.

Observing Mars

Observe the reddish color and features on the disk (such as Syrtis Major, Hellas, and the polar caps) as the planet rotates; notice how the features change with Mars' seasons. Observe seasonal dust storms. An orange filter is useful to accentuate contrast on Mars' surface; a blue filter may reveal the status of dust storms.

Table 19. Oppositions of Mars

Date of Opposition	Nearest to Earth	Diameter of Disk (arc sec)	Distance to Earth A.U.	Distance to Earth km (millions)
May 11, 1984	May 19	18	0.53	79.5
July 10, 1986	July 16	23	0.40	60.4
September 28, 1988	September 22	24	0.39	58.8

MERCURY

Mercury, the closest planet to the sun, is less than half the size of the earth. It has no moons, no atmosphere, and is an exceedingly inhospitable place. The temperature at midday is over 400°C (750°F).

Mercury never appears too far from the sun in the sky, and so is visible only briefly after sunset or before sunrise, as shown in the Graphic Timetables in Chapter 9.

From earth, we can see Mercury's phases (Fig. 131), but even the largest telescopes do not reveal any detail on its surface. The surface features are known only from NASA's Mariner 10 spacecraft, which flew close to Mercury three times in 1974 and 1975. The pictures (Fig. 132) revealed that Mercury has many craters and resembles our moon. The craters are flatter and have thinner rims on Mercury, because of the higher gravity, but those are subtle effects.

Transits of Mercury

Transits of Mercury are not as rare as those of Venus: by the year 2000, there will have been 11 in this century. The most recent one was on November 10, 1973, when Mercury took $5\frac{1}{2}$ hours to cross the sun's disk. The next two, on November 13, 1986, and November 6, 1993, will be visible from Europe but not from North America. Observers should note the exact times when Mercury first touches the sun, when it moves entirely within the solar disk, when it first touches the other side of the sun, and when it leaves the solar disk entirely.

Following those transits, the next will be a grazing transit — when Mercury will appear to just graze one edge of the solar disk — on November 15, 1999. This grazing transit will be visible from Antarctica and possibly from southern Australia. The next transit of Mercury visible from the U.S. and Canada will be on May 7, 2003 (just over a year before the next transit of Venus).

Observing Mercury

Observe the phases. For best viewing, locate Mercury on a favorable morning when it rises well ahead of the sun, and keep your telescope on it until after sunrise.

Fig. 131. Two phases of Mercury. Mercury was 7.1 arc seconds across at left and 5.4 arc seconds across at right. (New Mexico State University Observatory)

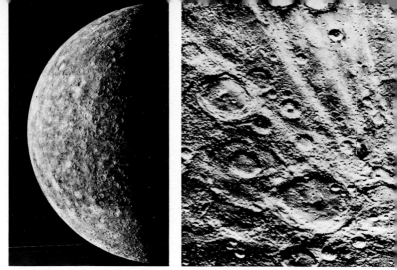

Fig. 132. *Left:* A montage of pictures of Mercury, taken from the Mariner 10 spacecraft. About $\frac{2}{3}$ of the portion seen is in Mercury's southern hemisphere. (NASA) *Right:* A Mariner 10 close-up of a field of rays radiating north (upward) from a crater off-camera to the south. The largest crater shown is 62 miles (100 km) across. (NASA)

URANUS and NEPTUNE

Uranus and Neptune reach magnitudes 6 and 8, respectively, at their brightest. They are each giant planets, about four times the earth's diameter, and about 15 times its mass. They appear as tiny greenish or bluish disks in earth-based telescopes; the color comes from methane in their atmospheres. They show as dots in binoculars or small telescopes.

Uranus' orbit is 19 A.U. in radius and the radius of Neptune's orbit is 30 A.U. These planets are so distant that Uranus never gets bigger than 4 seconds of arc, and Neptune gets no bigger than 2.5 arc seconds. Since the best resolution available from earth's surface is about $\frac{1}{2}$ second of arc, we just can't see much detail on the surface of either Uranus or Neptune.

A mnemonic for the names of Uranus' moons, in order of their distance from the planet — Miranda, Ariel, Umbriel, Titania, and Oberon — is "M-Auto." Uranus' moons are shown in Fig. 133. Neptune's moons (Fig. 134) are Triton, which is larger than our moon, and Nereid, which is more like an asteroid in size. Triton is covered in part by a layer of methane snow or ice and has a thin methane atmosphere. The Uranian satellites are coated with ordinary ice (frozen water) with some soil mixed in.

When Uranus occulted (that is, passed in front of and therefore hid) a star in 1977, the light from the star winked off and on a few times before and after the star was occulted by the planet itself.

Fig. 133. Uranus with its five known moons. (William Liller/The Cerro Tololo Inter-American Observatory)

This marked the discovery of a set of nine thin rings around Uranus, which have been studied further at subsequent occultations.

Voyager 2 will reach Uranus in 1986 and will continue on to Neptune in 1989. If the equipment is still working, we should learn a lot about these planets and their moons.

Observing Uranus and Neptune

Observe their greenish disks. (Charts that will help you find Uranus and Neptune appear in the January issue of *Sky & Telescope* each year.)

PLUTO

Tiny, distant Pluto is so far away and so small that we know little about it. Its orbit is very elliptical and has an average radius of 39 A.U. Pluto is now on the part of its orbit that is within the orbit of Neptune. Pluto will be within Neptune's orbit, closer than Neptune both to the sun and to the earth, until the year 2000.

A ground-based photograph taken in 1978 revealed that the image of Pluto was not quite round. A bulge at the edge of Pluto's image turned out to be a previously unsuspected moon (Fig. 135). Studies of the orbit of Pluto's moon have shown us that Pluto is only $\frac{1}{500}$ the mass of earth; other observations indicate that Pluto is less than $\frac{1}{3}$ the size of earth, even smaller than our moon.

Pluto — like Neptune's moon, Triton — has methane ice or snow on its surface and an atmosphere consisting of methane gas plus traces of other substances. Pluto is now about magnitude 13.7 and will be only slightly brighter in 1989, when it reaches *perihelion*, the closest point ("peri-") of its orbit to the sun ("helios"). Though Pluto is as bright as it has been for centuries, it is still

Fig. 134. (*left*) Neptune, with its inner moon, Triton. The other moon, Nereid, is farther out and too faint to be clearly visible. (Lick Observatory photo)

Fig. 135. (*right*) The bulge at the upper right of Pluto's image is its moon Charon, in this discovery photograph. Even though this picture is enlarged so much that film grains show, this is perhaps the best and clearest photograph ever taken of Pluto, which shows how poor our knowledge of this distant planet is. (James W. Christy/U.S. Naval Observatory photo)

much too faint to be seen in small telescopes. For keen observers, though, it may be seen as a point of light in 20-cm (8-inch) telescopes. No detail at all is detectable on its surface.

Observing Pluto

With a sufficiently large telescope, observe Pluto as a dot in the sky, changing position among the stars from night to night. A chart that will help you find Pluto appears in the January issue of *Sky & Telescope* each year.

Observing the Planets

By looking for a long time through a telescope, and waiting for instants when the "seeing" is especially good (that is, when the image is constant because the air is especially steady), the human eye can often see more detail than can be recorded with a camera. Thus excellent sketches can be made. You can follow not only the planets' rotation but also seasonal and weather changes on them.

The Association of Lunar and Planetary Observers, c/o Walter Haas, Box 3AZ, University Park, NM 88003, is an amateur society. Regional societies, such as the Amateur Astronomers Association, 1010 Park Avenue, New York, NY 10028, also may have special groups of members who are especially interested in planetary observing.

11

Comets and Asteroids

A bright comet looks spectacular in the sky. Comets are bodies that orbit in our solar system, passing between the planets. Also in the spaces between the nine major planets in our solar system are minor planets known as asteroids. In this chapter, we discuss comets and asteroids and how to observe them.

COMETS

Sometimes a fuzzy object becomes visible in the heavens. It may look like a smudge on the sky. If we are lucky, it will grow brighter over a period of weeks or months, and will form a tail. This tail may become so long that it extends across most of the sky (Fig. 136).

Such bodies are *comets,* large icy snowballs in the sky. Each comet we see began as one of the hundreds of billions of small bodies in a tremendous cloud surrounding the sun, far beyond the outermost planets. Sometimes a gravitational nudge from a passing star causes one of these bodies to move closer to the sun. Solar energy heats it, and the body gives off gas and dust that form a tail. The gas is pushed away from the comet by gas flowing outward from the sun — the *solar wind*. The dust is left behind in the comet's orbit by the pressure of the sun's radiation. Thus some comets have two visible tails — a dust tail trailing gracefully behind the comet and a gas tail whose irregularities show the different puffs of the solar wind (Fig. 137).

A comet bright enough to be seen with the naked eye or with binoculars can be a very beautiful sight (C.Pl. 69). If you plot its position carefully against the background of stars, you will see that it changes by about 1° per day, but the comet's motion is not apparent to the eye. A comet's tail results in part from its motion through the solar system, but a comet does not speed across our sky; its tail is not a result of any such apparent motion across the sky.

There may be a dozen or more comets in the sky at a single time, but most are so faint that they can be seen only with very large telescopes. Every few years, a comet can be seen with the naked eye, though most appear faint. Bright comets (1st or 2nd magnitude) might appear every decade or so. The appearance of a bright comet is usually unpredictable. Of the predictable comets, only the one known as Halley's Comet is spectacular.

Fig. 136. Comet Ikeya-Seki, observed in 1965 from Mt. Wilson overlooking Los Angeles. The photograph was a 32-second exposure on ASA 400 film at $f/1.6$. (William Liller)

The time comets take to complete their orbits depends on the size of their orbits. Encke's Comet returns every 3.3 years, but is faint. Halley's Comet returns about every 74–79 years, and is always bright enough to be seen with the naked eye. It has been seen on at least 27 occasions since 87 B.C., though it wasn't until early in the 18th century that Edmond Halley realized that the bright comets reported by previous observers were reappearances of the same comet. A comet loses less than 1% of its material at each passage near the sun, so it can reappear many times.

Fig. 137. Comet Kohoutek, photographed with the wide-field Schmidt telescope at the Palomar Observatory in 1974. The smooth dust tail and the kinky gas tail both show. (Palomar Observatory photo)

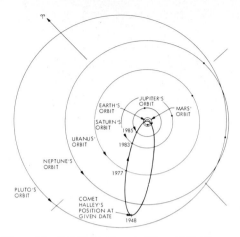

Fig. 138. The orbit of Halley's Comet. (Donald K. Yeomans, Jet Propulsion Laboratory)

Halley's Comet will again reach perihelion — the closest point of its orbit to the sun — on February 9, 1986 (Fig. 138). The relative locations of the sun, comet, and earth during this appearance will not be as favorable for observers on earth as they were during the 1910 appearance of Halley's Comet (Figs. 139, 140), so the comet will not be as bright this time. But it will still be a fascinating object to observe, particularly if you are far from city lights.

Fig. 141 shows the path of Halley's Comet during the months before its perihelion, and Fig. 142 shows a detailed view of its path in the sky near perihelion. Fig. 143 shows predictions of its brightness and the size of its tail for various times, as seen from different latitudes. For each 10° farther south you go than the latitude given, the comet will be about 10° higher in the sky.

Photographing a Comet

Occasionally, a comet may be bright enough to be photographed with an ordinary camera. No special equipment is required; a sturdy tripod and a lens that can be opened to a wide aperture (preferably about $f/1.4$, and at least $f/2$) are all that you will need. Use fast film, and be careful not to shake the camera. (Use a cable release, and raise the mirror of a reflex camera before snapping the shutter, if possible.) Both black-and-white and color film can produce pleasing results.

If you use a normal or wide-angle lens, you may be able to capture the comet at the top of the frame and the horizon or a tree on the bottom. Such a photograph gives a sense of scale, and may look more interesting than one in which the image of the comet is centered.

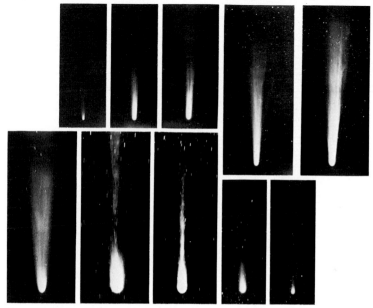

Fig. 139. Halley's Comet during its 1910 appearance. (Mt. Wilson and Las Campanas Observatories photo)

Fig. 140. The head of Halley's Comet on May 8, 1910. (Mt. Wilson and Las Campanas Observatories photo)

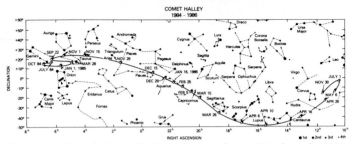

Fig. 141. The path of Halley's Comet, late 1984 through 1986. (Donald K. Yeomans, Jet Propulsion Laboratory)

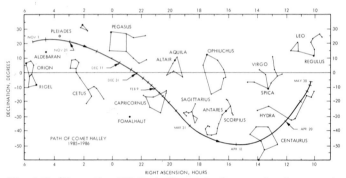

Fig. 142. The path of Halley's Comet in the months closest to its February 9, 1986, perihelion. (Donald K. Yeomans, Jet Propulsion Laboratory)

Take a wide range of exposures: try 1, 2, 4, 8, 15, 30, and 60 seconds. Film is cheap compared with the loss of a rare opportunity, so don't hesitate to take more pictures than you usually take at one time.

If you can place your camera piggyback on a telescope that tracks the stars, you will be able to take longer exposures. Again, take exposures over a wide range of times, with your lens wide open or almost wide open.

It can also be very interesting to draw the comet and its tail, instead of photographing it. The eye can detect very interesting detail.

If You Discover a Comet

Many new comets are discovered by amateur observers, usually with large binoculars (those with huge front lenses) or medium-

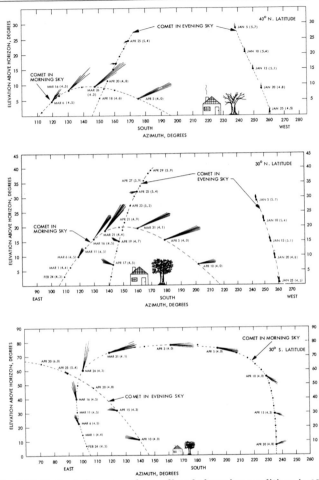

Fig. 143. Halley's Comet — the predicted observing conditions in 1986 for observers with binoculars, at 30° and 40° north and 30° south latitude. The approximate total visual magnitude is shown (in parentheses) for each day; whether the comet will be visible in the morning or in the evening is also indicated. Positions are given for the beginning of morning astronomical twilight or the end of evening astronomical twilight. (Astronomical twilight is defined as the time when the sun is 18° below the horizon.) (Donald K. Yeomans, Jet Propulsion Laboratory)

sized telescopes. The main background you must have to discover a comet is a detailed knowledge of what is in the sky, so that you can tell whether an object you see is a new comet or a well-known nebula.

To search for comets, the American comet observer John Bortle recommends sweeping your field of view back and forth at a rate of no more than 1° of sky per second. He recommends starting at the horizon after sunset, and sweeping back and forth 45° on either side of the sunset point, gradually moving your binoculars or telescope upward so as to slightly overlap your preceding area of observation. Continue until you are about 45° above the horizon — halfway up to the zenith. If you are searching before sunrise, begin 45° above the horizon and sweep back and forth while moving your field of view downward.

If you think that you have found a comet, note its position in the sky (right ascension and declination) by finding its position in relation to the stars or the Atlas Charts in Chapter 7. Also note its brightness (its approximate magnitude) by comparing it with that of plotted stars and record the time. Then watch for a while to note the direction in which the comet is moving and the rate of its motion. It is helpful to plot the position of the comet in the star atlas at different times, so that its motion can be accurately charted; the presence of motion is an important sign that you have in fact found a comet. Then send that information to the Central Bureau for Astronomical Telegrams of the International Astronomical Union, which is located at the Smithsonian Astrophysical Observatory, Cambridge, MA 02138. If you have access to a Telex II (TWX) machine, you can send a message to 710-320-6842; the answerback is ASTROGRAM CAM. Or you can send a telegram, giving this number and answerback as the address. The Bureau also has a telephone with an answering machine at (617) 864-5758, though they prefer to receive messages via Telex II (TWX) or telegram. Don't forget to include your name in the message, and to leave your address and telephone number.

The IAU Central Bureau names each comet after the first people — up to three observers, if there are near-simultaneous independent discoveries — to find it.

Articles by John Bortle in *The "Sky & Telescope" Guide to the Heavens* give details about searching for and observing comets.

ASTEROIDS

In between the major planets are thousands of minor planets ranging from one kilometer to hundreds of kilometers across; they are called *asteroids* or *minor planets*. The first asteroid, discovered on January 1, 1801, was at first thought to be a new planet; it was named Ceres. The numbers assigned to asteroids (in order of dis-

Table 20. The Brightest Asteroids

Asteroid	Brightest Visual Magnitude	Diam. (in km)	Asteroid	Brightest Visual Magnitude	Diam. (in km)
4 Vesta	5.1	555	15 Eunomia	7.9	261
2 Pallas	6.4	583	8 Flora	7.9	160
1 Ceres	6.7	1025	324 Bamberga	8.0	256
7 Iris	6.7	222	1036 Ganymed	8.1	40
433 Eros	6.8	20	9 Metis	8.1	168
6 Hebe	7.5	206	192 Nausikaa	8.2	99
3 Juno	7.5	249	20 Massalia	8.3	140
18 Melpomene	7.5	164			

Note: Brightness data from R. Shuort via J. U. Gunter; diameters based on Tucson Revised Index of Asteroid Data (TRIAD).

covery) are now given along with the names, so we call this asteroid 1 Ceres. Within a few years 2 Pallas, 3 Juno, and 4 Vesta were also discovered, and scientists realized that they were dealing with several little planets instead of a few major ones.

Many asteroids only 1 km in diameter are known. Over 200 of them with diameters greater than 100 km have been discovered, and half a dozen are known to exceed 300 km in diameter. Though most asteroids are in the *asteroid belt* located between the orbits of Mars and Jupiter, many other asteroids come very close to the earth. The asteroids in one group, the Apollo asteroids (named after the first of the group to be discovered, 1862 Apollo), have orbits that cross the earth's orbit. We know of about 3 dozen asteroids in this group. The orbit of an Aten asteroid (2062 Aten and those with similar orbits) not only crosses the earth's orbit but is even smaller than the earth's orbit. The orbit of 1566 Icarus, another unusual asteroid, takes it closer to the sun than Mercury. Icarus came within 6 million km of the earth in 1968, a near miss on an astronomical scale.

Vesta can be brighter than 6th magnitude, the limit of visibility to the naked eye. A small telescope can see a number of asteroids (Table 20). Some five dozen may become brighter than 10th magnitude at some time in their orbits. The motion of these asteroids with respect to the stars can be followed from night to night.

Sometimes an asteroid goes in front of — *occults* — a star. Then, by measuring the length of time that the star is hidden, we can calculate the size of the asteroid. The stars are so far away that light rays from them are nearly parallel, so the shadow an asteroid casts on the earth is almost exactly the size of the asteroid itself. Such occultations, for example, have shown us that the size of 2 Pallas is 558 x 526 x 532 km. To observe these occultations people had to be stationed, on the basis of last-minute observations, in the locations where the asteroid's shadow would most likely pass.

The magazine *Sky & Telescope* publishes the latest information about where asteroid occultations will take place. They also provide addresses from which special bulletins and lists are available for serious observers.

Asteroids show up as streaks on astronomical photographs made with telescopes that track the stars (Fig. 144). On photographs that track the asteroid, the stars are streaks, while the asteroid is a dot (Fig. 145).

Fig. 144. (*above*) An asteroid shows up as a trail in a photograph taken through a telescope that is tracking the stars. (Harvard College Observatory)

Fig. 145. (*below*) The asteroid 1967 Menzel. The telescope is tracking the asteroid, so the stars are trailed in this photo. (Harvard College Observatory)

12

Meteors and Meteor Showers

If you are outdoors and looking upward in the late evening around August 12th any year, you will probably see a "shooting star," a *meteor*, travel across the sky in less than a second (C.Pl. 75). You are actually seeing solid particles from space — larger than atoms but much smaller than asteroids (minor planets) — burn up in the earth's atmosphere. Each August 12th, the earth's orbit intersects the center of a stream of particles that makes up the Perseid meteor shower, which in some years consists of meteors that are visible at rates of up to one per minute.

When in space, the bodies are called *meteoroids*. Some of the meteoroids survive their fiery passage through the earth's atmosphere. Any part of a meteoroid that reaches the earth (or, indeed, any place where we can examine it, which would include the moon, airplanes, or spacecraft) is called a *meteorite*.

On any night with perfect observing conditions, you are likely to see a random meteor in the sky every 10 minutes or so. These are *sporadic meteors*.

Many times a year, the earth's orbit crosses a stream of particles, believed to be (in most cases) from a defunct comet. These particles make up *meteor showers* (Table 21). Some showers have more meteors per hour than others. Some, like the Perseids around August 12, the Geminids near December 14, and the Quadrantids near January 3rd, are of about the same strength each year, and we can expect to see approximately the same number of meteors at maximum. Others, like the Leonids around November 17, differ in strength from year to year. Still others, like the delta Aquarids from mid-July to late August or the Taurids in the fall, are spread out over many weeks.

The Leonid meteor shower is particularly spectacular about every 33 years. During a one-hour period of the 1966 Leonids, over 100 meteors were observed per minute, and during one 40-minute period, over 1000 meteors per minute were observed. Average magnitudes were 1.5 or 2, and some of the brighter trains lasted more than a minute. At maximum, some observers were able to see up to 40 Leonids per second! The 1998 and 1999 Leonid showers will be eagerly awaited.

The best way to observe a meteor shower is to lie back on a lawn chair or blanket on the grass and enjoy the sight. A meteor shower may appear in virtually any part of the sky, so to use a telescope or

Table 21. Major Meteor Showers

Shower	Date of Maximum	Time (E.S.T.)	Radiant (at maximum) r.a. dec. (2000.0)		
	1984, 1988		h	m	°
Lyrids	Apr. 21	11 p.m.	18	16	+34
Eta Aquarids	May 4	2 a.m.	22	24	0
Delta Aquarids	July 28	5 a.m.	22	36	−17
Perseids	Aug. 11	8 p.m.	03	04	+58
Orionids	Oct. 21	midnight	06	20	+15
South Taurids	Nov. 3	—	03	32	+14
Leonids	Nov. 17	7 a.m.	10	08	+22
Geminids	Dec. 13	7 p.m.	07	32	+32
Ursids	Dec. 22	1 a.m.	14	28	+76
	1985, 1989				
Quadrantids	Jan. 3	3 a.m.	15	28	+50

Notes: Since the year is actually $365\frac{1}{4}$ days long, we have listed the date and time starting in March for leap years (1984, 1988, etc.); add 6 hours for each year following in the four-year sequence.

Most showers are named after the constellation in which their radiant is located, or after the bright star their radiant is near. The Quadrantids were named after Quadrans Muralis (in the northern part of Boötes), a constellation suggested by J.E. Bode in 1801 that is not now accepted as a constellation.

even binoculars would simply limit your field of view. Instead, just look up, moving your eyes slowly around the sky. You will probably see a meteor out of the corner of your eye. Of course, the farther you are from streetlights, the better. If the moon is up, and particularly if it is more than half full, the sky will be too bright to see the meteor shower at its best. Sometimes you can wait until the moon sets, or observe before it rises.

Try to trace the paths of meteors in a certain shower back across the sky; all the paths will seem to converge in the same part of the sky, called the *radiant* (Fig. 146, p. 396). This is an effect of perspective, since the meteors are hitting the earth in parallel lines. Just as parallel railroad tracks seem to converge at a distant point on the horizon, the parallel meteor paths seem to come from a

Table 21 (contd.). Major Meteor Showers

Associated Comet	Avg. Hourly Rate (for a single observer)*	Avg. Duration of Maximum (between days of ¼ maximum)**
1861 I Thatcher	15	2
Halley	20	3
	20	7
1862 III Swift-Tuttle	50	5
Halley	25	2
Encke	15	†
Temple-Tuttle	15	—
	50	3
Tuttle	15	2
	40	1

* = Number of meteors (both shower and sporadic meteors) you can expect to see under ideal observing conditions — *i.e.*, no clouds, and a clear, dark sky in which 6th-magnitude stars are visible. Shower strength fluctuates from night to night and from year to year.
** = Background rate of sporadic meteors subtracted.
— = Broad maximum; duration of shower uncertain. †Meteors from a large number of small radiants near the constellation Taurus are visible for several weeks in November.

convergent point or area in the sky. During meteor showers, you can often see more meteors after midnight, because the earth is rotated then so that your side is plowing into the meteors rather than having them catch up from behind.

To tentatively identify a meteor with a certain shower, you can trace its path back to see if it came from a constellation that has a known, active radiant. On the night of a shower's maximum, most meteors will come from the shower, but on nights far from the shower's peak, determining whether a meteor came from a known radiant is the only way to identify meteors from spread-out showers.

Occasionally a sporadic meteor is extremely bright, as bright or brighter than even Venus. Such an object is called a *fireball;* a

Fig. 146. The Leonid meteor shower on November 17, 1966. The meteors appear to diverge from a radiant; two meteors appear as points because they are coming straight at us. Regulus is the bright star in the lower half of the frame. About 70 Leonid trails were recorded on the negative of this 3.5-min. exposure on Tri-X film at $f/3.5$. (Dennis Milon)

bright fireball may be a chip of a broken-up asteroid. Sometimes a fireball will leave a *train,* a path in the sky that remains for a few seconds. You may even be able to hear a sound, in which case there is more chance that a piece of the meteoroid is falling to earth as a meteorite. One may even come through your roof, as happened in Connecticut in 1982. Fireballs are so rare that you can't plan to see one. If you should see one, you should note as much information as you can, including the time; the brightness compared to nearby stars, planets, or even the moon; and, if possible, the altitude and azimuth of the beginning and ending points of the fireball's path among the stars. This information should be reported to the Smithsonian Institution, Museum of Natural History, Washington, D.C. 20560, in the United States; to the Herzberg Institute of Astrophysics, Ottawa, Ontario K1A 0R6, in Canada; or to suitable organizations in other countries. If you hear a sound, it is particularly important to report it quickly, for a meteorite may have landed nearby and could be picked up for scientific analysis.

Meteors ionize their paths in the earth's upper atmosphere. Both professionals and some amateur astronomers use radio astronomy to detect these paths; radio equipment needed to do so is relatively inexpensive and advertisements for suitable equipment can be found in many magazines for amateur radio operators and occasionally in *Sky & Telescope.*

Meteorites are particularly important to astronomers because aside from the moon rocks brought back between 1969 and 1972 by the Apollo missions and some dust brought back by Soviet spacecraft, meteorites provide the only extraterrestrial material we have to study in a laboratory. There are two basic types of meteorites (though, of course, experts make finer divisions): nickel-iron meteorites (called *irons*) and stony meteorites (called *stones*). Most meteorites found on the ground by accident are irons; they are very dense and appear quite different from ordinary rocks. On the other hand, when someone sees a meteorite fall and then searches for it, stony meteorites are found most of the time. This indicates that most meteorites are actually "stones," but that most of these are never discovered because they resemble normal (terrestrial) stones on the ground. The largest number of meteorites have been found in recent years in Antarctica, where they have accumulated undisturbed over long periods of time.

Nobody who is interested in meteorites should miss seeing the meteorites in museums (Fig. 147). You can also visit the Barringer Meteor Crater in southern Arizona (Fig. 148), which is 1.2 km (almost a mile) in diameter and resulted from the most recent large meteorite to hit the earth, about 25,000 years ago. Over a dozen even larger meteor craters are known on earth.

Fig. 147. The largest meteorite ever discovered was the 31,000-kg (34-ton) Ahnighito meteorite, brought back by Peary (the discoverer of the North Pole) from the Arctic in 1892. It is now on display at the American Museum of Natural History in New York. (Jay M. Pasachoff)

Fig. 148. The Barringer meteor crater near Winslow, Arizona, is 1.2 km (almost 1 mile) across and was formed about 25,000 years ago. (Meteor Crater, Northern Arizona)

To Observe Meteors

You will need a comfortable place to lie back outdoors, far from lights; an accurate watch, set to the nearest second (which formerly required receiving the station WWV on a short-wave radio but now can be done with a digital watch set from the radio or television); and perhaps a friend to take notes (which helps preserve your adaptation to darkness). When a meteor moves overhead, record the time of the event to the nearest second, the duration of the meteor (usually less than 2 seconds or so), the length of the trail, the meteor's magnitude (by comparing it to nearby stars), and the color. The data usually reported are the number of meteors seen by a single observer in an hour, and changes in the average hourly rate through the night; this information shows when the peak of the shower occurred. It may even be useful to measure and plot the number of meteors during 15-minute intervals. When possible, it is good to record whether the meteor seems to be coming from the radiant, or if it is a sporadic meteor. More advanced observers often do this by recording the path of each meteor and its magnitude on a star chart, using a different chart for each hour of observation.

Your data can be reported to the American Meteor Society, Department of Physics and Astronomy, SUNY, Geneseo, NY 14454, or to the British Meteor Society, 26 Adrian Street, Dover, Kent, CT17 9AT, England, and to the Meteor Section of the British Astronomical Association, 2 Hyde Road, Denchworth, Wantage, Oxfordshire, OX12 0DR, England.

13

Observing the Sun

Most of the stars are visible only at night, but one star — our
sun — is visible in the daytime. The sun is much closer to us than
any other star. It takes light only about 8 minutes to travel from
the sun to us; the nearest star is over 4 light-years away. As a result
of the sun's closeness to earth, its disk covers $\frac{1}{2}$° of sky and we can
see many details on its surface.

The Sun's Surface

The sun, like all stars, is a ball of hot gas. It has an interior and an
atmosphere. The surface of the sun sends us the light and heat that
provide energy for us on earth.

The energy is formed deep inside the sun's interior by the process
of nuclear fusion. In the sun's fusion process, groups of four hydrogen atoms are transformed into single helium atoms. Each helium
atom that results has slightly less mass (0.7%) than the sum of the
four hydrogens. At the turn of the century, Albert Einstein realized that mass can change into energy according to the equation
$E = mc^2$. Einstein's equation tells us that the little bit of mass that
disappears is transformed into a relatively large amount of energy.
(Since c — the speed of light — is such a large number, the amount
of energy that results adds up to a lot.) The sun is a good example
of a fusion reactor, a type that we are not yet able to build on
earth. The sun is a fusion reactor located 150 million kilometers
(93 million miles) from us.

The solar surface is called the *photosphere,* from the Greek word
photos, meaning "light." When we look at the sun using the pro-
tective filters and precautions described on p. 402, we can see that
it is largely uniform in brightness, that it becomes slightly darker
towards its edges, and that there often are a few dark areas on its
surface (Fig. 149). The dark areas — *sunspots* — are relatively cool
regions of the solar surface.

Each sunspot has a dark center called its *umbra* (plural,
umbrae) and a less dark region called its *penumbra* (Fig. 150).
Sunspots normally form in groups, and each spot can last for
weeks. Since the sun rotates about once every 25 days, we can
watch the sunspots form and rotate across the visible solar surface.

Sunspots are regions of the sun where the magnetic field is especially strong, perhaps 3000 times stronger than the average field of

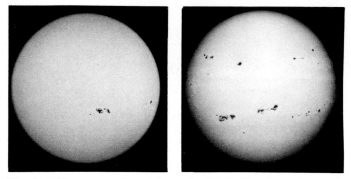

Fig. 149. The sun at solar minimum in 1974 (*left*) and at solar maximum in 1979 (*right*). On these white-light photographs, the sun is noticeably darker towards its limb; lighter areas called faculae become visible there. (*left:* William Livingston, The Kitt Peak National Observatory; *right:* Harold Zirin, Big Bear Solar Observatory, California Institute of Technology)

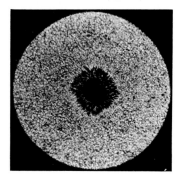

Fig. 150. A sunspot, showing its dark central umbra and its surrounding penumbra. The surface of the sun itself shows a salt-and-pepper effect known as granulation. Each granule is about 500 miles (700 km) across — roughly the distance from New York to Detroit. (Sacramento Peak Observatory)

the sun, of the earth, or of a toy magnet, which are all comparable in strength. Spots tend to occur in groups. Within each group of sunspots in one of the sun's hemispheres, the sunspots that lead in the direction the sun is rotating have north magnetic poles and the spots that trail have south magnetic poles. In the opposite hemisphere, the leading and trailing polarities are reversed.

The number of sunspots on the sun waxes and wanes over an 11-year period (Fig. 151). The vertical axis plots the *sunspot number,* which is not actually a direct count; it is a compound value that makes allowance for not only individual spots but also for the presence of sunspot groups. The sunspot number = $k(10g + f)$, where k is a correction factor to account for personal decisions of each individual observer (as to what constitutes an individual spot, for example), g is the number of groups, and f is the total number of spots. (Your value for k can be assigned only by a central registry, after they compare the numbers of spots you report with the established number.) An isolated spot is considered to be a group with one spot in it, so it adds 1 to each value of g and f.

After the 11-year cycle, the polarities in opposite hemispheres of the sun reverse; that is, in a hemisphere where the leading spots have had north magnetic poles, those spots now have south magnetic poles. Thus it takes two cycles, or 22 years, for the leading spots in each hemisphere to regain their original polarity, so the solar activity cycle is really 22 years long.

The next maximum of sunspots is expected in about 1990. As that time approaches, there will be more huge eruptions on the sun called *solar flares*. Flares send out particles, x-rays, and gamma rays that can hit the earth; these could cause surges in power lines, zap passengers in high-flying aircraft or astronauts on space shuttles, and contribute to creating the aurora. The flares also excite the aurora in a way that allows it to be seen by observers at latitudes more southerly than normal. Since the earth's magnetic field

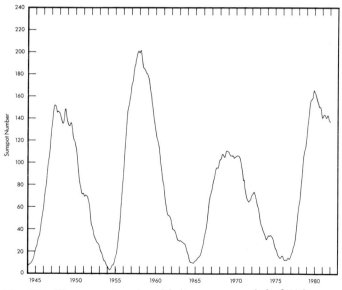

Fig. 151. The sunspot cycle, with its average period of 11.2 years. (World Data Center A for Solar-Terrestrial Physics)

guides the particles that can cause the aurora, and since the earth's magnetic field comes out of the earth at the magnetic poles, the aurora can most often be seen close to the magnetic poles. The northern magnetic pole is near Hudson Bay in Canada.

When we look at the center of the solar disk, we are looking through gas. The effect is the same as looking into the air on a foggy day; we can see only so far. When we observe the solar *limbs* — the parts of the sun near the edge — we are looking diagonally through the solar atmosphere and thus through more gas that obscures our view; we cannot see as close to the sun's center before the gas appears opaque. As a result, near the sun's edge we see higher levels of the sun's atmosphere. Since these higher levels are slightly cooler than the lower atmospheric levels we can see when we look toward the center of the sun's disk, the sun's surface looks a little darker toward its limb.

When we look at the sun through the earth's atmosphere, the atmosphere bends the different colors in sunlight by varying amounts. This effect — refraction — becomes most extreme at sunrise and sunset, and leads to the beautiful, elusive phenomenon known as the *green flash*. It lasts only a second or so, and can be seen only when our view of the setting sun is completely unobstructed; usually you must be looking over water, and there must not be any haze or clouds on the horizon. (It is even harder to see the green flash at sunrise, because you don't know where to look in advance.) As the sun nears the horizon, we wind up essentially with overlapping images of the sun. The colors are in the order ROY G. BIV (red, orange, yellow, green, blue, indigo, and violet), with the red image lowest. But the blue, indigo, and violet rays are scattered so much by the atmosphere that they don't reach us. The orange and yellow are absorbed by water vapor in the earth's atmosphere. So we are left with only red and green images; when the red image sets, we sometimes see a brief flash of green on the horizon (C.Pl. 72).

Safely Observing the Sun

Most of the time, the sun is too bright to look at safely with the naked eye. Except for certain short periods during total eclipses (see p. 405), the only time when we can safely look at the sun is when it is dimmed by haze, usually close to sunset.

The safest way to study the sun is not to look at it directly at all. You can use a telescope or binoculars to project an image of the sun onto a piece of cardboard. Stand with your back to the sun and look at the cardboard; do not look through the telescope at the sun at all. Looking at the sun through a telescope for even a second could blind you. If a finder telescope is mounted on the telescope you are using, make certain that it is securely capped.

Adjust the eyepiece of your telescope or binoculars so that it is behind its normal position. Then you can vary both the position of the eyepiece and the distance of the paper behind it to put the sun's image in focus. An image the width of your hand should show enough detail to reveal sunspots clearly. You can trace the outlines of the sunspots on a daily basis, and follow the way they change as the sun rotates and the sunspots evolve. Again, **never** look up through a telescope at the sun, not even to point or focus the telescope. (You can usually align the telescope fairly well by adjusting the shadows along the outside of the tube so that they are at a minimum.)

The sun is about one million times brighter than the full moon. So to observe the sun directly in order to see the sunspots, we need a filter that cuts out all but 1/1,000,000th of the sun's rays. A few filters of this type are made and distributed by telescope manufacturers. The filters normally go over the front end of the telescope, so that most of the sun's light never enters the telescope tube. (Filters that cut out light near the eyepiece are no longer thought to be safe for amateur astronomers; there is always the danger that they might slip or crack.)

Once you know how to observe the sun safely, you can make your own count of the sunspot number. The AAVSO (American Association of Variable Star Observers) in Cambridge, Massachusetts has a solar group that keeps track of sunspots.

The Chromosphere, Prominences, and Special Solar Filters

Many features of the sun's atmosphere are not visible when you look in white light — all the sun's rays together; they show up only when the light of the unique color (specific wavelength) of hydrogen is isolated from visible light. This light, represented by the so-called *H-alpha line,* falls in the red part of the spectrum. Professional solar astronomers and increasing numbers of amateurs have filters that pass only this red H-alpha line of hydrogen (C.Pls. 76 and 77).

When we look at the sun through an H-alpha filter, we are seeing the chromosphere. The *chromosphere* (from the Greek word for color, *chromos,* since it looks colorful during eclipses) is a thin layer just above the solar photosphere. The chromosphere is not really a layer; it is actually made up of spikes of gas called *spicules* that rise and fall with a period of 15 minutes each. The spikes are usually too delicate to see except with professional equipment used at exceptional sites.

H-alpha filters often come with fittings so they can be attached to the eyepiece of a telescope; in that location, they should be used with caution. When we look through an H-alpha filter, the solar surface looks mottled, and dark lines called *filaments* snake their

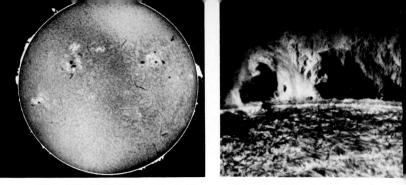

Fig. 152. (*left*) The sun's surface on July 25, 1981, photographed through an H-alpha filter that passes light in a band 0.7 angstroms wide. The central part of the image shows the solar disk, with dark filaments and bright active regions. Surrounding it is an image made at the same scale, but with the center of the disk occulted — hidden — by a device in the telescope. These photographs were taken by an amateur astronomer. (R.J. Poole/DayStar Filter Corp.)

Fig. 153. (*right*) A solar prominence. This is a quiescent (long-lasting) type; others may be eruptive. (Big Bear Solar Observatory, California Institute of Technology)

way across the solar surface (Fig. 152). When a filament is rotated so that it appears at the edge of the sun, it is then called a *prominence* (Fig. 153).

On any sunny day, you can see exciting things with a telescope pointed at the sun. Eyepiece projection or a special filter over the aperture reveals the sunspots. An H-alpha filter shows the filaments and prominences. Since the sun's surface changes from day to day, looking at the sun provides a changing image.

Some technical notes about equipment: if you are getting filters or other apparatus to observe the sun, it is best to check the advertisements in *Sky & Telescope* magazine to see what is currently available. At present, Celestron and Questar make high-quality solar filters that cut down the intensity of white light, though they are not cheap. DayStar makes H-alpha filters. A filter that passes a band of wavelengths narrower than 1 angstrom (0.0001 micrometer) is necessary to see the filaments on the solar disk; a filter that passes as much as 3 or 4 angstroms is sufficient to see the prominences around the solar limb.

Solar Eclipses: Nature's Spectacular

The most awesome sight we can see is a solar eclipse. The moon gradually covers the sun over a period of an hour or two, and then the crescent sun abruptly gives way to night. A few points of light on the edge of the sun — *Baily's beads* — are visible for a few seconds. A bright point of light on the edge of the moon, so bright that it glistens like a jewel and is called *the diamond-ring effect,* is the last Baily's bead (C.Pl. 82). It lasts 3 or 5 seconds and then is also covered by the moon. The pinkish chromosphere is visible for another few seconds; prominences may be visible somewhat longer. Scientists spread the sunlight out into its component colors (C.Pl. 80) for analysis.

The diamond-ring effect comes from the last bit of the solar photosphere shining through a valley on the edge of the moon. When the diamond ring is over, we see a halo of light around the sun — the *corona* (from the Latin word for "crown"). The corona is the tenuous layer of gas surrounding the sun at a temperature of 2,000,000°C. It always surrounds the sun, but is a million times fainter than the surface of the sun (the photosphere) and is normally fainter than the earth's blue sky. Normally the sky looks blue because light from the solar photosphere bounces around in it. For us to be able to see the chromosphere and corona, the sun must be up at a time when the sky is not illuminated, and that is exactly the situation we have during a total solar eclipse (Fig. 154).

In the hour or two before the total phase of the eclipse, when the sun is only partially covered by the moon (C.Pl. 78), even the remaining part of the sun is still too bright to look at safely. A few minutes before totality, the total amount of light from the photosphere is reduced enough that the eye-blink reflex that normally protects our eyes doesn't work. Even then, you can still hurt your eyes by looking at the remaining part of the sun without a special filter. Only when the diamond-ring effect is over — and nobody ever has trouble deciding when that is — can you take your special filter away from your eyes and stare at the sun.

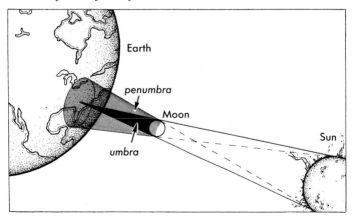

Fig. 154. From inside the umbra of the moon's shadow, we see a total eclipse of the sun. From inside the penumbra, we see a partial eclipse. In the drawing, the point of the umbra does not quite reach the earth, which means that the eclipse is annular, like the one in the southeastern U.S. on May 30, 1984. (Susan Eder)

Fig. 155. The 1980 total solar eclipse, photographed from India. We see the dark back side of the moon, surrounded by the spiky solar corona. The sun's diameter, barely hidden by the moon, is 1.4 million km, so we can see that the coronal streamers extend millions of kilometers into space. (Jay M. Pasachoff)

The corona is irregular in shape, with *streamers* extending millions of kilometers into space (Fig. 155, C.Pl. 79). The sun acts on the whole like a giant bar magnet, so we can sometimes see *polar tufts* — thin rays coming out of the sun's poles — as well as the streamers that appear at lower solar latitudes. The corona is expanding, forming the *solar wind* that extends throughout our solar system.

During the total part of the eclipse, when the corona is visible, the corona is perfectly safe to look at. It is then about the same brightness as the full moon. When the second diamond-ring effect occurs, marking the end of totality, you must again stop looking directly at the sun.

During totality, the sky is as dark as it is at night, or at least as dark as it is during evening twilight. If you look toward the horizon in any direction (C.Pl. 81), you can see beyond the darkest part of the moon's shadow. You will see a pinkish glow on the horizon that looks like a sunset extending 360° around the sky.

For the minute or two before totality begins or after it ends, narrow bands of shadow seem to race across the landscape. These *shadow bands* occur when inhomogeneous regions of the earth's upper atmosphere bend the narrow crescent of light from the partially eclipsed sun. Shadow bands can be seen especially well if you spread out a white sheet on the ground, perhaps with distances measured across it to help you determine the spacing between adjacent bands and the distance a band appears to travel each second (which can be converted into velocity in kilometers or miles per hour). It is also interesting to note the direction in which the bands are travelling, which is often different before and after totality.

Solar eclipses are not especially rare; partial ones, in which the central part of the moon's shadow never touches the earth, occur about 3 times every year. Total solar eclipses occur about every 18 months (Table 22). Since the moon and the sun appear approximately the same size in the sky — the sun is actually 400 times larger than the moon but happens to be 400 times farther away —

(*text continues on p. 408*)

Table 22. Solar Eclipses*

May 30, 1984, annular, lasting less than 11 seconds in the United States. The narrow path, only 4–11 km (2–7 miles) wide, crosses central Mexico and the southeastern U.S. Its center passes 20 km (12 miles) northwest of New Orleans, La.; 25 km (16 miles) northwest of Montgomery, Ala.; over Atlanta, Ga., Greenville, S.C., Greensboro, N.C., Petersburg, Va., and Assateague I., Md. The path ends in northwestern Africa. Partial phases will be visible over most of North America and Europe.

November 22, 1984, total. Totality in Irian Jaya, Indonesia, and Papua New Guinea will last just under 1 minute. The path is mostly over the Pacific Ocean, where the maximum duration is 1 minute 59 seconds and the maximum path width is 85 km (53 miles). Partial phases will be visible in Indonesia, Australia, New Zealand.

May 19, 1985, partial, maximum magnitude 0.84. Northeast Asia, Japan, northern North America, Iceland.

November 12, 1985, total. Very inaccessible — southernmost Pacific Ocean, down to area near Antarctica. Duration up to 1 min 59 sec.

April 9, 1986, partial, maximum magnitude 0.82. Visible in Indonesia, Australia, Papua New Guinea, and South Island of New Zealand.

October 3, 1986, annular, with a total midsection. Totality lasts only up to about 0.3 seconds in a path 2 km (1¼ mile) wide in the North Atlantic off Greenland and Iceland, with the sun only 5° in altitude. The partial phases will be visible in most of North America.

March 29, 1987, annular, with a total midsection. Totality lasts only up to about 8 seconds in a path only 5 km (3 miles) wide over the South Atlantic, off the coast of West Africa. Annularity enters West Africa near the equator and curves northeast to the Indian Ocean.

September 23, 1987, annular. Crosses China, Okinawa, and the Pacific; ends near Samoa. Partial phases visible in Asia and northeastern Australia.

March 17–18, 1988, total. Totality crosses Indonesia and the southern Philippines with a path 175 km (109 miles) wide and the sun about 65° in altitude, and then continues into the Pacific. Maximum totality is 3 minutes 46 seconds. Partial phases visible in eastern Asia, Indonesia, Micronesia, northwestern Australia, and western Hawaiian Islands.

September 11, 1988, annular. Annularity begins off Somalia, crosses the Indian Ocean, and passes south of Australia. Partial phases visible in eastern Africa, southern Asia, Indonesia, Australia, New Zealand, and Antarctica.

March 7, 1989, partial, maximum magnitude 0.83. Visible in Hawaii, northwestern North America, Greenland, and the Arctic.

August 31, 1989, partial, maximum magnitude 0.63. Visible in extreme southeastern Africa, Madagascar, and Antarctica.

January 26, 1990, annular. Begins in Antarctica and ends in South Atlantic. Partial phases visible in South Island of New Zealand and much of South America.

*Accessible total eclipses are listed in italics.

(*continued on page 408*)

Table 22 (contd.). Solar Eclipses

July 22, 1990, total. Begins in Finland and passes along the northern coasts of Europe and Asia. Path of totality is 208 km (130 miles) wide at maximum and sun is about 40° in altitude. Maxim..m totality is 2 minutes 33 seconds. Totality crosses Alaska's Aleutian Islands, with duration of 2 minutes 20 seconds. Seguam Island, Amlia Island, and Atka are in totality. Partial phases visible in northeastern Europe, northwestern North America, northern Asia, and Hawaii.

January 15–16, 1991, annular. The path of annularity crosses southwestern Australia, Tasmania, and New Zealand.

July 11, 1991, total. Crosses the "Big Island" of Hawaii in the state of Hawaii, with the central line passing very close to the Mauna Kea volcano and its observatory, where totality will last 4 minutes 8 seconds. The northern end of the path of totality will graze the southern coast of the Hawaiian island of Maui. The path of totality will pass over Mexico, where totality will last up to 6 minutes 54 seconds, and then over Central and South America. The maximum width of the path will be 258 km (161 miles).

the shadow is not more than about 200 miles (300 km) wide when it hits the earth. It traces a narrow path across the earth (Fig. 156), thousands of miles (kilometers) long but only tens or hundreds of miles (kilometers) wide. Only within that narrow path can you see the corona and experience the eclipse.

More and more people are taking long trips to see total solar eclipses; once you see one, it is hard to resist going to see another (and perhaps taking a friend along). Among the observations that you can make at an eclipse are: (1) timing the *contacts,* where first contact is when the moon first begins to cover the sun, second contact is when totality begins, third contact is when totality ends, and fourth contact is when the moon's image leaves the solar disk; (2) sketching or photographing the prominences; (3) sketching or photographing the corona; and (4) observing the shadow bands. Some people like to cover one eye with an eyepatch before totality, so the eye will be dark-adapted when totality begins. Many people like to observe the corona through binoculars during totality.

Photographing an eclipse is always interesting, and it is particularly nice that the choice of exposure is not critical. Each combination of lens opening (aperture) and exposure time gives a different effect, but all are pretty. The corona falls off rapidly in brightness the farther it is from the sun, so the longer the exposure time, the more corona you can see. But you also want to keep your exposure times short enough so that the moon's edge will not blur as the sun and moon move across the sky. Exposure times of up to about 1 second with a 500 mm telephoto lens, or 10 seconds with a "normal" 50 mm lens on a 35 mm camera, will do. It is important to use as steady a tripod as possible, to release the shutter carefully (preferably with a cable release), and to wait for the camera to stop vibrating after you advance the film (you may need to wait a sec-

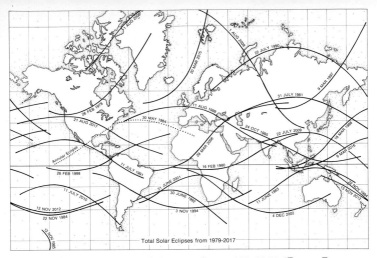

Fig. 156. Total eclipses of the sun, from 1979–2017. (Bryan Brewer, Earth View, Inc.)

ond or two). If possible, lock up the mirror on a reflex camera to reduce vibration.

Telephotos of at least 200 mm focal length and preferably 500 mm focal length give the best images of the corona. You can use normal or wide-angle lenses to take overall views that include the corona at the top of the scene and scenery in the foreground. You will need a special solar filter to photograph the partial phases of an eclipse; remove the filter to photograph the diamond ring, prominences, or corona. Then you must put it on again at the end of the eclipse, perhaps after having taken a quick picture of the second diamond ring. Recommended exposures appear in Table 23. A difficult tradeoff must be made between faster films like high-speed Ektachrome and slower but less grainy films like Koda-chrome; there is no "right answer" to this question. Usually slide film instead of print film should be used, because it captures a wider range of intensities; prints can be made from slides later on.

Some people see a partial eclipse, and wonder why others talk so much about a total eclipse. But the difference is like night and day. Since the photosphere is 1 million times brighter than the corona, even a 99% partial eclipse still leaves 1% of 1 million times or 10,000 times more light from the photosphere than there is from the corona. And so the sky is too bright to allow the corona to be seen during a partial eclipse. It is truly worth travelling to the zone of totality each time there is an eclipse. Seeing a partial eclipse and saying that you have seen an eclipse is like standing outside an opera house and saying that you have seen the opera; in both cases, you have missed the main event.

Because the orbits of the moon around the earth and the earth around the sun are elliptical, sometimes the moon appears too small to completely cover the sun during an eclipse. Then an *an-*

Fig. 157. Looking over someone's shoulder through a solar filter made of fogged and developed black-and-white film as he also looks through the solar filter, we see the annular eclipse of 1973. An entire circle of solar photosphere remained visible at all times. (Jay M. Pasachoff)

nulus — a ring — of photosphere remains visible (Fig. 157), and we have an *annular eclipse,* like the one visible from Mexico and the southeastern United States on May 30, 1984. Since the photosphere is always visible, special filters must be used throughout an annular eclipse, both for the eye and for a camera, or pinhole cameras could be used. You should use a telephoto lens of at least 200 mm focal length to take photos through filters.

On a technical note, the same special filters that were described on p. 404 can also be used at eclipses. Some cheaper materials, such as aluminized Mylar, can also be used for solar filters, and are available for this purpose from Tuthill, Inc.; look for advertisements in recent issues of *Sky & Telescope* or *Astronomy* magazines. The Mylar is inexpensive, but gives a bluish cast to the sun instead of a more pleasant orange one; it is also subject to potentially dangerous pinpricks.

You can make your own solar filter for observing an eclipse if you start a few days in advance. First completely fog a roll of black-and-white film, perhaps by simply unrolling it in the sunlight. Then develop it to maximum density. One or two thicknesses of this fogged and developed black-and-white film should cut down the solar intensity to a safe level; try two thicknesses first, and if the sun is not visible, try one thickness.

The silver in black-and-white film absorbs all sunlight across the entire spectrum, including the infrared (heat) radiation that is invisible to our eyes. Color film does not use silver, so it cannot be used to make a safe solar filter for eclipses or for any other observations of the sun. Also, gelatin photographic filters, such as Kodak's neutral-density Wratten filters, cannot be used for observing the sun with the naked eye, since they are effective only in the range for which photographic film is sensitive and pass infrared radiation that can be harmful to the eye.

No matter what filter you use, **never stare at the sun.** You will be able to see a great deal just by glancing through the filter at the sun for a few seconds. The partial phases do not change rapidly enough to make it interesting to look at the sun longer in any case.

Table 23. Solar Eclipse Photography

Do **not** look through your viewfinder at the sun—except during totality—unless the lens is covered with neutral-density filters.

Use film with the finest grain possible; for example, use Kodachrome 64 instead of any kind of Ektachrome for most purposes.

Mount your camera or telescope on as sturdy a tripod as possible. Exposure times for totality are not critical; most problems come from shaky mounts. Your camera will vibrate slightly each time you advance the film; wait a second or two to make sure the vibration has stopped before you take the next exposure. Use a cable release, and if possible, lock up the mirror on reflex cameras.

For close-ups of totality with a 35 mm camera, use a telephoto lens of at least 300 mm and preferably 500 mm.

Bracket your exposures widely—take a variety of exposures on either side of any given lens opening and shutter speed.

Make sure that you don't run out of film in the middle of your exposure sequence.

A sample eclipse sequence follows, assuming that you are using ASA/ISO 64 film (such as Kodachrome 64):

Partial phases: Every five minutes, take a bracketed series of exposures through a filter of neutral density 5, that is, a filter that passes only 1/100,000 (where the density 5 means that there are 5 zeroes) of the incoming sunlight. **Note:** Do *not* look through any viewfinders unless they are also covered with a neutral-density 5 filter; Wratten photographic neutral-density filters are *not safe* for the eye. If you have made your own filter out of exposed and developed photographic film, take a wide range of exposures to make certain that you have a good one. Remember that your meter reading will probably not be accurate since it will average the light available over the entire field of view, while you are photographing a bright object surrounded by darkness. Put in a fresh roll of film five minutes before totality.

Diamond ring: Take off the neutral-density filters, and carefully, without shaking the camera, take one or two exposures at about 1/30 second at $f/8$.

Prominences: As soon as the diamond ring disappears, take a series of exposures at $f/8$ for 1/60, 1/30, 1/15, and 1/8 second.

Corona: During totality, take a series of time exposures. If you are using an $f/8$ or $f/5.6$ lens, try 1/2 second, 1 second, 2 seconds, and 4 seconds. If any time is left, repeat the sequence. Then get ready to take the diamond-ring effect at the end of totality.

If you are using a wide-angle lens, open your lens as wide as possible (to $f/2.8$, $f/2$, or $f/1.4$, for example). Use fast film, such as ASA/ISO 200 or ASA/ISO 1000. Make certain that the eclipse is in the frame, near the top, and that some objects are visible in the foreground at the bottom of the frame. Take 1, 2, 4, and 8 second exposures.

At the end of the eclipse: Photograph the second diamond ring without looking through the viewfinder, then turn your camera so it is no longer facing the sun. (Otherwise, the sunlight focused through the lens could burn a hole in your shutter.) Add your neutral-density filters later for occasional photographs of the partial phases.

14

Coordinates, Time, and Calendars

As the earth rotates on its axis, the sky overhead seems to turn in the opposite direction. The stars appear to rise in the east, move overhead in the course of the night, and set in the west. It is convenient for many purposes to think of the stars as being fixed on a *celestial sphere* that rotates every 24 hours.

Positions in the Sky: Right Ascension and Declination

On the surface of the earth, we measure the positions of cities and towns in longitude and latitude. Longitude is measured by lines that extend from north pole to south pole, usually up to 180° E and 180° W of a zero-degree (0°) line that goes through Greenwich, England.

Latitude on earth is measured by parallel lines that mark the number of degrees north or south of the equator. The north pole is +90°; the longitude scale has no meaning there, since all longitudes converge at the poles. Similarly, the south pole is −90°.

Astronomers have set up a similar set of coordinates in the sky. The celestial equator is an imaginary line around the sky, above the earth's equator. The celestial poles are the imaginary points where extensions of the earth's axis into space would meet the celestial sphere; they lie above the earth's poles.

The astronomers' coordinate that is similar to longitude is called *right ascension.* A line that extends over the sky from the celestial north pole through your zenith to the celestial south pole is your *meridian.* It crosses the celestial equator perpendicularly, and at any one moment marks a line of constant right ascension. We measure right ascension along the celestial equator. Since the sky seems to turn overhead once every 24 hours, we measure right ascension in *hours.* The celestial equator is divided into 24 hours of right ascension; since the sky turns a full circle of 360° each 24 hours, each hour of right ascension equals 15°. Each *minute* of right ascension is $\frac{1}{60}$ of 15°, and equals 15′ (15 minutes of arc). Each *second* of right ascension is equal to $\frac{1}{60}$ of a minute, and so is equal to 15″ (15 seconds of arc). When we speak of "minutes" or "seconds," we must be clear as to whether we mean units of arc or of time.

On the Atlas Charts in Chapter 7, right ascension is marked on

Table 24. Angular Units for Coordinates in the Sky

right ascension	units of arc (degrees, minutes, or seconds)
24^h	360°
1^h	15°
4^m	1°
1^m	15′
4^s	1′
1^s	15″

the horizontal axis. As we look up when facing north, the sky seems to rotate counterclockwise around the north celestial pole, which is marked by the North Star, Polaris. Above Polaris in the sky, we see stars rising in the east (to our right), moving overhead, and setting in the west (to our left). But some stars are *circumpolar,* that is, they always stay above the horizon. They circle under Polaris from our west (left) to our east (right). When we look at them underneath Polaris, we have a view similar to that of the Atlas Charts in Chapter 7, on which north is at the top. The hours of right ascension passing overhead increase as time goes on (1^h, 2^h, 3^h, etc.), so right ascension increases from right to left on the Atlas Charts.

The vertical axis on the charts is *declination,* the number of degrees north (+) or south (−) of the celestial equator. Thus the celestial equator has declination 0°, the north celestial pole has declination +90°, and the south celestial pole has declination −90°. The regions of right ascension and declination shown on the Atlas Charts are set out in Table 14 and are outlined on the endpapers of this guide.

In order to fix the scale of right ascension, we must arbitrarily assume some particular line to be zero. We choose this zero point, 0^h of right ascension, to be the *vernal equinox.* The equinoxes are the two points where the celestial equator crosses the *ecliptic,* the path the sun follows across the sky in the course of the year. The vernal equinox is the one of these points that the sun crosses on its way north each year; the other is the *autumnal equinox.*

Although day and night are theoretically equal in length on the days of the equinoxes, that would be true only if the sun were a point, not a disk, and if the earth's atmosphere did not bend sunlight. However, the top of the sun actually rises a few minutes before the center of the sun's disk — the point used in calculations. Also, the earth's atmosphere bends sunlight, so we can see the sun for several minutes before the time sunrise would occur and after the time sunset would occur if the earth had no atmosphere.

To find a star's celestial coordinates (Fig. 158), we measure the

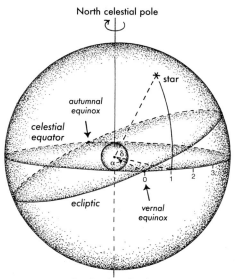

North celestial pole

South celestial pole

Fig. 158. Right ascension, noted by α (alpha), is measured in hours eastward around the celestial equator from the vernal equinox. Declination, noted by δ (delta), is measured in degrees north (+) or south (−) of the celestial equator. (Susan Eder)

number of hours around the celestial equator to its hour circle, and the number of degrees north or south of the celestial equator to its declination. Right ascension is often abbreviated either *r.a.* or with the Greek letter α (alpha). Declination is often abbreviated either *dec.* or with the Greek letter δ (delta). A star's right ascension is equal to the length of time that elapses after the vernal equinox crosses your meridian until the star crosses your meridian. The interval is measured in *sidereal time;* one rotation of the sky takes 24 hours of sidereal time.

The stars are essentially fixed in the sky, and so their right ascension and declination do not change measurably over short periods of time. The sun, moon, and planets, though, wander through the sky with respect to the stars; their right ascension and declination change during the course of a year.

Precession

The axis on which the earth spins is not perpendicular to the plane of the earth's orbit. It is, rather, tipped by $23\frac{1}{2}°$. The celestial coordinate system is complicated somewhat by the fact that the earth's axis wobbles as the earth spins, just as a spinning top wobbles. For the earth, gravity from the sun and moon cause the wobbling, the same way that the earth's gravity makes a top wobble. The wobbling is called *precession.*

The earth's axis actually traces out a huge curve in the sky over a 26,000-year period. Polaris is only temporarily the North Star; in about 13,000 years, Vega will be within a few degrees of the north pole. As a result of the precession of the earth's axis, the celestial coordinates of stars "precess" slightly with time; that is, they change their values from year to year. The vernal equinox moves slowly westward along the ecliptic at about 50″ per year. The effect is not large, but must be taken into account to plot the positions of astronomical objects accurately.

Stellar positions are usually given in one of a few "standard epochs," such as the year 1950 or 2000. Dates within a year are given as decimals, such as 1989.4 or 2000.0. Changing from the epoch in which positions are given to the current date is now easily done with a pocket calculator; the computers that operate many telescopes do it automatically. Useful formulae for precession are:

$$\text{new r.a.} = \text{r.a.} + [3.074^s + 1.336^s \sin{(\text{r.a.})} \tan{(\text{dec.})}] \times N,$$
$$\text{and new dec.} = \text{dec.} + 20.04'' \cos{(\text{r.a.})} \times N,$$

where N is the number of years since a standard epoch. In this *Field Guide,* we join the International Astronomical Union in adopting the year 2000.0 as the standard epoch. Thus N will be negative until the year 2000. (For 1985, for example, $N = -15$.) In any case, casual sky observers do not need to worry about calculating precession.

Since the constellation boundaries fixed by the International Astronomical Union in 1930 were defined to lie along right ascension and declination lines for epoch 1875.0, the constellation boundaries no longer coincide with round numbers in epoch 2000.0 coordinates, as is clear from the Atlas Charts in Chapter 7.

A smaller effect than precession, called *nutation,* is a "nodding" of the earth's axis caused chiefly by changes in the location of the moon's orbit. The coordinates thus change in a small ellipse, with axes of only 18.5″ and 13.7″, over a 19-year period. No amateur telescope would have to be set to that degree of accuracy.

Another small effect, the *aberration of starlight,* is a shift of up to 20.5″ resulting from the fact that the earth is moving through space. Just as you have to tilt your umbrella slightly forward to keep dry when moving rapidly through a rainstorm, since your motion makes the raindrops appear to slant toward you, aberration makes astronomers tilt their telescopes slightly forward as the earth orbits the sun.

Positions in most star catalogues, including the tables in this guide, take precession into account, but not nutation or aberration.

Time by the Stars

The stars return to the same point overhead where they were on the preceding night, but in that time, the earth has gone part (about $\frac{1}{365}$) of the way around the sun. So the earth has to rotate a little further for the sun to return to the same place in our sky (one *solar day*) compared to the length of time it takes the stars to return to the same place in our sky (one *sidereal day*). One day divided by 365 is about 4 minutes, so the solar day is longer than the sidereal day by about 4 minutes, actually 3^m56^s.

At about the vernal equinox each year, sidereal midnight (the beginning of a sidereal day) and solar "noon" coincide: 0 hours sidereal time = 12^h solar time. Then they drift apart by 4 minutes per day for another year. At the autumnal equinox, 12 hours sidereal time = 12^h solar time. The conversions between sidereal time and solar time are tabulated in Appendix Table A-9.

Your sidereal time is the number of hours since the vernal equinox has crossed your meridian. Your sidereal time is also equal to the right ascension of stars now crossing your meridian. Thus if you know your sidereal time, you know which stars are most favorably placed for observing, *i.e.,* those within a few hours of the current sidereal time, if they are in the right range of declination.

The Big Dipper and the North Star actually make a convenient sidereal clock in the sky. A line from Polaris through the Pointers sweeps counterclockwise around the sky once a day. When that line is pointed straight upward, the sidereal time is 11^h, since the right ascension of the Pointers is almost exactly 11^h. When the line from Polaris through the Pointers goes due left, toward due west on the horizon, the sidereal time is 6 hours later, or 17^h. When the Pointers are below Polaris so the line from Polaris through the Pointers goes straight downward, the sidereal time is 23^h.

Calendars and Time by the Sun

One solar day for any observer is the length of time that the sun takes to return to our meridian. It takes about $365\frac{1}{4}$ of those solar days to make a year. Instead of changing our clocks by $\frac{1}{4}$ day each year, we have three 365-day years followed by one 366-day *leap year,* a scheme worked out by the astronomer Sosigenes for Julius Caesar in 46 B.C.

But a year is 365.2422 days long, which is not exactly $365\frac{1}{4}$ days. In 1582, the time had slipped substantially from the proper season because of this difference; the Gregorian calendar skipped 10 days to compensate and made some new rules about leap year: the "century years" (1900, 2000, etc.) would not be leap years unless they are evenly divisible by 400. That is, 1800 and 1900 were not leap years, but 2000 will be a leap year.

Actual *apparent solar time* does not advance at a constant rate through the year, because the earth travels at varying speeds as it

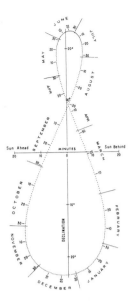

Fig. 159. The analemma, the figure-8 that shows how far mean solar time is ahead of or behind true solar time for a given day of the year. (*Sky & Telescope*)

traverses its elliptical orbit. As a result, it is more convenient for us to keep *mean solar time,* the average rate of solar time. Watches keep time at the same rate as mean solar time.

Your *local apparent solar time* is when the sun crosses your meridian. But your neighbors a few kilometers to the east or west have different meridians than yours, so their local time will be slightly different than yours. To compensate for these slight differences, we use a system of *standard time,* in which the solar time is the same in a band of earth longitudes 15° wide. The meridian of Greenwich is centered in one of these bands. The actual boundaries of the time zones often bend to follow political divisions between states, provinces, or countries.

The local apparent solar time minus the local mean solar time is called *the equation of time.* The difference can be as large as 16 minutes; we see its effect in Fig. 159.

The equation of time shows how far the real sun is ahead or behind the mean sun. As this value changes through the year, the sun also goes higher and lower in the sky. Thus the position of the sun at a given standard (mean) time each day appears to follow a figure-8 called the *analemma* (Fig. 159). C.Pl. 71 is a single multiple-exposure photograph, with an exposure of the sun taken at 8:30 a.m. E.S.T. every two weeks. The summer solstice is the highest point and the winter solstice is the lowest point; the lens was left open from dawn until shortly before the time for the standard exposure on those days and on the day of the crossover. The exposure of the foreground building was added on still another day, without the dense filter that normally allowed only the sun to be visible.

Astronomers often keep time in *Universal Time* (U.T.), basically the time for the meridian of Greenwich, with 0^h occurring at midnight. (The name "Greenwich mean time," GMT, is no longer commonly used, since it was ambiguous whether it began at noon or midnight.) A more accurate version of time calculated after the fact, incorporating small corrections, is called Ephemeris Time (E.T.), and is used for ephemeris calculations. E.T. is later than U.T. by about 1 minute; though we cannot say definitely in advance, the difference will be about 54.5 seconds on January 1, 1984, and will advance about 1 second per year. U.T. and E.T. are often kept on a 24-hour clock. To keep U.T. and E.T. aligned with time kept by atomic clocks, which keep time unaffected by changes in the earth's rotation, leap seconds are occasionally added at the end of December or June.

Time gets earlier as you move westward on earth. At a given instant, you subtract the following number of hours to go from U.T. to time zones in the U.S. and Canada:

Table 25. Time Zones

To convert from Universal Time (U.T.) to	To get Standard Time, subtract:	To get Daylight Saving Time, subtract:
Atlantic Time Zone	4 hours	3 hours
Eastern Time Zone	5 hours	4 hours
Central Time Zone	6 hours	5 hours
Mountain Time Zone	7 hours	6 hours
Pacific Time Zone	8 hours	7 hours
most of Alaska	9 hours	8 hours
Hawaii Time Zone	10 hours	—

Most states employ Daylight Saving Time, in which you add one hour to the Standard Time for the summer months; Arizona, Hawaii, and parts of Indiana have Standard Time all year. To remember to add one hour in the spring and subtract it again in the fall, think of "spring ahead, fall back."

Halfway around the world from Greenwich, in the middle of the Pacific, the date changes by one at the *International Date Line,* so that the same hours on the same day don't keep endlessly circling the world. The time gets earlier by one hour each time you cross a time zone as you go from east to west; as you cross the International Date Line from east to west, one day is added.

Appendix Tables

Glossary

Bibliography

Telescope Information

Index

Table A-1. The Constellations

Abbre-viation	Latin Name	Pronun-ciation	Genitive	Translation
And	Andromeda	ăn-drŏm'ə-də	Andromedae	Andromeda
Ant	Antlia	ănt'lē-ə	Antliae	Pump
Aps	Apus	ā'pəs	Apodis	Bird of Paradise
Aqr	Aquarius	ə-kwâr'ē-əs	Aquarii	Water Bearer
Aql	Aquila	ăk'wə-lə	Aquilae	Eagle
Ara	Ara	ā'rə	Arae	Altar
Ari	Aries	âr'ēz, âr'ē-ēz'	Arietis	Ram
Aur	Auriga	ô-rī'gə	Aurigae	Charioteer
Boo	Boötes	bō-ō'tēz	Boötis	Herdsman
Cae	Caelum	sē'ləm	Caeli	Chisel
Cam	Camelopardalis	kə-mĕl'ō-pär'-də-lĭs	Camelopardalis	Giraffe
Cnc	Cancer	kăn'sər	Cancri	Crab
CVn	Canes Venatici	kā'nēz vī-năt'ə-sī'	Canum Venaticorum	Hunting Dogs
CMa	Canis Major	kā'nĭs mā'jər	Canis Majoris	Big Dog
CMi	Canis Minor	kā'nĭs mī'nər	Canis Minoris	Little Dog
Cap	Capricornus	kăp'rĭ-kôr'nəs	Capricorni	Goat
Car	Carina	kə-rī'nə	Carinae	Ship's Keel
Cas	Cassiopeia	kăs'ē-ə-pē'ə	Cassiopeiae	Cassiopeia
Cen	Centaurus	sĕn-tôr'əs	Centauri	Centaur
Cep	Cepheus	sē'fyōōs'	Cephei	Cepheus
Cet	Cetus	sē'təs	Ceti	Whale
Cha	Chamaeleon	kə-mēl'yən	Chamaeleonis	Chameleon
Cir	Circinus	sûr'sə-nəs	Circini	Compass
Col	Columba	kə-lŭm'bə	Columbae	Dove
Com	Coma Berenices	kō'mə bĕr'ə-nī'sēz	Comae Berenices	Berenice's Hair
CrA	Corona Australis	kə-rō'nə ôstrā'lĭs	Coronae Australis	Southern Crown
CrB	Corona Borealis	kə-rō'nə bôr'ē-ăl'ĭs	Coronae Borealis	Northern Crown
Crv	Corvus	kôr'vəs	Corvi	Crow
Crt	Crater	krā'tər	Crateris	Cup
Cru	Crux	krŭks	Crucis	Southern Cross
Cyg	Cygnus	sĭg'nəs	Cygni	Swan
Del	Delphinus	dĕl-fī'nəs	Delphini	Dolphin
Dor	Dorado	də-rä'dō	Doradus	Swordfish
Dra	Draco	drā'kō	Draconis	Dragon
Equ	Equuleus	ĭ-kwōō'lē-əs	Equulei	Little Horse
Eri	Eridanus	ĭ-rĭd'n-əs	Eridani	River Eridanus
For	Fornax	fôr'năks'	Fornacis	Furnace
Gem	Gemini	jĕm'ə-nī'	Geminorum	Twins
Gru	Grus	grŭs	Gruis	Crane
Her	Hercules	hûr'kyə-lēz'	Herculis	Hercules
Hor	Horologium	hôr'ə-lō'jē-əm	Horologii	Clock
Hya	Hydra	hī'drə	Hydrae	Water Snake
Hyi	Hydrus	hī'drəs	Hydri	Water Snake
Ind	Indus	ĭn'dəs	Indi	Indian

Table A-1 (contd.). The Constellations

Abbre-viation	Latin Name	Pronun-ciation	Genitive	Translation
Lac	Lacerta	lə-sûr′tə	Lacertae	Lizard
Leo	Leo	lē′ō	Leonis	Lion
LMi	Leo Minor	lē′ō mī′nər	Leonis Minoris	Little Lion
Lep	Lepus	lē′pəs	Leporis	Hare
Lib	Libra	lī′brə	Librae	Scales
Lup	Lupus	lōō′pəs	Lupi	Wolf
Lyn	Lynx	lĭngks	Lyncis	Lynx
Lyr	Lyra	lī′rə	Lyrae	Harp
Men	Mensa	mĕn′sə	Mensae	Table
Mic	Microsco-pium	mī′krə-skō′-pē-əm	Microscopii	Microscope
Mon	Monoceros	mə-nŏs′ər-əs	Monocerotis	Unicorn
Mus	Musca	mŭs′kə	Muscae	Fly
Nor	Norma	nôr′mə	Normae	Level
Oct	Octans	ŏk′tănz′	Octantis	Octant
Oph	Ophiuchus	ŏf′ē-yōō′kəs	Ophiuchi	Ophiuchus
Ori	Orion	ō-rī′ən	Orionis	Orion
Pav	Pavo	pä′vō	Pavonis	Peacock
Peg	Pegasus	pĕg′ə-səs	Pegasi	Pegasus
Per	Perseus	pûr′sē-əs	Persei	Perseus
Phe	Phoenix	fē′nĭks	Phoenicis	Phoenix
Pic	Pictor	pĭk′tər	Pictoris	Easel
Psc	Pisces	pī′sēz	Piscium	Fish
PsA	Piscis Austrinus	pī′sis ôs-trī′nəs	Piscis Austrini	Southern Fish
Pup	Puppis	pŭp′ĭs	Puppis	Ship's Stern
Pyx	Pyxis	pĭk′sĭs	Pyxidis	Ship's Compass
Ret	Reticulum	rĭ-tĭk′yə-ləm	Reticuli	Net
Sge	Sagitta	sə-jĭt′ə	Sagittae	Arrow
Sgr	Sagittarius	săj′ə-târ′ē-əs	Sagittarii	Archer
Sco	Scorpius	skôr′pē-əs	Scorpii	Scorpion
Scl	Sculptor	skŭlp′tər	Sculptoris	Sculptor
Sct	Scutum	skyōō′təm	Scuti	Shield
Ser	Serpens	sûr′pənz	Serpentis	Serpent
Sex	Sextans	sĕks′təns	Sextantis	Sextant
Tau	Taurus	tôr′əs	Tauri	Bull
Tel	Telescopium	tĕl′ə-skō′pē-əm	Telescopii	Telescope
Tri	Triangulum	trī-ăng′gyə-ləm	Trianguli	Triangle
TrA	Triangulum Australe	trī-ăng′gyə-ləm ô-strā′lē	Trianguli Australis	Southern Triangle
Tuc	Tucana	tōō-kä′nə	Tucanae	Toucan
UMa	Ursa Major	ûr′sə mā′jər	Ursae Majoris	Big Bear
UMi	Ursa Minor	ûr′sə mī′nər	Ursae Minoris	Little Bear
Vel	Vela	vē′lə	Velorum	Ship's Sails
Vir	Virgo	vûr′gō	Virginis	Virgin
Vol	Volans	vō′länz	Volantis	Flying Fish
Vul	Vulpecula	vŭl-pĕk′yə-lə	Vulpeculae	Little Fox

Table A-2. The Brightest Stars, to Magnitude 3.5

This list includes 287 stars, down to and including visual magnitude 3.5. The first column is the star's Flamsteed number, the second is the abbreviation of its Greek-letter Bayer designation and its constellation. For the few stars that have neither a Flamsteed number nor a Bayer designation, the appropriate numbers in the Smithsonian Astrophysical Observatory (SAO) catalogue are given. Positions are given for epoch 2000.0.

V is the *visual magnitude* — the magnitude measured through a standard yellowish filter that simulates the range of sensitivity of the human eye. (For variable stars, the maximum brightness is given.) B − V is a *color index*, the magnitude seen through a standard blue filter minus the visual magnitude; that gives the star's temperature. Then follows the star's absolute magnitude (M_v, the magnitude the star would have if it were at a distance of 10 parsecs = 32.6 light-years) and its spectral type. In the spectral type column, Roman numerals indicate the star's luminosity class: V indicates normal dwarfs like the sun; IV indicates subgiants; III, giants; II, bright giants; and I, supergiants. The notation Md means that the star (Mira — omicron Ceti) is a dwarf M star. W marks Wolf-Rayet stars — stars that show broad emission lines (from gas they are ejecting) in addition to absorption lines in their spectrum; the C in WC stands for strong radiation from carbon and oxygen.

Under **notes**, an *m* means that the star is part of a binary or multiple system or has a close optical neighbor, a *v* means that the star is variable, and a question mark means that variability is questionable or suspected.

Another popular name for the star, if any, appears in the last column.

Name and Catalogue Numbers		Position (2000.0) r.a. h m s	dec. ° ′ ″	Magnitudes V	B − V	M_v	Spectral Type	Distance (l-y)	Notes	Other Name
21	α And	0 08 23	+29 05 26	2.1	−0.11	0.3	A0	72	m	Alpheratz
11	β Cas	0 09 11	+59 08 59	2.3	0.3	1.9	F2 IV	42	m	Caph
88	γ Peg	0 13 14	+15 11 01	2.8	−0.2	−3.0	B2 IV	490	mv	Algenib
	β Hyi	0 25 45	−77 15 16	2.8	0.6	3.8	G1 IV	21		
	α Phe	0 26 17	−42 18 22	2.4	1.1	0.2	K0 III	78	m	Ankaa
31	δ And	0 39 20	+30 51 40	3.3	1.3	−0.2	K3 III	160	m	
18	α Cas	0 40 30	+56 32 15	2.2	1.2	−0.9	K0 II	120	mv?	Schedar
16	β Cet	0 43 35	−17 59 12	2.0	1.0	0.2	K0 III	68		Deneb Kaitos
24	η Cas	0 49 06	+57 48 58	3.4	0.6	4.6	G0 V	19	m	

Table A-2 (contd.). The Brightest Stars, to Magnitude 3.5

Name and Catalogue Numbers	Position (2000.0) r.a. h m s	dec. ° ' "	Magnitudes V	B − V	M$_v$	Spectral Type	Distance (l-y)	Notes	Other Name
27 γ Cas	0 56 42	+60 43 00	2.5	−0.2	−4.6	B0 IV	780	mv	
β Phe	1 06 05	−46 43 07	3.3	0.9	0.3	G8 III	130	m	
31 η Cet	1 08 35	−10 10 56	3.5	1.2	−0.1	K2 III	120	m	
43 β And	1 09 44	+35 37 14	2.1	1.6	−0.4	M0 III	88	m	Mirach
37 δ Cas	1 25 49	+60 14 07	2.7	0.1	2.1	A5 V	62	mv?	Ruchbah
γ Phe	1 28 22	−43 19 06	3.4	1.6	−4.4	K5 V	910		
α Eri	1 37 43	−57 14 12	0.5	−0.2	−1.6	B5 IV	85		Achernar
52 τ Cet	1 44 04	−15 56 15	3.5	0.7	5.7	G8 V	12	m	
2 α Tri	1 53 05	+29 34 44	3.4	0.5	2.2	F6 IV	59	m	
45 ε Cas	1 54 24	+63 40 13	3.4	−0.2	−2.9	B3 III	520		
6 β Ari	1 54 38	+20 48 29	2.6	0.1	2.1	A5 V	46		Sheratan
α Hyi	1 58 46	−61 34 12	2.9	0.3	2.6	F0 V	36		
57 γ¹ And	2 03 54	+42 19 47	2.2	1.2	−0.1	K2 III	121	m	Almach
13 α Ari	2 07 10	+23 27 45	2.0	1.2	−0.1	K2 III	85		Hamal
4 β Tri	2 09 33	+34 59 14	3.0	0.1	0.3	A5 III	110		
68 o Cet	2 19 21	−2 58 39	3.0	1.4	0.9	Md	95	mv	Mira
1 α UMi	2 31 50	+89 15 51	2.0	0.6	−4.6	F8 I	—	mv	Polaris
86 γ Cet	2 43 18	+3 14 09	3.5	0.1	1.4	A2 V	75	m	
θ¹ Eri	2 58 16	−40 18 17	3.2	0.1	1.7	A3 V	55	m	Acamar
92 α Cet	3 02 17	+4 05 23	2.5	1.6	−0.5	M2 III	130		Menkar
23 γ Per	3 04 48	+53 30 23	2.9	0.7	0.3	G8 III	110	m	
25 ρ Per	3 05 11	+38 50 25	3.4	1.7	−0.5	M4 III	200	v	
26 β Per	3 08 10	+40 57 21	2.1	−0.1	−0.2	B8 V	95	mv	Algol
33 α Per	3 24 19	+49 51 40	1.8	0.5	−4.6	F5 I	620	m	Mirfak

Table A-2 (contd.). The Brightest Stars, to Magnitude 3.5

Name and Catalogue Numbers	Position (2000.0) r.a. h m s	dec. ° ′ ″	Magnitudes V	B − V	M_v	Spectral Type	Distance (l-y)	Notes	Other Name
39 δ Per	3 42 55	+47 47 15	3.0	−0.1	−2.2	B5 III	330	m	
γ Hyi	3 47 15	−74 14 20	3.2	1.6	−0.4	M0 III	160		
25 η Tau	3 47 29	+24 06 18	2.9	−0.1	−1.6	B7 III	240	m	Alcyone
44 ζ Per	3 54 08	+31 53 01	2.9	0.1	−5.7	B1 I	1,110	m	Atik
45 ε Per	3 57 51	+40 00 37	2.9	−0.2	0.1	B0 V	130	m	
34 γ Eri	3 58 02	−13 30 31	3.0	1.6	−0.4	M0 III	140	m	Zaurak
35 λ Tau	4 00 41	+12 29 25	3.5	−0.1	−1.7	B3 V	330	v	
α Ret	4 14 26	−62 28 26	3.4	0.9	−2.1	G6 II	390	m	
78 θ² Tau	4 28 40	+15 52 15	3.4	0.2	0.5	A7 III	120	m	
α Dor	4 34 00	−55 02 42	3.3	−0.1	−0.6	A0 III	190	m	
87 α Tau	4 35 55	+16 30 33	0.9	1.5	−0.3	K5 III	68	m	Aldebaran
1 π³ Ori	4 49 50	+6 57 41	3.2	0.5	3.8	F6 V	25	m	
3 ι Aur	4 56 59	+33 09 58	2.7	1.5	−2.3	K3 II	270		
7 ε Aur	5 01 58	+43 49 24	3.0	0.5	−8.5	F0 I	4,570	mv	
2 ε Lep	5 05 28	−22 22 16	3.2	1.5	−0.3	K5 III	160		
10 η Aur	5 06 31	+41 14 04	3.2	−0.2	−1.7	B3 V	200		
67 β Eri	5 07 51	−5 05 11	2.8	0.1	0.0	A3 III	90	m	Cursa
5 μ Lep	5 12 56	−16 12 20	3.3	−0.1	−0.8	B9 III	220		
19 β Ori	5 14 32	−8 12 06	0.1	0.0	−7.1	B8 I	910	m	Rigel
13 α Aur	5 16 41	+45 59 53	0.1	0.8	0.3	G8 III	42	m	Capella
28 η Ori	5 24 29	−2 23 50	3.4	−0.2	−3.5	B1 V	750	mv	
24 γ Ori	5 25 08	+6 20 59	1.6	−0.2	−3.6	B2 III	360	m	Bellatrix
112 β Tau	5 26 18	+28 36 27	1.7	−0.1	−1.6	B7 III	130	m	Elnath
9 β Lep	5 28 15	−20 45 35	2.8	0.8	−2.1	G2 II	320	m	Nihal

424

Table A-2 (contd). The Brightest Stars, to Magnitude 3.5

Name and Catalogue Numbers	Position (2000.0) r.a. h m s	dec. ° ′ ″	Magnitudes V	B − V	M_v	Spectral Type	Distance (l-y)	Notes	Other Name
34 δ Ori	5 32 00	−0 17 57	2.2	−0.2		O9 II		mv	Mintaka
11 α Lep	5 32 44	−17 49 20	2.6	0.2	−4.7	F0 I	950	m	Arneb
39 λ Ori	5 35 08	+9 56 02	3.4	−0.2	2.0	O8		m	Meissa
44 ι Ori	5 35 26	−5 54 36	2.8	−0.2	−6.0	O9 III	1,860	m	
46 ε Ori	5 36 13	−1 12 07	1.7	−0.2	−6.2	B0 I	1,210	m	Alnilam
123 ζ Tau	5 37 39	+21 08 33	3.0	−0.2	−3.0	B2 IV	490		
α Col	5 39 39	−34 04 27	2.6	−0.1	−0.2	B8 V	120		Phact
50 ζ Ori	5 40 46	−1 56 34	1.8	−0.2	−5.9	O9 I	1,110	m	Alnitak
53 κ Ori	5 47 45	−9 40 11	2.1	−0.2	0.4	B0 I	68		Saiph
β Col	5 50 58	−35 46 06	3.1	1.2	−0.1	K2 III	140		Wazn
58 α Ori	5 55 10	+7 24 26	0.5	1.9	−5.6	M2 I	310	mv	Betelgeuse
34 β Aur	5 59 32	+44 56 51	1.9	0.0	0.6	A2 IV	72	mv	Menkalinan
37 θ Aur	5 59 43	+37 12 45	2.6	−0.1	2.2	A0	82	m	
7 η Gem	6 14 53	+22 30 24	3.3	1.6	−0.5	M3 III	190	mv	Propus
1 ζ CMa	6 20 19	−30 03 48	3.0	−0.2	−1.7	B3 V	290		Furud
2 β CMa	6 22 42	−17 57 22	2.0	−0.2	−4.8	B1 II	720	mv	Mirzam
13 μ Gem	6 22 58	+22 30 49	2.9	1.6	−0.5	M3 III	150	m	
α Car	6 23 57	−52 41 44	−0.7	0.2	−8.5	F0 I	1,170		Canopus
24 γ Gem	6 37 43	+16 23 57	1.9	0.0	0.0	A0 IV	85	m	Alhena
ν Pup	6 37 46	−43 11 45	3.2	−0.1	−1.2	B8 III	250		
27 ε Gem	6 43 56	+25 07 52	3.0	1.4	−4.5	G8 I	690	m	Mebsuta
9 α CMa	6 45 09	−16 42 58	−1.5	0.0	1.4	A1 V	9	m	Sirius
31 ξ Gem	6 45 17	+12 53 44	3.4	0.4	0.7	F5 III	75		
α Pic	6 48 11	−61 56 29	3.3	0.2	2.1	A5 V	52		

Table A-2 (contd.). The Brightest Stars, to Magnitude 3.5

Name and Catalogue Numbers	Position (2000.0) r.a. h m s	dec. ° ' "	Magnitudes V	B−V	M_v	Spectral Type	Distance (l-y)	Notes	Other Name
τ Pup	6 49 56	−50 36 53	2.9	1.2	0.2	K0 III	82		
21 ε CMa	6 58 38	−28 58 20	1.5	−0.2	−4.4	B2 II	490	m	Adhara
22 σ CMa	7 01 43	−27 56 06	3.5	1.7	−5.7	M0 I	1,500	m	
24 o² CMa	7 03 01	−23 50 00	3.0	−0.1	−6.8	B3 I	2,810		
25 δ CMa	7 08 23	−26 23 36	1.9	0.7	−8.0	F8 I	3,070		
π Pup	7 17 09	−37 05 51	2.7	1.6	−0.3	K5 III	130		
31 η CMa	7 24 06	−29 18 11	2.4	−0.1	−7.0	B5 I	2,480	m	Aludra
3 β CMi	7 27 09	+8 17 21	2.9	−0.1	−0.2	B8 V	140	m	Gomeisa
σ Pup	7 29 14	−43 18 05	3.3	1.5	−0.3	K5 III	170		
66 α Gem	7 34 36	+31 53 18	1.6	0.0	1.2	A1 V	46	m	Castor
10 α CMi	7 39 18	+5 13 30	0.4	0.4	2.6	F5 IV	11	m	Procyon
78 β Gem	7 45 19	+28 01 34	1.1	1.0	0.2	K0 III	36	m	Pollux
7 ξ Pup	7 49 18	−24 51 35	3.3	1.2	−4.5	G3 I	750	m	
χ Car	7 56 47	−52 58 56	3.5	−0.2	−3.0	B2 IV	590		
ζ Pup	8 03 35	−40 00 12	2.3	−0.3		O5			
15 ρ Pup	8 07 33	−24 18 15	2.8	0.4	−2.0	F6 II	300	mv	
γ Vel	8 09 32	−47 20 12	1.8	−0.2		WC7		m	
ε Car	8 22 31	−59 30 34	1.9	1.3	−2.1	K0 II	200		Avior
1 o UMa	8 30 16	+60 43 05	3.4	0.8	−0.9	G4 II	230	m	Muscida
δ Vel	8 44 42	−54 42 30	2.0	0.0	0.6	A0 V	68	m	
11 ε Hya	8 46 47	+6 25 07	3.4	0.7	0.6	G0 III	110	m	
16 ζ Hya	8 55 24	+5 56 44	3.1	1.0	0.2	K0 III	120		
9 ι UMa	8 59 12	+48 02 29	3.1	0.2	2.4	A7 V	49	m	Talitha
λ Vel	9 08 00	−43 25 57	2.2	1.7	−4.4	K5 I	490	m	

Table A-2 (contd.). The Brightest Stars, to Magnitude 3.5

Name and Catalogue Numbers	Position (2000.0) r.a. h m s	dec. ° ′ ″	Magnitudes V	B − V	M_v	Spectral Type	Distance (l-y)	Notes	Other Name
236693	9 10 58	−58 58 01	3.4	−0.2	−3.0	B2 IV	620		a Car
β Car	9 13 12	−69 43 02	1.7	0.0	−0.6	A0 III	85		Miaplacidus
ι Car	9 17 05	−59 16 31	2.3	0.2	−4.7	F0	820		Aspidiske
40 α Lyn	9 21 03	+34 23 33	3.1	1.6	−0.4	M0 III	170		
κ Vel	9 22 07	−55 00 38	2.5	−0.2	−3.0	B2 IV	390		
30 α Hya	9 27 35	−8 39 31	2.0	1.4	−0.2	K3	85	m	Alphard
237067	9 31 13	−57 02 04	3.1	1.6	−0.3	K5	150	v?	N Vel
25 θ UMa	9 32 51	+51 40 38	3.2	0.5	2.2	F6 IV	46	m	
17 ε Leo	9 45 51	+23 46 27	3.0	0.8	−2.0	G0 II	310		
ν Car	9 47 06	−65 04 18	3.0	0.3	−2.0	A7 II	320	m	
32 α Leo	10 08 22	+11 58 02	1.4	−0.1	−0.6	B7 V	85	m	Regulus
ω Car	10 13 44	−70 02 16	3.3	−0.1	−1.0	B7 IV	230		
36 ζ Leo	10 16 41	+23 25 02	3.4	0.3	0.6	F0 III	120	m	Adhafera
250905	10 17 05	−61 19 56	3.4	1.5	−4.4	K5 I	910	m	q Car
33 λ UMa	10 17 06	+42 54 52	3.5	0.0	0.6	A2 IV	120		Tania Borealis
41 γ¹ Leo	10 19 58	+19 50 30	2.3	1.1	0.2	K0 III		m	Algieba
34 μ UMa	10 22 20	+41 29 58	3.1	2.0	−0.4	M0 III	160		Tania Australis
251006	10 32 01	−61 41 07	3.3	−0.1	−1.7	B3 V	310	v	p Car
θ Car	10 42 57	−64 23 39	2.8	−0.2	−4.1	B0 V	750		
μ Vel	10 46 46	−49 25 12	2.7	0.9	0.3	G5 III	98	m	
ν Hya	10 49 37	−16 11 37	3.1	1.3	−0.1	K2 III	130		
48 β UMa	11 01 50	+56 22 56	2.4	0.0	1.2	A1 V	62		Merak
50 α UMa	11 03 44	+61 45 03	1.8	1.1	0.2	K0 III	75	m	Dubhe
52 ψ UMa	11 09 40	+44 29 54	3.0	1.1	0.0	K1 III	120		

Name and Catalogue Numbers	Position (2000.0) r.a. h m s	dec. ° ' "	Magnitudes V	B − V	M_v	Spectral Type	Distance (l-y)	Notes	Other Name
68 δ Leo	11 14 06	+20 31 25	2.6	0.1	1.9	A4 V	52	m	Zosma
70 θ Leo	11 14 14	+15 25 46	3.3	0.0	1.4	A2 V	78		Chertan
54 ν UMa	11 18 29	+33 05 39	3.5	1.4	−0.2	K3 III	150	m	Alula Borealis
λ Cen	11 35 47	−63 01 11	3.1	0.0	−0.8	B9 III	190	m	
94 β Leo	11 49 04	+14 34 19	2.1	0.1	1.7	A3 V	39	m	Denebola
64 γ UMa	11 53 50	+53 41 41	2.4	0.0	0.6	A0 V	75		Phecda
δ Cen	12 08 22	−50 43 20	2.6	−0.1	−2.5	B2 V	330	m	
2 ε Crv	12 10 07	−22 37 11	3.0	1.3	−0.1	K2 III	100		
δ Cru	12 15 09	−58 44 55	2.8	−0.2	−3.0	B2 IV	260		
69 δ UMa	12 15 26	+57 01 57	3.3	0.1	1.7	A3 V	65	m	Megrez
4 γ Crv	12 15 48	−17 32 31	2.6	−0.1	−1.2	B8 III	190		Gienah
α¹ Cru	12 26 36	−63 05 56	1.4	0.1	−3.9	B1 IV	360	m	Acrux
α² Cru	12 26 37	−62 05 58	1.9		−3.3	B3	360	m	
7 δ Crv	12 29 52	−16 30 55	3.0	−0.1	0.2	B9 V	120	m	Algorab
γ Cru	12 31 10	−57 06 47	1.6	1.6	−0.5	M3 III	88	m	Gacrux
9 β Crv	12 34 23	−23 23 48	2.7	0.9	−2.1	G5 II	290		
α Mus	12 37 11	−69 08 07	2.7	−0.2	−2.3	B3 IV	330	m	
γ Cen	12 41 31	−48 57 34	2.2	0.0	−0.6	A0 III	110	m	
29 γ Vir	12 41 40	−1 26 57	2.8	0.4	2.6	F0 V	36	m	Porrima
β Mus	12 46 17	−68 06 29	3.1	−0.2	−1.7	B3 V	290	m	
β Cru	12 47 43	−59 41 19	1.3	−0.2	−5.0	B0 III	420	mv	Mimosa
77 ε UMa	12 54 02	+55 57 35	1.8	0.0	0.4	A0	62	v	Alioth
43 δ Vir	12 55 36	+3 23 51	3.4	1.6	−0.5	M3 III	150	m	
12 α² CVn	12 56 02	+38 19 06	2.9	−0.1	1.4	A0	65	mv	Cor Caroli

Table A-2 (contd.). The Brightest Stars, to Magnitude 3.5

Name and Catalogue Numbers	Position (2000.0) r.a. h m s	dec. ° ′ ″	Magnitudes V	B − V	M_v	Spectral Type	Distance (l-y)	Notes	Other Name
47 ε Vir	13 02 11	+10 57 33	2.8	0.9	0.2	G9 III	100	m	Vindemiatrix
46 γ Hya	13 18 55	−23 10 17	3.0	0.9	0.3	G5 III	100	m	
ι Cen	13 20 36	−36 42 44	2.8	0.0	1.4	A2 V	52		
79 ζ UMa	13 23 56	+54 55 31	2.3	0.0	1.4	A2 V	59	m	Mizar
67 α Vir	13 25 12	−11 09 41	1.0	−0.2	−3.5	B1 V	260	mv	Spica
79 ζ Vir	13 34 42	−0 35 46	3.4	0.1	1.7	A3 V	75		
ε Cen	13 39 53	−53 27 58	2.3	−0.2	−3.5	B1 V	490		
85 η UMa	13 47 32	+49 18 48	1.9	−0.2	−1.7	B3 V	110	m	Alkaid
ν Cen	13 49 30	−41 41 16	3.4	−0.2	−2.5	B2 V	490		
μ Cen	13 49 37	−42 28 25	3.0	−0.2	−1.7	B3 V	290	mv	
8 η Boo	13 54 41	+18 23 51	2.7	0.6	2.7	G0 IV	32	m	Muphrid
ζ Cen	13 55 32	−47 17 17	2.6	−0.2	−3.0	B2 IV	360		
β Cen	14 03 49	−60 22 22	0.6	−0.2	−5.1	B1 II	460	m	
49 π Hya	14 06 22	−26 40 56	3.3	1.1	−0.1	K2 III	150		
5 θ Cen	14 06 41	−36 22 12	2.1	1.0	1.7	K0 III	46	m	Menkent
16 α Boo	14 15 40	+19 10 57	0.0	1.2	−0.2	K2 III	36		Arcturus
27 γ Boo	14 32 05	+38 18 30	3.0	0.2	0.5	A7 III	100	mv?	Seginus
η Cen	14 35 30	−42 09 28	2.3	−0.2	−2.9	B3 III	360		
α² Cen	14 39 35	−60 50 13	1.4		5.8	K1 V	4	m	
α¹ Cen	14 39 37	−60 50 02	0.0	0.7	4.4	G2 V	4	m	Rigil Kentaurus
α Lup	14 41 56	−47 23 17	2.3	−0.2	−4.4	B1 III	690	m	
α Cir	14 42 28	−64 58 43	3.2	0.2	2.6	F0 V	46	m	
36 ε Boo	14 44 59	+27 04 27	2.4	1.0	−0.9	K0 II	150	m	Izar
7 β UMi	14 50 42	+74 09 19	2.1	1.5	−0.3	K4 III	95	m	Kochab

Table A-2 (contd.). The Brightest Stars, to Magnitude 3.5

Name and Catalogue Numbers	Position (2000.0) r.a. h m s	dec. ° ′ ″	Magnitudes V	B − V	M_v	Spectral Type	Distance (l-y)	Notes	Other Name
9 α² Lib	14 50 53	−16 02 30	2.8	0.2	4.7	A	72	m	Zubenelgenubi
β Lup	14 58 32	−43 08 02	2.7	−0.2	−2.5	B2 V	360		
κ Cen	14 59 10	−42 06 15	3.1	−0.2	−2.5	B2 V	420	m	
42 β Boo	15 01 57	+40 23 26	3.5	1.0	0.3	G8 III	140		Nekkar
20 σ Lib	15 04 04	−25 16 55	3.3	1.7	−0.5	M4 III	170		
ζ Lup	15 12 17	−52 05 57	3.4	0.9	0.3	G8 III	140	m	
49 δ Boo	15 15 30	+33 18 53	3.5	1.0	0.3	G8 III	140	m	
27 β Lib	15 17 00	−9 22 58	2.6	−0.1	−0.2	B8 V	120		Zubeneschamali
γ Tra	15 18 55	−68 40 46	2.9	0.0	0.6	A0 V	91		
13 γ UMi	15 20 44	+71 50 02	3.1	0.1	−1.1	A3 II	230		Pherkad
δ Lup	15 21 22	−40 38 51	3.2	−0.2	−3.0	B2 IV	590		
ε Lup	15 22 41	−44 41 21	3.4	−0.2	−2.3	B3 IV	460	m	
12 ι Dra	15 24 56	+58 57 58	3.3	1.2	−0.1	K2 III	160	m	Edasich
5 α CrB	15 34 41	+26 42 53	2.2	0.0	0.6	A0 V	78	v	Alphecca
γ Lup	15 35 08	−41 10 00	2.8	−0.2	−1.7	B3 V	260	m	
24 α Ser	15 44 16	+6 25 32	2.7	1.2	−0.1	K2 III	85	m	Unukalhai
β Tra	15 55 08	−63 25 50	2.9	0.3	3.0	F2 V	33		
6 π Sco	15 58 51	−26 06 50	2.9	−0.2	−3.5	B1 V	620	m	
η Lup	16 00 07	−38 23 48	3.4	−0.2	−2.5	B2 V	490		
7 δ Sco	16 00 20	−22 37 18	2.3	−0.1	−4.1	B0 V	550		
8 β¹ Sco	16 05 26	−19 48 19	2.6	−0.1		B0 V		m	Graffias
1 δ Oph	16 14 21	−3 41 39	2.7	1.6	−0.5	M1 III	140	m	Yed Prior
2 ε Oph	16 18 19	−4 41 33	3.2	1.0	0.3	G8 III	100	m	Yed Posterior
20 σ Sco	16 21 11	−25 35 34	2.9	0.1	−4.4	B1 III	590	mv	

Table A-2 (contd.). The Brightest Stars, to Magnitude 3.5

Name and Catalogue Numbers	Position (2000.0) r.a. h m s	dec. ° ′ ″	Magnitudes V	B − V	M_v	Spectral Type	Distance (l-y)	Notes	Other Name
14 η Dra	16 23 59	+61 30 51	2.7	0.9	0.3	G8 III	85	m	
21 α Sco	16 29 24	−26 25 55	1.0	1.8	−4.7	M1 I	330	mv	Antares
27 β Her	16 30 13	+21 29 22	2.8	0.9	0.3	G8 III	100	m	Kornephoros
23 τ Sco	16 35 53	−28 12 58	2.8	−0.3	−4.1	B0 V	780		
13 ζ Oph	16 37 09	−10 34 02	2.6	0.0	−4.4	O9 V	550		
40 ξ Her	16 41 17	+31 36 10	2.8	0.7	3.0	G0 IV	31		
α Tra	16 48 40	−69 01 39	1.9	1.4	−0.1	K2 III	55	m	Atria
26 ε Sco	16 50 10	−34 17 36	2.3	1.2	−0.1	K2 III	65		
μ¹ Sco	16 51 52	−38 02 51	3.0	−0.2	−3.0	B1 V	520	mv	
27 κ Oph	16 57 40	+9 22 30	3.2	1.2	−0.1	K2 III	120	v?	
ζ Ara	16 58 37	−55 59 24	3.1	1.6	−0.3	K5 III	140		
22 ζ Dra	17 08 47	+65 42 53	3.2	−0.1	−1.9	B6 III	320		
35 η Oph	17 10 23	−15 43 30	2.4	0.1	1.4	A2 V	59	m	Sabik
η Sco	17 12 09	−43 14 21	3.3	0.4	0.6	F2 III	68		
64 α¹ Her	17 14 39	+14 23 25	3.2	1.4	−0.9	M5 II	220	mv	Rasalgethi
65 δ Her	17 15 02	+24 50 21	3.1	0.1	0.9	A3 IV	91	m	
67 π Her	17 15 03	+36 48 33	3.2	1.4	−2.3	K3 II	390		
42 θ Oph	17 22 00	−24 59 58	3.3	−0.2	−3.0	B2 IV	590		
β Ara	17 25 18	−55 31 47	2.9	1.5	−4.4	K3 I	780		
γ Ara	17 25 24	−56 22 39	3.3	−0.1	−4.4	B1 III	1,080	m	
23 β Dra	17 30 26	+52 18 05	2.8	1.0	−2.1	G2 II	270	m	Rastaban
34 υ Sco	17 30 46	−37 17 45	2.7	−0.2	−5.7	B3 I	1,570		Lesath
α Ara	17 31 50	−49 52 34	3.0	−0.2	−1.7	B3 V	190	m	
35 λ Sco	17 33 36	−37 06 14	1.6	−0.2	−3.0	B2 IV	270	m	Shaula

Table A-2 (contd.). The Brightest Stars, to Magnitude 3.5

Name and Catalogue Numbers	Position (2000.0) r.a. h m s	dec. ° ′ ″	Magnitudes V	B − V	M_v	Spectral Type	Distance (l-y)	Notes	Other Name
55 α Oph	17 34 56	+12 33 36	2.1	0.2	0.3	A5 III	62		Rasalhague
θ Sco	17 37 19	−42 59 52	1.9	0.4	−5.6	F0 I	910		
κ Sco	17 42 29	−39 01 48	2.4	−0.2	−3.0	B2 IV	390		
60 β Oph	17 43 28	+4 34 02	2.8	1.2	−0.1	K2 III	120		Cebalrai
86 μ Her	17 46 27	+27 43 15	3.4	0.8	3.9	G5 IV	26	m	
ι¹ Sco	17 47 35	−40 07 37	3.0	0.5	−8.4	F2 I	5,550	m	
209318	17 49 51	−37 02 36	3.2	1.2	−0.1	K2 III	150	m	G Sco
33 γ Dra	17 56 36	+51 29 20	2.2	1.5	−0.3	K5 III	100	m	Eltanin
64 ν Oph	17 59 01	−9 46 25	3.3	1.0	0.2	K0 III	140		
10 γ Sgr	18 05 48	−30 25 26	3.0	1.0	0.2	K0 III	120		Alnasl
η Sgr	18 17 38	−36 45 42	3.1	1.6	−2.4	M3 II	430	m	
19 δ Sgr	18 21 00	−29 49 42	2.7	1.4	−0.1	K2 III	82	m	Kaus Media
58 η Ser	18 21 18	−2 53 56	3.3	0.9	1.7	K0	52	m	
20 ε Sgr	18 24 10	−34 23 05	1.9	0.0	−0.3	B9 IV	85	m	Kaus Australis
22 λ Sgr	18 27 58	−25 25 18	2.8	1.0	−0.1	K2 III	98		Kaus Borealis
3 α Lyr	18 36 56	+38 47 01	0.0	0.0	0.5	A0 V	26	m	Vega
27 φ Sgr	18 45 39	−26 59 27	3.2	−0.1	−1.2	B8 III	250		
10 β Lyr	18 50 05	+33 21 46	3.5	0.0	−0.6	B7 V	300	mv	Sheliak
34 σ Sgr	18 55 16	−26 17 48	2.0	−0.2	−2.0	B3 IV	210	m	Nunki
14 γ Lyr	18 58 56	+32 41 22	3.2	−0.1	−0.8	B9 III	190	m	Sulafat
38 ζ Sgr	19 02 37	−29 52 49	3.0	0.1	0.6	A2 IV	78	m	Ascella
17 ζ Aql	19 05 24	+13 51 48	3.0	0.0	0.2	B9 V	100	m	
16 λ Aql	19 06 15	−4 52 57	3.4	−0.1	0.0	B8 V	98		
40 τ Sgr	19 06 56	−27 40 13	3.3	1.2	0.0	K1 III	130		

Table A-2 (contd.). The Brightest Stars, to Magnitude 3.5

Name and Catalogue Numbers	Position (2000.0) r.a. h m s	dec. ° ' "	Magnitudes V	B − V	M_v	Spectral Type	Distance (l-y)	Notes	Other Name
41 τ Sgr	19 09 46	−21 01 25	2.9	0.4	−2.0	F2 II	310	m	
57 δ Dra	19 12 33	+67 39 41	3.1	1.0	0.2	G9 III	120	m	Altais
30 δ Aql	19 25 30	+3 06 53	3.4	0.3	2.1	F0 IV	52	m	
6 β Cyg	19 30 43	+27 57 35	3.1	1.1	−2.3	K3 II	390	m	Albireo
18 δ Cyg	19 44 58	+45 07 51	2.9	0.0	0.6	A0 III	160	m	
50 γ Aql	19 46 15	+10 36 48	2.7	1.5	−2.3	K3 II	280	m	Tarazed
53 α Aql	19 50 47	+8 52 06	0.8	0.2	2.2	A7 IV	17	m	Altair
12 γ Sge	19 58 45	+19 29 32	3.5	1.6	−0.3	K5 III	177		
65 θ Aql	20 11 18	−0 49 17	3.2	−0.1	−0.8	B9 III	200	m	
9 β Cap	20 21 01	−14 46 53	3.1	0.8	4.0	F8 V	100	m	Dabih
37 γ Cyg	20 22 14	+40 15 24	2.2	0.7	−4.6	F8 I	750	m	Sadr
α Pav	20 25 39	−56 44 06	1.9	−0.2	−2.3	B3 IV	230	m	Peacock
α Ind	20 37 34	−47 17 29	3.1	1.0	0.2	K0 III	120	m	
50 α Cyg	20 41 26	+45 16 49	1.3	0.1	−7.5	A2 I	1,830	m	Deneb
β Pav	20 44 57	−66 12 12	3.4	0.2	1.2	A5 IV	91		
3 η Cep	20 45 17	+61 50 20	3.4	0.9	3.2	K0 IV	46	m	
53 ε Cyg	20 46 13	+33 58 13	2.5	1.0	0.2	K0 III	82	m	
64 ζ Cyg	21 12 56	+30 13 37	3.2	1.0	−2.1	G8 II	390	m	
5 α Cep	21 18 35	+62 35 08	2.4	0.2	1.9	A7 IV	46	m	Alderamin
8 β Cep	21 28 39	+70 33 39	3.2	−0.2	−3.6	B2 III	750	mv	Alfirk
22 β Aqr	21 31 33	−5 34 16	2.9	0.8	−4.5	G0 I	980	m	Sadalsuud
8 ε Peg	21 44 11	+9 52 30	2.4	1.5	−4.4	K2 I	520	m	Enif
49 δ Cap	21 47 02	−16 07 38	2.9	0.3	2.0	A	49	mv	Deneb Algedi
γ Gru	21 53 56	−37 21 54	3.0	−0.1	−1.2	B8 III	230		

Table A-2 (contd.). The Brightest Stars, to Magnitude 3.5

Name and Catalogue Numbers	Position (2000.0) r.a. h m s	dec. ° ' "	Magnitudes V	B – V	Mᵥ	Spectral Type	Distance (l-y)	Notes	Other Name
34 α Aqr	22 05 47	− 0 19 11	3.0	1.0	−4.5	G2 I	950	m	Sadalmelik
α Gru	22 08 14	−46 57 40	1.7	−0.1	−1.1	B5 V	68	m	Al Naïr
21 ζ Cep	22 10 51	+58 12 05	3.4	1.6	−4.4	K1 I	720		
α Tuc	22 18 30	−60 15 35	2.9	1.4	−0.2	K3 III	110		
42 ζ Peg	22 41 28	+10 49 53	3.4	−0.1	0.0	B8 V	160	m	Homam
β Gru	22 42 40	−46 53 05	2.1	1.6	−2.4	M3 II	170		
44 η Peg	22 43 00	+30 13 17	2.9	0.9	−0.9	G2 II	170	m	Matar
ε Gru	22 48 33	−51 19 01	3.5	0.1	1.4	A2 V	82		
48 μ Peg	22 50 00	+24 36 06	3.5	0.9	0.2	K0 III	150		Sadalbari
76 δ Aqr	22 54 39	−15 49 15	3.3	0.1	−0.2	A2 III	98		Skat
24 α Psa	22 57 39	−29 37 20	1.2	0.1	2.0	A3 V	22		Fomalhaut
53 β Peg	23 03 46	+28 04 58	2.4	1.7	−1.4	M2 II	180	mv	Scheat
54 α Peg	23 04 46	+15 12 19	2.5	−0.0	0.2	B9 V	100		Markab
35 γ Cep	23 39 21	+77 37 57	3.2	1.0	2.2	K1 IV	52		Errai

The information in this table was kindly extracted for us from *Sky Catalogue 2000.0*, edited by Alan Hirshfeld and Roger W. Sinnott (copyright 1982 by Sky Publishing Corp.), and is reprinted with permission.

Table A-3. Properties of the Principal Spectral Types

Spectral Type	Apparent Color	Intrinsic Brightness (M_v)	Surface Temperature (K)	Primary Absorption Lines in Spectrum	Examples
O	blue	less than −0.2	25,000–40,000	Strong lines of ionized helium and highly ionized metals; hydrogen lines weak	ζ Orionis (O9.5)
B	blue	−0.2–0.0	11,000–25,000	Lines of neutral helium prominent; hydrogen lines stronger than in type O	Spica (B1) Rigel (B8)
A	blue to white	0.0–0.3	7,500–11,000	Strong lines of hydrogen, ionized calcium, and other ionized metals; weak helium lines	Vega (A0) Sirius (A1) Deneb (A2)
F	white	0.3–0.6	6,000–7,500	Hydrogen lines weaker than in type A; ionized calcium strong; lines of neutral metals becoming prominent	Canopus (F0) Procyon (F5) Polaris (F8)
G	white to yellow	0.6–1.1	5,000–6,000	Numerous strong lines of ionized calcium and other ionized and neutral metals; hydrogen lines weaker than in type F	Sun (G2) Capella (G8)
K	orange to red	1.1–1.5	3,500–5,000	Numerous strong lines of neutral metals	Arcturus (K2) Aldebaran (K5)
M	red	greater than 1.5	3,000–3,500	Numerous strong lines of neutral metals; strong molecular bands (primarily titanium oxide)	Antares (M1) Betelgeuse (M2)

Note: The number after the letter in each spectral type (see last column, above) indicates a further subdivision within each type. For example, Sirius (type A1) is hotter than Deneb (type A2).
From *Sky Catalogue 2000.0* by Alan Hirshfeld and Roger W. Sinnott, courtesy of Sky Publishing Corp.

Table A-4. The Nearest Stars

No. (Rank)	Name	Position (2000.0) r.a. h m	dec. ° '	Distance (l-y)	Spectral Type	Magnitude V	M_v	Luminosity ($L_{sun} = 1$)
1	Sun				G2 V	−26.72	4.85	1.0
2	Proxima Cen	14 32	−62 49	4.24	dM5e	11.05	15.49	0.00006
	α (alpha) Cen A	14 42	−60 59	4.34	G2 V	−0.01	4.37	1.6
	α (alpha) Cen B				K0 V	1.33	5.71	0.45
3	Barnard's star	17 58	+04 36	5.97	M5 V	9.54	13.22	0.00045
4	Wolf 359	10 59	+06 54	7.76	dM8e	13.53	16.65	0.00002
5	BD + 36°2147 (Lalande 21185)	11 06	+35 51	8.22	M2 V	7.50	10.50	0.0055
6	L 726-8 = A	01 39	−17 49	8.42	dM6e	12.52	15.46	0.00006
	UV Cet = B				dM6e	13.02	15.96	0.00004
7	Sirius A	06 46	−16 45	8.64	A1 V	−1.46	1.42	23.5
	Sirius B				DA	8.3	11.2	0.003
8	Ross 154 (V 1216 Sgr)	18 52	−23 46	9.46	dM5e	10.45	13.14	0.00048
9	Ross 248 (HH And)	23 42	+44 21	10.37	dM6e	12.29	14.78	0.00011
10	ε (epsilon) Eri	03 34	−09 22	10.76	K2 V	3.73	6.14	0.30
11	Ross 128 (FI Vir)	11 50	+00 41	10.96	dM5	11.10	13.47	0.00036
12	61 Cyg A	21 08	+38 50	11.09	K5 V	5.22	7.56	0.082
	61 Cyg B				K7 V	6.03	8.37	0.039
13	ε (epsilon) Ind	22 06	−56 37	11.22	K5 V	4.68	7.00	0.14
14	BD + 43°44 A (GX And)	00 21	+43 39	11.25	M1 V	8.08	10.39	0.0061
	+ 43°44 B (GQ And) (Groombridge 39 AB)				M6 Ve	11.06	13.37	0.00039
15	L 789-6	22 41	−15 12	11.25	dM7e	12.18	14.49	0.00014

Notes: r.a. = right ascension; dec. = declination; l-y = light-years; V = visual magnitude; M_v = absolute magnitude. Under Spectral Type, the Roman numerals indicate the star's luminosity class: V = normal dwarfs, like the sun. Also, d = dwarf; e = emission lines present in spectrum. Courtesy of W. Gliese (private communication, 1979).

Table A-5. Selected Bright Planetary Nebulae

Planetary Nebula	Name and Constellation	r. a. (2000.0) h m s	dec. ° ′ ″	Magnitude (visual)	Diameter (arc sec)
NGC 7293	Helix in Aquarius	22 29 38	−20 50 09	6.5	770
NGC 6853	Dumbbell in Vulpecula, M27	19 59 36	22 43 15	7.6	350
NGC 3132	Eight-burst in Antlia	10 07 01	−40 25 41	8.2	67
NGC 2392	Eskimo in Gemini	07 29 10	20 54 43	8.3	45
NGC 7009	Saturn in Aquarius	21 04 11	−11 22 48	8.4	48
NGC 246	in Cetus	00 47 03	−11 52 22	8.5	230
NGC 6543	Cat's Eye in Draco	17 58 33	66 37 59	8.8	22
NGC 6826	Blinking Planetary in Cygnus	19 44 48	50 31 31	8.8	25
NGC 7662	in Andromeda	23 25 54	42 32 06	8.9	31
NGC 3242	Ghost of Jupiter in Hydra	10 24 47	−18 38 24	9.0	37
NGC 6720	Ring Nebula in Lyra, M57	18 53 34	33 01 41	9.3	71
NGC 1535	in Eridanus	04 14 15	−12 44 29	9.3	18
NGC 6572	in Ophiuchus	18 12 07	06 51 14	9.6	11
NGC 6210	in Hercules	16 44 30	23 48 01	9.7	19
NGC 6818	Little Gem in Sagittarius	19 43 57	−14 09 11	9.9	25

Courtesy of Yervant Terzian, Cornell University.

Table A-6. Properties of the Planets

Intrinsic and Rotational Properties of the Planets

Name	Equatorial Radius (km)	Equatorial Radius ÷ Earth's	Mass ÷ Earth's	Mean Density (g/cm³)	Oblateness	Surface Gravity (Earth = 1)	Sidereal Rotation Period	Inclination of Equator to Orbit (degrees)	Apparent Magnitude at 1982 Opposition	Apparent Equatorial Diameter (arc sec)
Mercury	2,439	0.3824	0.0553	5.43	0.0	0.38	58.646^d	0	−1.8	5.5
Venus	6,052	0.9489	0.8150	5.24	0.0	0.89	243.01^dR	177.3	−4.3	30.5
Earth	6,378.140	1	1	5.515	0.0034	1	$23^h56^m04.1^s$	23.45		
Mars	3,397.2	0.5326	0.1074	3.93	0.005	0.38	$24^h37^m22.662^s$	25.19	−1.2	8.9
Jupiter	71,398	11.194	317.89	1.36	0.061	2.54	9^h50^m to $>9^h55^m$	3.12	−2.0	98.4
Saturn	60,000	9.41	95.17	0.71	0.109	1.07	$10^h39.9^m$	26.73	+0.5	82.8
Uranus	26,145	4.1	14.56	1.30	0.03	0.8	12^h to $24^h \pm 4^h$	97.86	+5.8	32.9
Neptune	24,300	3.8	17.24	1.8	0.03	1.2	$18^h12^m \pm 24^m$	29.56	+7.7	31.1
Pluto	1,500–1,800	0.4	0.02	0.5–0.8	?	?	$6^d9^h17^m$	118 ?	+13.7	0.1

R = retrograde; d = days; h = hours; m = minutes; s = seconds.

Orbital Properties of the Planets

Name	Semimajor Axis (A.U.)	Semimajor Axis (10⁶ km)	Sidereal Period (years)	Sidereal Period (days)	Synodic Period (days)	Eccentricity of Orbit	Inclination to Ecliptic
Mercury	0.3871	57.9	0.24084	87.96	115.9	0.2056	7°0′26″
Venus	0.7233	108.2	0.61515	224.68	584.0	0.0068	3°23′40″
Earth	1	149.6	1.00004	365.26		0.0167	0°0′14″
Mars	1.5237	227.9	1.8808	686.95	779.9	0.0934	1°51′09″
Jupiter	5.2028	778.3	11.862	4337	398.9	0.0483	1°18′29″
Saturn	9.5388	1427.0	29.456	10,760	378.1	0.0560	2°29′17″
Uranus	19.1914	2871.0	84.07	30,700	369.7	0.0461	0°48′26″
Neptune	30.0611	4497.1	164.81	60,200	367.5	0.0100	1°46′27″
Pluto	39.5294	5913.5	248.53	90,780	366.7	0.2484	17°09′03″

Table A-7. Planetary Satellites

	Satellite	Semimajor Axis of Orbit (km)	Sidereal Period d h m	Orbital Eccentricity	Orbital Inclination (degrees)	Diameter (km)	Visible Magnitude*
Earth	the Moon	384,500	27 07 43	0.055	18–29	3476	−12.7
Mars	Phobos	9,378	0 07 39	0.015	1.1	27 × 21 × 18	11.3
	Deimos	23,459	1 06 18	0.00052	0.9 – 2.7v	15 × 12 × 10	12.4
Jupiter	1979 J1	127,000	0 07 04			40	
XVI	Metis	129,000	0 07 06	0		20	
XIV	Adrastea	134,000	0 07 09			20	
V	Amalthea	180,000	0 11 57	0.003	0.4	270 × 165 × 153	14.1
XV	Thebe	222,000	0 16 11			80	
I	Io	422,000	1 18 28	0.000	0	3632 ± 60	5.0
II	Europa	671,000	3 13 14	0.000	0.5	3126 ± 60	5.3
III	Ganymede	1,070,000	7 03 43	0.001	0.2	5276 ± 60	4.6
IV	Callisto	1,885,000	16 16 32	0.01	0.2	4820 ± 60	5.6
XIII	Leda	11,110,000	240	0.147	26.7	10	20
VI	Himalia	11,470,000	251	0.158	27.6	170	14.7
X	Lysithea	11,710,000	260	0.130	29.0	20	18.4
VII	Elara	11,740,000	260	0.207	24.8	80	16.4
XII	Ananke	20,700,000	617R	0.169	147	20	18.9
XI	Carme	22,350,000	629R	0.207	164	30	18.0
VIII	Pasiphae	23,330,000	735R	0.378	145	40	17.7
IX	Sinope	23,370,000	758R	0.275	153	30	18.3
Saturn							
17	Atlas	137,670	14 27	0.002	0.3	80 × 60 × 40	17
16	Inner F-ring shepherd	139,353	14 43	0.003	0.0	140 × 100 × 80	16
15	Outer F-ring shepherd	141,700	15 05	0.004	0.05	110 × 90 × 70	16

Table A-7 (contd.). Planetary Satellites

		Satellite	Semimajor Axis of Orbit (km)	Sidereal Period d	h	m	Orbital Eccentricity	Orbital Inclination (degrees)	Diameter (km)	Visible Magnitude*
Saturn contd.	10	Janus	151,422		16	40	0.007	0.1	220 × 200 × 160	15
	11	Epimetheus	151,472		16	40	0.009	0.3	140 × 120 × 100	
	1	Mimas	185,600		22	37	0.020	1.5	390	12.9
	2	Enceladus	238,100	1	08	52	0.005	0.0	510	11.7
	3	Tethys	294,700	1	22	15	0.000	1.9	1,050	10.3
	13	Telesto	294,700	1	22	15			34 × 28 × 26	19
	14	Calypso	294,700	1	22	15			34 × 22 × 22	19
	4	Dione	377,500	2	17	36	0.002	0.0	1,120	10.4
	12	Dione B	378,060	2	17	45			36 × 22 × 30	19
	5	Rhea	527,200	4	12	16	0.001	0.4	1,530	9.7
	6	Titan	1,221,600	15	21	51	0.029	0.3	5,150	8.3
	7	Hyperion	1,483,000	21	06	45	0.104	0.4	400 × 250 × 220	14.2
	8	Iapetus	3,560,100	79	03	43	0.028	14.7	1,440	11.2
	9	Phoebe	12,950,000	549	03	33	0.163	159	200	16.3
Uranus	5	Miranda	130,000	1	09	56	0.01	3.4	600	16.5
	1	Ariel	191,000	2	12	29R	0.003	0	1410 ± 105	14.4
	2	Umbriel	260,000	4	03	27R	0.004	0	1160 ± 90	15.3
	3	Titania	436,000	8	16	56R	0.002	0	1670 ± 90	14.0
	4	Oberon	583,000	13	11	07R	0.001	0	1690 ± 90	14.2
Neptune	3		75,000							
		Triton	354,000	5	21	03R	0.000	160.0	3200 ± 400	13.6
		Nereid	5,570,000	365	5		0.76	27.4	600	18.7
Pluto	3	Charon	20,000?	6	9	17	0?	105?	1000	16–17

Notes: R = retrograde; v = variable; * = magnitude given for mean opposition distance.

Table A-8. Planetary Longitudes

Year	Date		Julian Day 2440000+	Sun ☉	Mercury ☿	Venus ♀	Mars ♂	Jupiter ♃	Saturn ♄
1983	Dec	31	5700	279	280	239	204	266	224
1984	Jan	10	5710	289	270	251	209	268	225
1984	Jan	20	5720	300	275	263	214	270	225
1984	Jan	30	5730	310	286	275	219	272	226
1984	Feb	9	5740	320	300	288	224	274	226
1984	Feb	19	5750	330	316	300	228	276	227
1984	Feb	29	5760	340	333	312	232	278	227
1984	Mar	10	5770	350	351	324	235	279	227
1984	Mar	20	5780	360	12	337	238	281	227
1984	Mar	30	5790	10	29	349	239	282	226
1984	Apr	9	5800	20	37	2	239	282	226
1984	Apr	19	5810	29	34	14	238	283	225
1984	Apr	29	5820	39	28	26	236	283	224
1984	May	9	5830	49	26	39	233	283	223
1984	May	19	5840	58	32	51	229	283	222
1984	May	29	5850	68	45	63	226	282	222
1984	Jun	8	5860	78	61	76	224	281	221
1984	Jun	18	5870	87	81	88	222	280	221
1984	Jun	28	5880	97	103	100	222	279	221
1984	Jul	8	5890	106	123	113	223	277	220
1984	Jul	18	5900	116	140	125	226	276	221
1984	Jul	28	5910	125	152	137	231	274	221
1984	Aug	7	5920	135	161	150	235	274	221
1984	Aug	17	5930	144	163	162	240	273	222
1984	Aug	27	5940	154	156	174	246	273	223
1984	Sep	6	5950	164	149	187	252	273	223
1984	Sep	16	5960	174	155	199	258	273	224
1984	Sep	26	5970	183	171	211	265	274	225
1984	Oct	6	5980	193	190	223	271	275	225
1984	Oct	16	5990	203	207	236	278	276	226
1984	Oct	26	6000	213	223	248	285	277	227
1984	Nov	5	6010	223	239	260	292	279	228
1984	Nov	15	6020	233	253	272	300	281	230
1984	Nov	25	6030	243	265	284	307	283	231
1984	Dec	5	6040	253	271	296	315	285	232
1984	Dec	15	6050	264	262	307	323	287	233
1984	Dec	25	6060	274	254	319	330	289	234
1985	Jan	4	6070	284	261	330	338	292	235
1985	Jan	14	6080	294	273	341	345	294	236
1985	Jan	24	6090	304	287	352	353	297	237
1985	Feb	3	6100	315	303	1	0	299	237
1985	Feb	13	6110	325	320	10	8	302	238
1985	Feb	23	6120	335	338	17	15	304	239
1985	Mar	5	6130	345	357	21	23	306	239
1985	Mar	15	6140	355	13	23	31	308	239
1985	Mar	25	6150	5	19	20	38	310	239

Table A-8 (contd.). Planetary Longitudes

Year	Date	Julian Day 2440000+	Sun ☉	Mercury ☿	Venus ♀	Mars ♂	Jupiter ♃	Saturn ♄
1985	Apr 4	6160	15	13	14	45	312	239
1985	Apr 14	6170	24	7	8	52	313	238
1985	Apr 24	6180	34	8	5	59	314	237
1985	May 4	6190	44	17	7	66	316	236
1985	May 14	6200	53	30	11	73	317	235
1985	May 24	6210	63	47	18	80	317	234
1985	Jun 3	6220	73	67	26	86	317	234
1985	Jun 13	6230	82	90	36	92	317	233
1985	Jun 23	6240	92	109	46	99	317	232
1985	Jul 3	6250	101	125	56	106	316	232
1985	Jul 13	6260	111	137	67	112	315	232
1985	Jul 23	6270	120	145	78	119	314	232
1985	Aug 2	6280	130	144	89	126	312	232
1985	Aug 12	6290	139	137	101	132	311	233
1985	Aug 22	6300	149	132	113	139	310	233
1985	Sep 1	6310	159	141	124	145	309	234
1985	Sep 11	6320	168	158	136	151	308	234
1985	Sep 21	6330	178	177	148	157	307	235
1985	Oct 1	6340	188	195	161	164	307	236
1985	Oct 11	6350	198	211	173	170	307	237
1985	Oct 21	6360	208	226	186	176	307	237
1985	Oct 31	6370	218	240	198	182	308	238
1985	Nov 10	6380	228	251	211	188	309	239

Year	Date	Julian Day 2440000+	Sun ☉	Mercury ☿	Venus ♀	Mars ♂	Jupiter ♃	Saturn ♄
1985	Nov 20	6390	238	256	223	195	311	240
1985	Nov 30	6400	248	245	236	201	312	242
1985	Dec 10	6410	258	239	249	207	314	243
1985	Dec 20	6420	269	247	261	213	316	244
1985	Dec 30	6430	279	260	274	220	318	245
1986	Jan 9	6440	289	275	286	226	320	246
1986	Jan 19	6450	299	291	299	232	322	247
1986	Jan 29	6460	309	307	311	238	324	247
1986	Feb 8	6470	319	325	324	243	327	248
1986	Feb 18	6480	329	343	336	249	329	249
1986	Feb 28	6490	339	358	349	255	332	250
1986	Mar 10	6500	349	1	1	260	335	250
1986	Mar 20	6510	359	353	14	266	337	250
1986	Mar 30	6520	9	348	26	271	339	250
1986	Apr 9	6530	19	352	39	276	341	250
1986	Apr 19	6540	29	2	51	281	343	250
1986	Apr 29	6550	39	16	64	286	345	250
1986	May 9	6560	48	33	76	290	347	249
1986	May 19	6570	58	53	88	293	349	248
1986	May 29	6580	68	76	100	294	350	247
1986	Jun 8	6590	77	95	112	295	352	246
1986	Jun 18	6600	87	110	124	295	353	245
1986	Jun 28	6610	96	122	135	294	353	244

Table A-8 (contd.). Planetary Longitudes

Year	Date	Julian Day 2440000+	Sun ⊙	Mercury ☿	Venus ♀	Mars ♂	Jupiter ♃	Saturn ♄
1986	Jul 8	6620	106	126	147	291	353	244
1986	Jul 18	6630	115	123	158	285	353	244
1986	Jul 28	6640	125	117	169	283	353	243
1986	Aug 7	6650	134	116	180	282	352	243
1986	Aug 17	6660	144	126	190	282	351	244
1986	Aug 27	6670	154	144	200	283	350	244
1986	Sep 6	6680	163	164	209	285	348	244
1986	Sep 16	6690	173	182	217	288	348	245
1986	Sep 26	6700	183	199	225	293	346	246
1986	Oct 6	6710	193	214	229	298	345	246
1986	Oct 16	6720	203	227	231	304	344	247
1986	Oct 26	6730	213	236	229	310	343	248
1986	Nov 5	6740	223	239	224	316	343	249
1986	Nov 15	6750	233	228	219	323	343	250
1986	Nov 25	6760	243	224	215	329	343	251
1986	Dec 5	6770	253	233	217	336	344	253
1986	Dec 15	6780	263	247	222	343	345	253
1986	Dec 25	6790	273	263	229	350	347	255
1987	Jan 4	6800	284	279	237	357	348	256
1987	Jan 14	6810	294	295	247	4	350	257
1987	Jan 24	6820	304	312	257	11	352	258
1987	Feb 3	6830	314	329	268	18	354	258
1987	Feb 13	6840	324	342	279	25	356	259

Year	Date	Julian Day 2440000+	Sun ⊙	Mercury ☿	Venus ♀	Mars ♂	Jupiter ♃	Saturn ♄
1987	Feb 23	6850	334	343	291	32	358	260
1987	Mar 5	6860	344	333	302	39	0	261
1987	Mar 15	6870	354	330	314	46	3	261
1987	Mar 25	6880	4	336	326	52	5	262
1987	Apr 4	6890	14	347	338	59	8	262
1987	Apr 14	6900	24	2	350	66	11	262
1987	Apr 24	6910	34	19	2	73	13	262
1987	May 4	6920	43	39	14	79	15	261
1987	May 14	6930	53	62	25	86	17	261
1987	May 24	6940	63	81	38	92	19	260
1987	Jun 3	6950	72	95	50	98	21	259
1987	Jun 13	6960	82	105	62	105	23	258
1987	Jun 23	6970	91	106	74	112	25	257
1987	Jul 3	6980	101	102	87	118	26	256
1987	Jul 13	6990	110	97	99	124	28	255
1987	Jul 23	7000	120	100	111	130	28	255
1987	Aug 2	7010	129	112	124	137	29	255
1987	Aug 12	7020	139	130	136	143	30	255
1987	Aug 22	7030	149	151	148	150	30	255
1987	Sep 1	7040	158	170	161	156	30	255
1987	Sep 11	7050	168	186	173	163	29	255
1987	Sep 21	7060	178	201	186	169	28	256
1987	Oct 1	7070	188	214	198	175	27	256

Table A-8 (contd.). Planetary Longitudes

Year	Date	Julian Day 2440000+	Sun ☉	Mercury ☿	Venus ♀	Mars ♂	Jupiter ♃	Saturn ♄
1987	Oct 11	7080	197	222	211	182	26	257
1987	Oct 21	7090	207	222	223	188	24	258
1987	Oct 31	7100	217	211	235	195	22	259
1987	Nov 10	7110	227	209	248	201	21	260
1987	Nov 20	7120	238	220	261	208	21	261
1987	Nov 30	7130	248	235	273	214	20	262
1987	Dec 10	7140	258	251	286	220	20	263
1987	Dec 20	7150	268	267	298	227	20	264
1987	Dec 30	7160	278	282	310	234	20	266
1988	Jan 9	7170	288	299	322	241	20	267
1988	Jan 19	7180	299	315	335	247	21	267
1988	Jan 29	7190	309	327	347	254	22	268
1988	Feb 8	7200	319	326	359	261	24	269
1988	Feb 18	7210	329	315	11	267	26	270
1988	Feb 28	7220	339	314	22	274	28	271
1988	Mar 9	7230	349	322	34	281	30	271
1988	Mar 19	7240	359	334	45	287	32	272
1988	Mar 29	7250	9	349	55	294	35	272
1988	Apr 8	7260	19	6	65	301	37	273
1988	Apr 18	7270	29	26	74	307	39	273
1988	Apr 28	7280	38	48	82	314	42	273
1988	May 8	7290	48	67	87	321	44	273
1988	May 18	7300	58	80	90	327	47	272

Year	Date	Julian Day 2440000+	Sun ☉	Mercury ☿	Venus ♀	Mars ♂	Jupiter ♃	Saturn ♄
1988	May 28	7310	67	86	90	334	49	272
1988	Jun 7	7320	77	85	84	340	52	271
1988	Jun 17	7330	86	79	78	347	54	270
1988	Jun 27	7340	96	77	73	353	56	269
1988	Jul 7	7350	105	84	72	358	58	268
1988	Jul 17	7360	115	97	76	3	60	268
1988	Jul 27	7370	124	116	81	7	61	267
1988	Aug 6	7380	134	138	89	10	63	266
1988	Aug 16	7390	144	157	97	12	64	266
1988	Aug 26	7400	153	174	107	13	65	266
1988	Sep 5	7410	163	188	117	13	66	266
1988	Sep 15	7420	173	199	128	11	66	266
1988	Sep 25	7430	183	207	139	7	66	267
1988	Oct 5	7440	192	205	150	2	66	267
1988	Oct 15	7450	202	194	162	0	66	268
1988	Oct 25	7460	212	194	174	359	65	268
1988	Nov 4	7470	222	206	186	359	63	269
1988	Nov 14	7480	232	222	198	0	62	270
1988	Nov 24	7490	242	238	211	2	60	271
1988	Dec 4	7500	253	254	223	6	59	272
1988	Dec 14	7510	263	270	235	11	57	273
1988	Dec 24	7520	273	286	248	16	57	275
1989	Jan 3	7530	283	301	260	21	56	276

444

Table A-8 (contd.). Planetary Longitudes

Year	Date	Julian Day 2440000+	Sun ☉	Mercury ☿	Venus ♀	Mars ♂	Jupiter ♃	Saturn ♄
1989	Jan 13	7540	293	312	273	27	56	277
1989	Jan 23	7550	303	308	286	32	56	278
1989	Feb 2	7560	314	297	298	38	56	279
1989	Feb 12	7570	324	298	311	44	57	280
1989	Feb 22	7580	334	307	323	50	58	281
1989	Mar 4	7590	344	320	336	56	59	282
1989	Mar 14	7600	354	335	348	62	60	282
1989	Mar 24	7610	4	353	1	68	62	283
1989	Apr 3	7620	14	12	13	74	64	283
1989	Apr 13	7630	23	33	25	80	66	284
1989	Apr 23	7640	33	52	38	86	68	284
1989	May 3	7650	43	64	50	92	70	284
1989	May 13	7660	52	67	63	99	72	284
1989	May 23	7670	62	63	75	105	75	284
1989	Jun 2	7680	72	58	87	111	77	283
1989	Jun 12	7690	81	59	100	118	79	283
1989	Jun 22	7700	91	68	112	124	82	282
1989	Jul 2	7710	100	83	124	130	84	281
1989	Jul 12	7720	110	102	137	136	86	280
1989	Jul 22	7730	119	124	149	142	89	279
1989	Aug 1	7740	129	144	161	149	91	278
1989	Aug 11	7750	138	160	172	155	92	278
1989	Aug 21	7760	148	175	184	161	94	278
1989	Aug 31	7770	158	185	196	167	96	278
1989	Sep 10	7780	168	191	208	174	97	277
1989	Sep 20	7790	177	187	219	180	99	277
1989	Sep 30	7800	187	177	231	187	100	277
1989	Oct 10	7810	197	179	242	194	100	278
1989	Oct 20	7820	207	192	253	200	101	278
1989	Oct 30	7830	217	209	264	207	101	279
1989	Nov 9	7840	227	226	274	213	101	280
1989	Nov 19	7850	237	242	284	220	100	281
1989	Nov 29	7860	247	258	292	227	99	282
1989	Dec 9	7870	257	273	300	234	98	283
1989	Dec 19	7880	268	287	305	241	97	284
1989	Dec 29	7890	278	296	308	248	95	285

Note: All values are degrees of celestial longitude (positions along the ecliptic).
Data prepared by William D. Stahlman and Owen Gingerich.

Table A-9. Local Sidereal Time at 00:00 Local Standard Time (computed for the year 1985)

Date	Jan	Feb	Mar	Apr	May	Jun	Jul	Aug	Sep	Oct	Nov	Dec
1	6:42	8:45	10:35	12:37	14:35	16:38	18:36	20:38	22:40	0:39	2:41	4:39
2	6:46	8:49	10:39	12:41	14:39	16:42	18:40	20:42	22:44	0:43	2:45	4:43
3	6:50	8:52	10:43	12:45	14:43	16:46	18:44	20:46	22:48	0:47	2:49	4:47
4	6:54	8:56	10:47	12:49	14:47	16:50	18:48	20:50	22:52	0:51	2:53	4:51
5	6:58	9:00	10:51	12:53	14:51	16:53	18:52	20:54	22:56	0:54	2:57	4:55
6	7:02	9:04	10:55	12:57	14:55	16:57	18:56	20:58	23:00	0:58	3:01	4:59
7	7:06	9:08	10:59	13:01	14:59	17:01	19:00	21:02	23:04	1:02	3:05	5:03
8	7:10	9:12	11:03	13:05	15:03	17:05	19:04	21:06	23:08	1:06	3:09	5:07
9	7:14	9:16	11:07	13:09	15:07	17:09	19:08	21:10	23:12	1:10	3:12	5:11
10	7:18	9:20	11:10	13:13	15:11	17:13	19:11	21:14	23:16	1:14	3:16	5:15
11	7:22	9:24	11:14	13:17	15:15	17:17	19:15	21:18	23:20	1:18	3:20	5:19
12	7:26	9:28	11:18	13:21	15:19	17:21	19:19	21:22	23:24	1:22	3:24	5:23
13	7:30	9:32	11:22	13:25	15:23	17:25	19:23	21:26	23:28	1:26	3:28	5:27
14	7:34	9:36	11:26	13:28	15:27	17:29	19:27	21:29	23:32	1:30	3:32	5:30
15	7:38	9:40	11:30	13:32	15:31	17:33	19:31	21:33	23:36	1:34	3:36	5:34
16	7:42	9:44	11:34	13:36	15:35	17:37	19:35	21:37	23:40	1:38	3:40	5:38
17	7:45	9:48	11:38	13:40	15:39	17:41	19:39	21:41	23:43	1:42	3:44	5:42
18	7:49	9:52	11:42	13:44	15:43	17:45	19:43	21:45	23:47	1:46	3:48	5:46
19	7:53	9:56	11:46	13:48	15:46	17:49	19:47	21:49	23:51	1:50	3:52	5:50
20	7:57	9:59	11:50	13:52	15:50	17:53	19:51	21:53	23:55	1:54	3:56	5:54

Table A-9 (contd.). Local Sidereal Time at 00:00 Local Standard Time
(computed for the year 1985)

Date	Jan	Feb	Mar	Apr	May	Jun	Jul	Aug	Sep	Oct	Nov	Dec
21	8:01	10:03	11:54	13:56	15:54	17:57	19:55	21:57	23:59	1:58	4:00	5:58
22	8:05	10:07	11:58	14:00	15:58	18:00	19:59	22:01	0:03	2:01	4:04	6:02
23	8:09	10:11	12:02	14:04	16:02	18:04	20:03	22:05	0:07	2:05	4:08	6:06
24	8:13	10:15	12:06	14:08	16:06	18:08	20:07	22:09	0:11	2:09	4:12	6:10
25	8:17	10:19	12:10	14:12	16:10	18:12	20:11	22:13	0:15	2:13	4:16	6:14
26	8:21	10:23	12:14	14:16	16:14	18:16	20:15	22:17	0:19	2:17	4:19	6:18
27	8:25	10:27	12:17	14:20	16:18	18:20	20:18	22:21	0:23	2:21	4:23	6:22
28	8:29	10:31	12:21	14:24	16:22	18:24	20:22	22:25	0:27	2:25	4:27	6:26
29	8:33		12:25	14:28	16:26	18:28	20:26	22:29	0:31	2:29	4:31	6:30
30	8:37		12:29	14:32	16:30	18:32	20:30	22:33	0:35	2:33	4:35	6:34
31	8:41		12:33		16:34		20:34	22:36		2:37		6:37

Compared with the sidereal time for dates in 1985:

From January 1, 1984 through February 28, 1984: subtract 3 minutes.

From February 29, 1984 through December 31, 1984: add 1 minute.

1985: see table.

From January 1, 1986 through December 31, 1986: subtract 1 minute.

From January 1, 1987 through December 31, 1987: subtract 2 minutes.

From January 1, 1988 through February 28, 1988: subtract 3 minutes.

From February 29, 1988 through December 31, 1988: add 1 minute.

1989: same as 1985; see table.

Copyright 1980 by Astronomical Data Service, Colorado Springs, Colorado U.S.A.

447

Glossary

Aberration of starlight: The tiny apparent displacement of stars resulting from the motion of the earth through space.

Absolute magnitude (M): The magnitude a celestial object would appear to have if it were at a distance of 10 parsecs.

Absolute visual magnitude (M_V): The absolute magnitude of an object measured through a special yellowish filter that approximates the visual range of the human eye.

Absorption nebula: A nebula seen in silhouette as it absorbs light from behind; also called a dark nebula.

Altitude: Angular distance (usually measured in degrees) above the horizon.

Analemma: The figure-8 representing the equation of time and the variation of the sun's altitude in the sky during the course of a year.

Angstrom: A unit of wavelength or distance, equivalent to 1/10,000 micrometer or 1/10,000,000,000 meter.

Annular eclipse: A solar eclipse in which a ring — an annulus — of solar photosphere remains visible.

Aphelion: The farthest point from the sun in an object's orbit around it.

Apparent magnitude (m): Magnitude as seen by an observer.

Apparent solar time: Time determined by the actual position of the sun in the sky; corresponds to time on most sundials.

Asterism: A noticeable pattern of stars that makes up part of one or more constellations; not a constellation itself.

Asteroid: A minor planet, smaller than any major planet in our solar system; not one of the satellites (moons) of a major planet such as the earth or Jupiter.

Astronomical unit (A.U.): The average distance from the earth to the sun, which equals 149,598,770 kilometers.

Autumnal equinox: The intersection of the ecliptic and the celestial equator that the sun passes each year on its way to southern (negative) declinations.

Baily's beads: A chain of several bright "beads" of white light, visible just before or after totality at a solar eclipse. The effect occurs when bits of photosphere shine through valleys at the moon's edge. See also **diamond-ring effect.**

Bayer designations: The Greek letters assigned to the stars in a constellation, usually in order of brightness, by Johann Bayer in his sky atlas (1603).

Belts: Dark bands in the clouds on giant planets such as Jupiter; compare with **zones.**

Binary star: A double star; a system containing two or more stars. In an *eclipsing binary,* one star goes behind the other periodically, changing the total amount of light we see.

Black hole: A region of space in which mass is packed so densely that (according to Einstein's general theory of relativity) nothing, not even light, can escape.

Cassini's division: The major division in Saturn's rings, which separates the A-ring from the B-ring.

Celestial equator: The imaginary great circle that lies above the earth's equator on the celestial sphere.

Celestial longitude: Longitude measured (in degrees) along the ecliptic to the east from the vernal equinox.

Celestial poles: The points in the sky where the earth's axis, extended into space, intersects with the celestial sphere.

Celestial sphere: The imaginary sphere surrounding the earth, with the stars and other astronomical objects attached to it.

Cepheid variable: A star that varies in the manner of delta Cephei. The absolute magnitudes of these variable stars can be calculated from their periods of variation; by comparing the absolute and apparent magnitudes, the distances to these stars and the galaxies they are in can be determined.

Chromosphere: A layer in the sun and many other stars just above the photosphere. During eclipses, the solar chromosphere glows reddish from hydrogen emission.

Circumpolar: Refers to a star, asterism, or constellation that is close enough to the celestial pole that, from the latitude at which you are observing, it never appears to set.

Comet: A body — probably resembling a "dirty snowball," between 0.1 and 100 km across — that travels through the solar system in an elliptical orbit of random inclination to the ecliptic. A comet grows a tail if it comes close enough to the sun.

Conjunction: The alignment of two celestial bodies that occurs when they reach the same celestial longitude. The bodies then appear approximately closest to each other in the sky. See also **inferior conjunction, superior conjunction.**

Constellation: One of the 88 parts into which the sky is divided; also refers to the historical, mythological, or other figures that represented earlier divisions of the sky.

Contact(s): The stage(s) of an eclipse, occultation, or transit when the edges of the apparent disks of astronomical bodies seem to touch. At a solar eclipse, first contact is when the advancing edge of the sun first touches the moon; second contact is when the advancing edge of the sun touches the other side of the moon, beginning totality; third contact is when the trailing edge of the sun touches the trailing edge of the moon, ending totality; and fourth contact marks the end of the eclipse.

Corona: The outermost layer of the sun and many other stars; a faint halo of extremely hot (million-degree) gas.

Crepe ring: Saturn's inner ring, also known as the C-ring, which extends inward to the planet from the brightest ring (the B-ring).

Crescent: One of the phases of the moon or the inner planets (Venus and Mercury) as seen from earth, caused by the relative angles of sunlight and the observer's viewpoint. From spacecraft, crescent phases of the earth, Mars, Jupiter, and Saturn have also been seen.

Declination: The celestial coordinate analogous to latitude, usually measured in degrees, minutes, and seconds of arc north (+) or south (−) of the celestial equator.

Diamond-ring effect: An effect created as the total phase of a solar eclipse is about to begin, when the last Baily's bead — a remaining bit of photosphere — glows so intensely by contrast with the sun's faint corona that it looks like the jewel on a ring. Also refers to the equivalent phase at the end of totality.

Double star: A system containing two or more stars. In a true double, the stars are physically close to each other; in an *optical double,* they lie in approximately the same direction from the earth and thus appear close to each other, but are actually far apart. See also **binary star.**

Earthshine: Sunlight reflected off the earth, which lights the side of the moon that does not receive direct sunlight.

Eclipse, lunar: The passage of the moon into the earth's shadow.

Eclipse, solar: The passage of the moon's shadow across the earth. See also **annular eclipse, contact(s), penumbra, umbra.**

Ecliptic: The apparent path the sun follows across the sky during the year; the same path is also followed approximately by the moon and planets.

Ejecta blanket: Chunks of rock, usually extending from one side of a crater, that were ejected during the crater's formation.

Elongation: Angular distance in celestial longitude from the sun in the sky.

Emission lines: Extra radiation at certain specific wavelengths in a spectrum, compared with neighboring wavelengths (colors).

Emission nebula: A gas cloud that receives energy from a hot star, allowing it to give off radiation in emission lines such as those of hydrogen. The characteristic reddish radiation of many emission nebulae is mostly from the hydrogen-alpha line.

Encke's division: A thin division in the A-ring of Saturn.

Ephemeris Time: The official system of mean solar time, used to calculate data for tables of changing astronomical phenomena (ephemerides). Ephemeris Time differs only slightly from Universal Time.

Equation of time: The variation of local apparent solar time minus local mean solar time over the year.

Equinox: One of the two intersections of the ecliptic and the celestial equator: see **autumnal equinox, vernal equinox.**

Filament: A dark region snaking across the sun; a prominence seen in projection against the solar disk.

Fireball: An extremely bright meteor, usually with an apparent magnitude brighter than -5; some fireballs are as bright as magnitude -20.

Flamsteed number: The number assigned to a star in a given constellation, in order of right ascension, in the 1725 catalogue of John Flamsteed.

Galactic: Pertaining to our galaxy, the Milky Way Galaxy.

Galactic cluster: An irregular grouping of stars of a common and possibly recent origin. Also called an **open cluster.**

Galactic equator, galactic poles: The equator and poles in a coordinate system in which the equator is placed along the plane of our galaxy, the Milky Way Galaxy.

Galaxy: A giant collection of stars, gas, and dust. Our galaxy, the Milky Way Galaxy, contains 1 trillion times the mass of our sun.

Giant: A star brighter and larger than most stars of its color and temperature. Stars become giants (normally **red giants**) when they use up all the hydrogen in their cores and leave the "main sequence" part of their life cycle. See also **supergiant.**

Gibbous: A phase of a moon or planet in which more than half of the side we see is illuminated. Remember: it "gibb" ("gives") us more light.

Globular cluster: A spherical grouping of stars of a common origin; globular clusters and the stars in them are very old.

Graben: On the surface of the earth, the moon, or other planets or moons, a long and narrow region between two faults that has subsided.

Half moon: The first-quarter or third-quarter phase, when half the visible side of the moon is illuminated.

Hour angle: The sidereal time elapsed since an object was on the meridian or, if the hour angle is negative, before the object reaches the meridian. (The hour angle equals the difference between the right ascension of an object and of your meridian.)

Hour circle: A line along which right ascension is constant, lying on a great circle that passes through the celestial poles and the object.

Hubble's Law: The relationship between the velocity and distance of galaxies and other distant objects; it shows that the universe is expanding.

Hydrogen-alpha line: The strongest spectral line of hydrogen in the visible part of the spectrum. It falls in the red, so that an emission hydrogen-alpha line is red; an absorption hydrogen-alpha line is the absence of that wavelength of red.

Inferior conjunction: The conjunction in which a planet whose orbit is inside that of the earth passes between the earth and sun.

Intrinsic brightness: The amount of energy (usually light) an object gives off; its true brightness, independent of the effects of distance or dimming by intervening material.

Ionized hydrogen: Hydrogen that has lost its electron; ionized hydrogen gas, commonly found in stars and nebulae, has free protons and free electrons.

Julian day: The number of days since noon on 1 January 4713 B.C. Variable-star observers and other astronomers commonly calculate the interval between dates of events by subtracting Julian days, eliminating the necessity to keep track of leap years and other calendar details.

Libration: The turning of the visible face of the moon, which allows us to see different amounts of the lunar surface around the limb (edge).

Light-year: The distance that light travels in a year, which equals 9,460,000,000,000 km or 63,240 A.U. (astronomical units).

Limb: The edge of the apparent disk of an astronomical body, such as the sun, moon, or a planet.

Magnitude: A logarithmic scale of brightness, in which each change of five magnitudes is equivalent to a change by a factor of 100. Adding one magnitude corresponds to a decrease in brightness by a factor of 2.512 .... See also **absolute magnitude and apparent magnitude.**

Main-sequence star: A star in the prime of its life, when hydrogen inside it is undergoing nuclear fusion; such stars form a band — the main sequence — across a graph of stellar temperatures vs. stellar brightness.

Maxima: The times when a variable star reaches its maximum brightness.

Mean solar time: Time as kept by a fictitious "mean" sun that travels at a steady rate across the sky throughout the year.

Meridian: The great circle passing through the celestial poles and your zenith.

Messier Catalogue: The list of 103 nonstellar, deep-sky objects compiled by Charles Messier in the 1770s, and subsequently expanded to 109 or 110 objects.

Meteor: A meteoroid streaking across the sky; a shooting star.

Meteorite: The part of a meteoroid that survives its passage through the earth's atmosphere.

Meteoroid: A small chunk of rock or metal in the solar system, often spread throughout a comet's orbit; sometimes a chip off an asteroid.

Meteor shower: The appearance of many meteors during a short period of time, as the earth passes through a comet's orbit.

Minima: The times of a variable star's minimum brightness.

Mira variable: A long-period variable star, like the star omicron Ceti (called "Mira").

Nebula: A region of gas or dust in a galaxy that can be observed optically. See also **emission nebula, absorption nebula,** and **reflection nebula.**

NGC: The prefix used before numbers assigned to nonstellar objects in the *New General Catalogue,* published by J.L.E. Dreyer in 1888.

Neutron star: A small (20-km diameter), dense (a billion tons per cubic cm) star, resulting from the collapse of a dying star to the point where only the fact that its neutrons resist being pushed still closer together prevents further collapse.

Nova: A newly visible star, or one that suddenly increases drastically in brightness.

Nutation: A small nodding motion of the earth's axis of rotation with a period of 19 years; this motion is superimposed on precession.

Oblate: A nonspherical shape formed by rotating an ellipse around its narrower axis; the equatorial diameter of an oblate body (such as Jupiter) is greater than its polar diameter.

Occultation: The hiding of one celestial body by another.

Open cluster: An irregular grouping of stars of a common and possibly recent origin. Also called a **galactic cluster.**

Opposition: The point in a planet's orbit at which its celestial longitude is 180° from that of the sun. A planet at opposition is visible all night long.

Parsec: The distance from which 1 A.U. appears to subtend (cover) 1 second of arc; 1 parsec equals 3.261633 . . . light-years.

Penumbra: At an eclipse, the part of the earth or moon's shadow from which part of the solar disk is visible. Also refers to the outer, less dark portion of a sunspot.

Perihelion: The nearest point to the sun in an object's orbit around it.

Photosphere: The visible surface of the sun or of another star.

Planetary nebula: A shell of gas ejected by a dying star that contains about as much mass as the sun.

Polar tufts: Small spikes visible in the solar corona near the sun's poles, formed by gas following the sun's magnetic field.

Position angle: The angle, centered at the brighter component of a double star, that an observer follows counterclockwise from north around to the fainter component.

Precession: The slow drifting of the orientation of the earth's axis over a period of 26,000 years. Also refers to its effect on the location of the equinoxes, and thus on the coordinate system of right ascension and declination used to plot positions of stars and other objects.

Prominence: Gas suspended above the solar photosphere by the sun's magnetic field; ordinarily visible at the solar limb (edge). A prominence glows reddish during eclipses because of its characteristic hydrogen-alpha radiation.

Proper motion: Apparent angular motion across the sky, shown as a change in an object's position with respect to the background stars.

Pulsar: A rotating neutron star that gives off sharp pulses of radio waves with a period ranging from about 0.001 to 4 seconds.

Quasar: A "quasi-stellar object" with an extremely large redshift; presumably a powerful event going on in the central region of a galaxy. According to Hubble's Law, quasars must be among the most distant objects in the universe.

Radiant: The location on the celestial sphere from which meteors in a given shower appear to radiate, because of perspective.

Red giant: A swollen star; a stage occurring at the end of a star's main-sequence period of life. See also **giant.**

Reflecting telescope: A telescope that uses a mirror in the principal stage of forming an image.

Reflection nebula: A dust cloud that reflects a star's light to us.

Refracting telescope: A telescope that uses a lens in the principal stage of forming an image.

Retrograde motion: The apparent backward (westward) loop in a planet's motion across the sky over a lengthy period of time. Copernicus explained it as a projection effect caused when the earth overtakes another planet as they both orbit the sun.

Revolution: The orbiting of a planet or other object around the sun or another central body. (Compare with **rotation.**)

Right ascension: The angle of an object around the celestial equator, measured in hours, minutes, and seconds eastward from the vernal equinox.

Rotation: The spinning of a planet or other object on its axis. (Compare with **revolution.**)

Separation: The angular distance (measured in degrees, minutes, and seconds of arc) between components of a double star.

Shadow bands: Light and dark bands that appear to sweep across the ground in the minutes before and after totality at a solar eclipse; caused by irregularities in the earth's upper atmosphere.

Sidereal time: Time by the stars; technically, the hour angle of the vernal equinox, which is equal to the right ascension of objects on your meridian.

Solar flare: An explosive eruption on the sun reaching temperatures of millions of degrees. Note: a flare is not a prominence.

Solstice(s): The positions of the sun when it reaches its northernmost declination (in northern-hemisphere summer) or southernmost declination (in northern-hemisphere winter).

Spectral line: A wavelength of the spectrum at which the intensity is greater than (an *emission line*) or less than (an *absorption line*) neighboring values.

Spectral type: One of several temperature classes — OBAFGKM, in decreasing order of temperature — into which stars are placed, based on analyses of their spectra.

Spectrum (pl. spectra): The radiation from an object, spread out into its component colors, wavelengths, or frequencies.

Star cloud: One of several regions of the Milky Way where great numbers of stars appear.

Streamers: Large-scale structures in the sun's corona, usually near the solar equator, shaped by the sun's magnetic field.

Sunspots: Relatively dark regions on the solar photosphere, corresponding to areas with exceptionally high magnetic fields.

Supergiant: A star brighter and larger than even giants of the same color and temperature. Only the most massive stars become supergiants, after passing through the giant stage.

Superior conjunction: The conjunction in which a planet whose orbit is inside that of the earth passes on the far side of the sun with respect to the earth.

Supernova: The explosion and devastation of a very massive star.

Supernova remnant: Gas left over from a supernova that can be seen in the sky or detected from its radio or x-ray emission. (The Crab Nebula, for example, can be detected all three ways.)

Surface brightness: The brightness of a unit area of an object's surface. For spread-out objects such as nebulae, the surface brightness determines the amount of contrast the object has against the background sky, and whether the object's surface is bright enough to make an image on your retina. Even though the object's total brightness may be high, it still may be hard to see if it is spread out enough so that its surface brightness is low.

Synodic: Related to the alignment of three bodies, often the earth, the sun, and a third body, such as the moon or a planet.

Terminator: The edge of the lighted region of a moon or planet; the line between day and night.

Train: A path left in the sky by a meteor.

Transient Lunar Phenomena (TLP's): Changes, such as emissions of gas, observed on the moon.

Transit: The passage of an inner planet (Mercury or Venus) across the sun's disk as seen from earth, or of a moon (such as one of Jupiter's Galilean satellites) across its planet's disk. Also, the passage of an object across an observer's meridian.

Umbra: At an eclipse, the part of the moon or earth's shadow from which the solar disk is entirely hidden. Also refers to the inner, darker portion of a sunspot. See also **penumbra.**

Universal Time (U.T.): Solar time at the meridian of Greenwich, England.

Variable star: A star whose apparent brightness changes over time.

Vernal equinox: The intersection of the ecliptic and the celestial equator that the sun passes on its way to northern (positive) declinations.

Zenith: The point directly overhead (wherever an observer is), 90° above the horizon.

Zodiac: Traditionally, a set of 12 constellations through which the sun, moon, and planets pass in the course of a year. Actually, that band of the sky contains many more parts of constellations, and because of precession, the sun is no longer in the constellations associated with its "traditional" dates at those times.

Zones: Bright bands in the cloud layers of the giant planets (Jupiter, Saturn, Uranus, and Neptune).

Bibliography

Sky Atlases and General References

Hirshfeld, Alan, and Sinnott, Roger W. 1982 (vol. 1), 1985 (vol. 2). *Sky Catalogue 2000.0.* Cambridge, Mass.: Sky Publishing Corp. Volume 1 is a list of stars; volume 2 provides lists of double stars, variable stars, galaxies, clusters, nebulae, and other objects. All positions are precessed to epoch 2000.0.

Meeus, Jean. 1983. *Astronomical Tables of the Sun, Moon, and Planets.* Richmond, Va.: Willmann-Bell. Lists of astronomical phenomena, including planetary oppositions and conjunctions, eclipses, transits, etc.

Pasachoff, Jay M. 1985. *Contemporary Astronomy.* 3rd ed. Philadelphia: Saunders College Publishing. A nonmathematical, well-illustrated survey.

――― 1983. *Astronomy: From the Earth to the Universe.* 2nd ed. Philadelphia: Saunders College Publishing. Another introductory college text by the author of this *Field Guide.*

Satterthwaite, Gilbert E.; Moore, Patrick; and Inglis, Robert G.; eds. 1978. *Norton's Star Atlas and Reference Handbook.* 17th ed. Cambridge, Mass.: Sky Publishing Corp. The old standard, updated.

Scovil, Charles E., ed. 1980. *AAVSO Variable Star Atlas.* Cambridge, Mass.: Sky Publishing Corp. Includes over 2000 variable stars brighter than magnitude 9.5, along with comparison stars and many deep-sky objects.

Tirion, Wil. 1982. *Sky Atlas 2000.0.* Cambridge, Mass.: Sky Publishing Corp. and New York: Cambridge University Press. Twenty-six larger-scale versions of the Atlas Charts used in Chapter 7 of this *Field Guide,* with stars that are one-half magnitude fainter.

Vehrenberg, Hans. 1983. *Atlas of Deep Sky Splendors.* 4th ed. Cambridge, Mass.: Sky Publishing Corp. Color and black-and-white photographs, at a uniform scale, of the most interesting parts of the sky. Includes close-ups of Messier and other objects.

Observing Handbooks

Bishop, Roy L., ed. *Observer's Handbook* (published annually). Toronto: Royal Astronomical Society of Canada. A popular guide to sky objects and events, available from the Society at 124 Merton St., Toronto, Ontario, Canada M4S 2Z2.

Burnham, Robert, Jr. 1978. *Burnham's Celestial Handbook.* 3 vols. New York: Dover. Detailed constellation-by-constellation discussions, with photographs of a wide variety of objects.

Chartrand, Mark R., III. 1982. *Skyguide.* New York: Western Publishing Co., Golden Press. Observing hints and constellation maps, with beautiful illustrations by Helmut K. Wimmer.

Gallant, Roy A. 1980. *Our Universe.* Washington, D.C.: National Geographic Society. A picture atlas of astronomy for children; includes an observation kit.

Jones, Kenneth Glyn, ed. 1981 (vols. 1–4), 1982 (vol. 5). *Webb Society Deep-Sky Observer's Handbook.* Hillside, N.J.: Enslow Publishers. Detailed observing notes and sketches of telescope fields of view.

Kirby-Smith, H. T. 1976. *U.S. Observatories: A Directory and Travel Guide.* New York: Van Nostrand Reinhold.

Mallas, John H., and Kreimer, Evered. 1978. *The Messier Album.* Cambridge, Mass.: Sky Publishing Corp. Discussions of each Messier object, with black-and-white photographs. Includes historical background and observing notes by Owen Gingerich.

Muirden, James. 1979. *Astronomy With Binoculars.* New York: Harper and Row, Thomas Y. Crowell.

Newton, Jack. 1977. *Deep Sky Objects: A Photographic Guide for the Amateur.* Toronto: Gall Publications. Black-and-white, wide-field photographs. Write to the publisher at 1293 Gerrard St. East, Toronto, Ontario, Canada M4L 1Y8.

Rey, H. A. 1962. *The Stars: A New Way to See Them.* Boston: Houghton Mifflin. Nontraditional constellation outlines, drawn to resemble actual objects more than the usual ways that stars are connected.

Roth, Gunter D. 1975. *Astronomy — A Handbook.* New York: Springer-Verlag.

Sherrod, R. Clay. 1981. *A Complete Manual of Amateur Astronomy.* Englewood Cliffs, N.J.: Prentice-Hall. Detailed recommendations.

Sidgwick, J. B. 1980. *Amateur Astronomer's Handbook.* 4th ed., revised by James Muirden. Hillside, N.J.: Enslow Publishers.
——— 1982. *Observational Astronomy For Amateurs.* 4th ed., revised by James Muirden. Hillside, N.J.: Enslow Publishers.

Whitney, Charles A. 1981. *Whitney's Star Finder.* 3rd ed. New York: Alfred A. Knopf.

Sky Bulletins

Astronomical Calendar (yearly). The changing sky and astronomical events such as eclipses. Available from Guy Ottewell, Dept. of Physics, Furman University, Greenville, S.C.

29613. Also available from the same source: *The Astronomical Companion,* a companion volume that serves as a general reference, and *The View from the Earth,* a children's version of the *Astronomical Calendar.*

Graphic Timetable of the Heavens (yearly). A more detailed version of the Graphic Timetables in this guide, available as a full-color poster from Scientia, Inc., 1815 Landrake Rd., Baltimore, Md. 21204.

Sky Calendar (monthly). Easy-to-use diagrams of the moon's phases and its daily changes in position against the starry background, plus diagrams of planetary conjunctions with bright stars, with the moon, and with other planets. Available by subscription from Abrams Planetarium, Michigan State University, East Lansing, Mich. 48824.

Skywatcher's Almanac (yearly). Computer-generated information on the visibility of the sun, moon, and other objects, tailored for the individual observer at a given latitude. Available from the Astronomical Data Service, 3922 Leisure Lane, Colorado Springs, Colo. 80917.

Magazines

Astronomy (monthly). Monthly sky features and articles summarizing different fields of astronomy in lay terms. *Odyssey* (a monthly magazine for children) and *Deep Sky* (a quarterly magazine for adult observers) are also available from the same publisher, at 625 E. St. Paul Ave., P.O. Box 92788, Milwaukee, Wisc. 53202.

Sky & Telescope (monthly). The standard journal for amateur observers; includes popular articles on astronomical topics and sky events, in addition to regular monthly features. For subscription information, write to 49 Bay State Rd., Cambridge, Mass. 02238.

For Calculators and Computers

Burgess, Eric. 1982. *Celestial BASIC: Astronomy on Your Computer.* Berkeley, Calif.: Sybex. Simple programs that will enable you to calculate astronomical information, including changes in celestial coordinates, planetary positions, dates of lunar eclipses, calendars, and so on.

Duffett-Smith, Peter. 1981. *Practical Astronomy With Your Calculator.* 2nd ed. New York: Cambridge University Press.

Meeus, Jean. 1982. *Astronomical Formulas for Calculators.* 2nd ed. Formulas for calendars, conjunctions, eclipses, lunar and planetary motion, equation of time, precession and nutation, etc. Available from Willmann-Bell, Inc., P.O. Box 3125, Richmond, Va. 23235.

Amateur Societies

American Association of Variable Star Observers (AAVSO), 187 Concord Ave., Cambridge, Mass. 02138.

American Meteor Society, Dept. of Physics and Astronomy, SUNY, Geneseo, N.Y. 14454.

Association of Lunar and Planetary Observers (ALPO), 8930 Raven Dr., Waco, Tex. 76710.

The Astronomical League, the umbrella group of amateur societies. For their newsletter, *The Reflector,* write Carol J. Beaman, Editor, 6804 Alvina Rd., Rockford, Ill. 61103.

Astronomical Society of the Pacific, 1290 24th Ave., San Francisco, Calif. 94122. This national and international group publishes a monthly magazine called *Mercury.*

British Astronomical Association, Burlington House, Piccadilly, London W1V 0NL, England.

British Meteor Society, 26 Adrian St., Dover, Kent, CT17 9AT, England.

Royal Astronomical Society of Canada, 124 Merton St., Toronto, Ontario, Canada M4S 2Z2.

Western Amateur Astronomers, 496 Drake Dr., Santa Rosa, Calif. 95405.

Professional Society

American Astronomical Society, 1816 Jefferson Pl., N.W., Washington, D.C. 20036. For a booklet on "A Career in Astronomy," write to Education Officer, AAS, Sharp Laboratory, University of Delaware, Newark, Del. 19711.

Telescope Information

The human eye collects light very efficiently, but all the light it collects must pass through its tiny pupil. A telescope gathers much more light in a given time and so allows you to see fainter objects. Further, the larger the lens or mirror the telescope has, the finer the detail that you can see. Telescopes can also magnify, but this is usually less important than their light-gathering power or resolution. It is not useful to magnify a faint or blurry image.

Anyone who is planning to buy a telescope should pay special attention to the sturdiness of the mount, for a telescope on a flimsy mount is useless. A mount not only supports the telescope but also may track astronomical objects across the sky as the earth rotates. With a mount that isn't tracking, an object drifts out of sight from the center of a 1° field of view in only two minutes. The effect of the earth's motion shows up in photographs of stars taken without a tracking mount. When the shutter is left open for several minutes to capture the images of fainter stars, the stars' images show up as curved trails rather than as points of light (C.Pl. 70).

Many telescopes come on *equatorial mounts,* in which one axis — the polar axis — points up at an angle so that it is directed at a celestial pole. Another, perpendicular, axis allows the telescope to point anywhere in the sky. A single motor on the polar axis allows the telescope to track stars, planets, or galaxies so they won't drift out of view.

Fig. 160. Tripod-mounted photographs taken with a standard 35 mm camera and a 50 mm lens at $f/2$. A 30-second exposure (*left*) shows the constellation Orion, including the nebula in its belt. A 2-minute exposure (*right*) shows more stars, though they are more obviously trailed. The bright star Sirius is visible at lower left. Its distorted shape shows the effect of being at the edge of the field of view. (Jay M. Pasachoff)

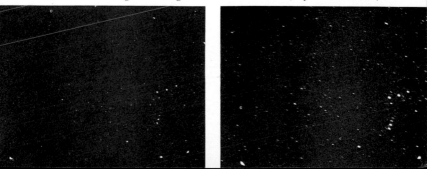

The alternative — an *altazimuth mount* — has one axis that points straight up, around which rotation around the compass (azimuth) takes place, and another axis that points from side to side, around which up-and-down rotation (altitude) takes place. Astronomical objects move across the sky at a slant, so varying motions on both axes are necessary for an altazimuth mount.

Types of Telescopes

A *reflecting telescope* uses a mirror to collect and focus light, while a *refracting telescope* uses a lens. The larger the telescope, the better the resolution it can provide, so the more magnification images can tolerate. Large telescopes allow you to see fainter objects and more detail, but you can enjoy studying the planets and other objects with a small telescope. After all, Galileo discovered the phases of Venus and the moons of Jupiter with a reflecting telescope only about $1\frac{1}{2}$ inches (4 cm) across. You can increase the magnification of a small telescope by using an eyepiece with a shorter focal length.

The traditional Newtonian telescope is still very popular and typically uses a mirror 6 inches across and a tube 48 inches long, although it is available in other sizes as well. Other telescopes that are now popular with amateurs are neither pure reflectors nor refractors; in these *compound telescopes*, the light passes through a lens and is then focused by mirrors inside the telescope's tube. Since the light bounces back and forth between the mirrors, the tube can be short and therefore more portable. Most telescopes of this type are Schmidt-Cassegrains: Schmidt telescopes have wide

Fig. 161. A reflecting telescope of the Newtonian type. Isaac Newton invented the method of using a small diagonal flat mirror to reflect the focused light out to the side, where it can be viewed without blocking the incoming light. (Meade Instruments Corp.)

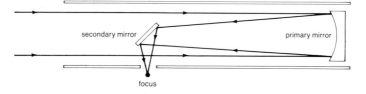

Fig. 162. A refracting telescope, in which the light is focused by a lens. (Meade Instruments Corp.)

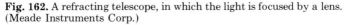
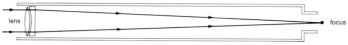

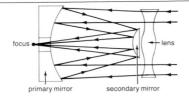

Fig. 163. A compound telescope of the Schmidt-Cassegrain type can show a wide field in good focus because of the combination of lens and mirrors used. (Meade Instruments Corp.)

fields; Cassegrain telescopes are those in which the image is directed through a hole in the main mirror to your pupil. Most of these telescopes come on equatorial mounts.

Dobsonian telescopes use thin mirrors, wooden cases with plastic bearings, and altazimuth mounts to obtain large apertures (mirror sizes) relatively inexpensively. These reflectors are designed for visual observing rather than for photography, since they do not track the stars.

Anyone evaluating telescopes should look at advertisements in recent issues of *Sky & Telescope* and *Astronomy* magazines to see what types and models are currently available. An article on "Selecting Your First Telescope" is available from the Astronomical Society of the Pacific, 1290 24th Ave., San Francisco, Calif. 94122.

The following manufacturers make telescopes, mounts, and filters that are popular with amateurs:

Bausch & Lomb, Bushnell Division, 2828 E. Foothill Blvd., Pasadena, Calif. 91107. *Criterion 4000* 4-in. system and traditional Newtonian telescopes.

Celestron International, 2835 Columbia St., Torrance, Calif. 90503. Binoculars and telescopes, including 5-in., 8-in., and 14-in. Schmidt-Cassegrain and Schmidt telescopes, a 5.5-in. wide-field Schmidt-Newtonian telescope, and refractors.

Coulter Optical Co., P.O. Box K, Idyllwild, Calif. 92349. Dobsonian telescopes.

DayStar Filter Corp., P.O. Box 1290, Pomona, Calif. 91769. Solar hydrogen-alpha filters.

Edmund Scientific, 101 East Gloucester Pike, Barrington, N.J. 08007. Newtonian telescopes, refractors, and the inexpensive wide-field *Astroscan 2001* telescope.

Meade Instruments, 1675 Toronto Way, Costa Mesa, Calif. 92626. 4-in., 8-in., and 10-in. Schmidt-Cassegrain telescopes and mounts; Newtonian telescopes.

Questar Corp., Box C, New Hope, Pa. 18938. Expensive but high-quality 3.5-in. compact, Maksutov-type folded telescopes.

Unitron Corp., 175 Express St., Plainview, N.Y. 11803. Refractors.

Index

The index provides references to objects, definitions, and general discussions of topics in astronomy; not all individual objects are indexed. The index should be used in conjunction with the tables and the glossary, the contents of which are not separately indexed. Page numbers for the Color Plates are shown in **boldface** type and page numbers for figures are *italicized*.

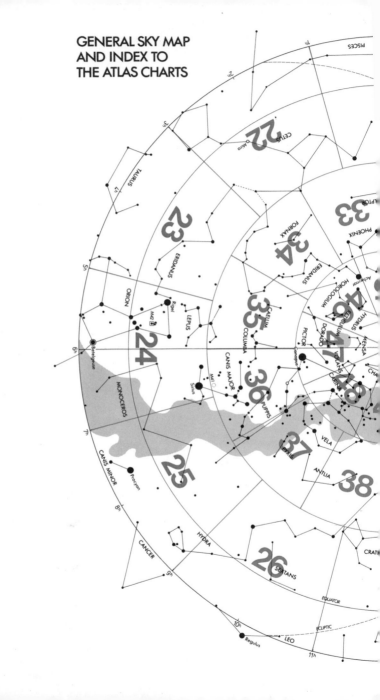

GENERAL SKY MAP
AND INDEX TO
THE ATLAS CHARTS

PISCES

CETUS

o Mira

22

TAURUS

23

ERIDANUS

ORION

Rigel

M42

LEPUS

24

Betelgeuse

MONOCEROS

CANIS MINOR

Procyon

CANCER

25

HYDRA

26

SEXTANS

LEO

Regulus

FORNAX

PHOENIX

ERIDANUS

34

33

CAELUM

COLUMBA

35

CANIS MAJOR

M41

Sirius

PUPPIS

36

VELA

ANTLIA

37

38

EQUATOR

ECLIPTIC

CRATER

Achernar

HOROLOGIUM

DORADO

PICTOR

Canopus

RETICULUM

HYDRUS

MENSA

VOLANS

CARINA

46

47

48

CHA

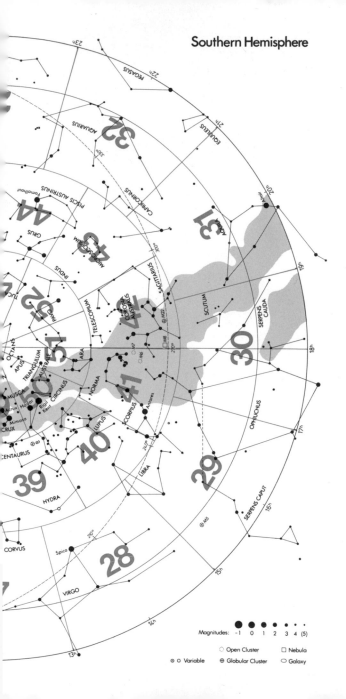

Southern Hemisphere